Sustainable Development Goals Series

The **Sustainable Development Goals Series** is Springer Nature's inaugural cross-imprint book series that addresses and supports the United Nations' seventeen Sustainable Development Goals. The series fosters comprehensive research focused on these global targets and endeavours to address some of society's greatest grand challenges. The SDGs are inherently multidisciplinary, and they bring people working across different fields together and working towards a common goal. In this spirit, the Sustainable Development Goals series is the first at Springer Nature to publish books under both the Springer and Palgrave Macmillan imprints, bringing the strengths of our imprints together.

The Sustainable Development Goals Series is organized into eighteen subseries: one subseries based around each of the seventeen respective Sustainable Development Goals, and an eighteenth subseries, "Connecting the Goals", which serves as a home for volumes addressing multiple goals or studying the SDGs as a whole. Each subseries is guided by an expert Subseries Advisor with years or decades of experience studying and addressing core components of their respective Goal.

The SDG Series has a remit as broad as the SDGs themselves, and contributions are welcome from scientists, academics, policymakers, and researchers working in fields related to any of the seventeen goals. If you are interested in contributing a monograph or curated volume to the series, please contact the Publishers: Zachary Romano [Springer; zachary.romano@springer.com] and Rachael Ballard [Palgrave Macmillan; rachael.ballard@palgrave.com].

Emilio José García • Brenda Vale

Wealth and the Built Environment

Springer

Emilio José García
School of Architecture and Planning
The University of Auckland
Auckland, New Zealand

Brenda Vale
Wellington School of Architecture
Victoria University of Wellington
Wellington, New Zealand

ISSN 2523-3084 ISSN 2523-3092 (electronic)
Sustainable Development Goals Series
ISBN 978-3-031-81956-8 ISBN 978-3-031-81954-4 (eBook)
https://doi.org/10.1007/978-3-031-81954-4

This Springer imprint is published by the registered company Springer Nature Switzerland AG
The registered company address is: Gewerbestrasse 11, 6330 Cham, Switzerland

Contents

Part II On Architecture and Wealth

About the Authors

Emilio José García is an architect and urban designer. Since 2013 he has been working as a Senior Lecturer in Sustainability and Resilience at the School of Architecture and Planning at the University of Auckland, New Zealand. He has been teaching, practising and researching in Argentina, Mexico and New Zealand while participating in international competitions in America, Europe and Asia. He won a Holcim Award for Sustainable Construction in 2008. He is interested in developing new areas of study that link social and environmental science with architecture and urban design. The book "Unravelling Sustainability and Resilience in the Built Environment" (Routledge: 2017), co-authored with Brenda Vale, explores what sustainability and resilience mean when applied to the built environment. The book uses a morphological approach to measure resilience in the built environment. His latest book "Collapsing Gracefully: making a Built Environment That Is Fit for the Future" (Springer: 2021), co-authored with Brenda and Robert Vale, challenges design approaches based on the myth that economic growth and technological development can solve social and environmental crises while proposing ways to avoid them.

Brenda Vale is an architect and academic. Along with Robert Vale, they wrote one of the first books on sustainable design and architecture, "The Autonomous House", in 1975. Following their design of several award-winning sustainable commercial buildings in the UK, they designed and built the UK's first autonomous house in 1993 and the first zero-emissions settlement in 1998. They have received international recognition, including the United Nations and European Solar Energy Society awards. They developed the Australian government's National Australian Built Environment Rating System (NABERS), which has now been implemented. Other books include "Time to Eat the Dog? The Real Guide to Sustainable Living", which analysed the impact on the environment of a western life-style and things people do every day, and "Living Within a Fair Share Ecological Footprint", which has chapters written by many of their former and existing postgraduate students. The book "Architecture on the Carpet" explores the links between architecture and construction toys such as Meccano and Lego over the last hundred years. Recently, Brenda has co-authored many environmental books, including "The Environmental Impact of cities. Death by Democracy and Capitalism" (2023), "Heating with Wolves, Cooling with Cacti: Thermo-bio-architectural Framework (ThBA)" (2021) and "Everyday Lifestyles and Sustainability: The Environmental Impact of Doing the Same Things Differently" (2018).

Part I

Wealth and Its Effect on the Built Environment

The first six chapters set out to explore links between wealth and the built environment, in terms of what is built, the cost of building and what is admired. It argues that unless the issue of wealth is addressed, the goals as set out in the 17 SDGs will be hard to achieve. Part I also looks at why changing how we think about the built environment if serious climate change and earth overshoot are to be avoided relates to wealth and its acquisition. Is wealth, as expressed through architecture and the built environment, the real barrier to climate change mitigation?

1 A Problem for the Sustainable Development Goals

Introduction

The environmental problems that humanity inflicts on planet Earth and all flora and fauna inhabiting it are well documented. Also well discussed is the issue of inequality within that human population as "71 percent of the world's population live in countries where inequality has grown" (UN, 2020).

What is less often discussed is how the accumulation of wealth affects these environmental and inequality issues. People born between 1981 and 1996, commonly described as Millennials, will be the wealthiest generation ever after inheriting assets from their parents (Kelly, 2023). If a business-as-usual mindset prevails, this wealth will be unevenly distributed among the population. A tiny group will receive more than anyone; another group will get less but more than enough, and half of the population will only inherit poverty. Since everything produced, traded and consumed by people is made out of nature, it is this accumulated or inherited wealth that leads to over consumption and environmental issues.

Moreover: "Resources of the Earth are not consumed by the growing population of developing countries for satisfying their basic needs, but by the decreasing population of developed countries for meeting their boundless demands" (Toth & Szigeti, 2016). Even though this explains why socioeconomic and environmental inequalities are inextricably linked, it still presents a significant challenge: The abundance of wealth can be detrimental to the environment and everyone in it: "As a tree that is heavy laden with fruit, breaks her own boughs, with their own greatness they ruin themselves" (Burton, 2021:559). How will it be possible for the wealthiest generation ever to deal with socioeconomic and environmental crises if the inherited fortune is a curse? Will this wealth inhibit the achievement of sustainable development as set out in the United Nations Sustainable Development Goals, which are supposed to "…end poverty and inequality, protect the planet, and ensure that all people enjoy health, justice and prosperity" (WHO, 2024)?

The Sustainable Development Goals

The United Nations 17 sustainable development goals (SDGs) were adopted in 2015 by the members of the United Nations and were supposed to be a plan for moving the world to "peace and prosperity for people and the planet, now and into the future" (United Nations, n.d.-a). There are 17 goals as shown in Fig. 1.1, each of which are subdivided into measurable targets. These 17 goals present a wish list, which, if achieved, would make planet Earth a wonderful place to live in for all humanity. The problem lies in how, and what it would take in terms of change to current ways of living, to achieve them, especially how this is tied up with current attitudes to wealth.

E. J. García, B. Vale, *Wealth and the Built Environment*, Sustainable Development Goals Series,
https://doi.org/10.1007/978-3-031-81954-4_1

Fig. 1.1 The 17 Sustainable Development Goals (United Nations, n.d.-a) (Reproduced with permission of the UN with the proviso: "The content of this publication has not been approved by the United Nations and does not reflect the views of the United Nations or its officials or Member States")

Figure 1.1 shows that the SDGs cover a wide range of global concerns, from poverty and hunger, to gender equity, to justice and climate action. It reveals that whatever sustainable development is supposed to be, it is not only about the environmental problems facing humanity, such as climate change mitigation and pollution of the seas and other waters. However, climate change (related to SDG 13, Climate Action), can potentially affect all humanity and the agriculture on which humanity relies, so in turn would affect SDG 15, Life on Land, since sustainable agriculture must exist within global climates. Other goals are more targeted towards those that currently have little, such as No Poverty (SDG 1) and Zero Hunger (SDG 2), two goals that may be difficult to achieve without climate change mitigation to avoid disruption to current agricultural practices.

This is one problem with having a series of goals without any sense of how they are nested together, making this reminiscent of the early diagram of the three pillars of sustainability that had sustainability occurring at the overlap of three circles labelled environment, economy and society. However, unless these circles are nested so that the economy sits within society, which sits within the environment, there can never be sustainability for humanity. It is necessary to see what the environment can sustain for the people that live within it and then the level of economy that can be achieved for that number of people within that environment. How much development can be achieved through that level of economic activity will only then become visible.

There is another problem: The SDGs fail to consider the issue of wealth and who has it when it comes to planning for eliminating poverty and inequality, both admirable targets. There is no sense in the 17 SDGs as to which needs to come first. Should it be Climate Action (SDG 13) so that people do not suffer the effects of droughts and floods on agriculture, which would go against achieving zero hunger, or should the emphasis be on Decent Work and Economic Growth (SDG 8), with all the problems of which economies need to grow, and which perhaps should not, given the resources of a finite planet?

What seems to be missing from the 17 SDGs is consideration of why human society is in its current problematic condition and the extent to which the idea of wealth and wealth accumulation is behind this. This book aims to address this by looking at wealth as a problem for the SDGs, with a focus on its effect on the built environment.

The Built Environment

The reason for the focus on the built environment is that everyone lives in it, and it is the arena where the inequalities and impact of wealth are often clearly visible. In the built environment, social, economic and environmental inequalities take shape because in a finite system, if someone takes a lot, someone else is left with very little. This idea echoes the German playwright Bertolt Brecht's (1934) poem *Alfabet* from his book for children:

> Reicher Mann und armer Mann
> Standen da und sahn sich an.
> Und der Arme sagte bleich:
> Wär ich nicht arm, wärst du nicht reich.

The poor man says to the rich man "If I was not poor, you would not be rich". The capitalist answer to this is to point to China where allowing some people to become very rich has reduced poverty in a very short time through the expansion of economic activities (Zitelmann, 2019). However, this has raised the issue of inequality in Chinese society, especially between urban and rural China. "China's inequality level used to be less than Europe's in the late 1970s, close to the most egalitarian Nordic countries, while it is now approaching US levels" (Piketty et al., 2019). Since SDG 10 is concerned with reducing inequalities, there seems to be a conundrum. Without redistribution of the wealth created in the name of development, there cannot be a reduction in inequality. It is wealth and who has it that sits at the centre of this problem.

This highlights the urgent need to start questioning the impact of wealth. The growth of unfair wealth accumulation makes the gap between the haves and have-nots so large that they become parallel realities. This increases the lack of awareness of each other's problems, together with a deep feeling of injustice or both. Unfortunately, an unequal society is not just divided by each part ignoring the other. This book attempts to explain how wealth has built an idea of development and progress, which is geared toward more wealth for a few through the design and consumption of products that benefit the wealthy to the detriment of the rest. From this point of view, the mix of greed and wealth can be considered a social-ecological hazard that increases its negative impact beyond what is built, can be seen, or perceived. It seems that the socio-economic reality of people, especially their wealth, has not been properly acknowledged or questioned in the paradigms and ideas that influence the decisions taken by both political leaders and the designers and planners of the built environment. Wealthy people who benefit from the current situation will want to keep those privileges because they do not see anything wrong with doing this. This has been made very clear at the 28th United Nations Climate Change conference COP28, where Conference president Sultan Al Jaber challenged the scientific basis of the phasing-out of fossil fuels that would lead the world to holding to a 1.5 °C rise in global temperature (Friedman, 2023). Sultan Al Jaber is also the CEO of the United Arab Emirates state oil company. This suggests a conflict of interest created by wealth, or at least the desire to hang on to creating more wealth. The problem, as noted above, is that wealth is inherently considered a development goal indicating progress and, therefore, appears intrinsic to sustainable development.

The same bias exists regarding the link between cities, wealth and sustainable development. One important reason to study the built environment and those who create it is that cities should contain the key to unlocking and solving all our problems, from economic prosperity to climate change. Jane Jacobs was probably one of the most famous advocates of the idea that cities play a crucial role in forming a nation's wealth and progress. Jacobs stressed that nations are much more abstract than cities, proposing that cities and their hinterland should be viewed like countries with their currencies. Using naturalistic analogies, Jacobs attempted to defend the diversity within cities and the contribution made by small businesses and firms to their development.

She stated that "Economic life develops by grace of innovating" (Jacobs, 1986:39), an essential quality of what she called "creative cities" (Jacobs et al., 2016:253). According to her hypothesis, the economy of creative cities relies on their capacity to innovate and replace imports with products. This means creating your own things instead of buying them from other countries. She explains that wealth is created by saving the difference and exporting your products.

Many years later, Richard Florida (2002) used similar ideas in his book *The Rise of the Creative Class,* calling for the creation of cities that provide opportunities to attract and retain talented people. In this context, "talented" and creative did not mean innovative bakers or skilful seamstresses as Jacobs hints, but primarily people who could create start-ups and clusters of technological development like Silicon Valley. The assumption is once again that technological innovation equals economic development. Florida seems to have missed Jacobs' initial intentions for creative cities as places "where many, many ordinary people can try out their hopes, insights, ambitions, and skills, and make something of them" (Jacobs et al., 2016:254). People move to cities attracted by the idea of progress, but many end up living in the deprived and excluded areas of the city because that is all they can afford. Living in cities has benefited a few people extraordinarily but has been hard for the rest. How many workers can start a business with a capital investment if they do not first have the chance to make this capital? How many people can have access to affordable tertiary studies to exploit their creativity and create a professional network? Do cities create opportunities for wealthy opportunists by depriving others of the same privilege? What is the role of the designers of buildings for these wealth-creating cities?

Later, Florida (2017:xxii) admitted he was naïve: "I realized I had been overly optimistic to believe that cities and the creative class could by themselves bring forth a better and more inclusive kind of urbanism". He observed that the creative class was only a homogeneous group of wealthy people and luxury businesses that gentrified the city and displaced the non-wealthy, increasing gentrification, unaffordability, and the decline of the middle class, leading to the "new urban crisis". Still, Jacobs had stated that the problems created by cities are opportunities to solve those problems and import those solutions to other places. "Life keeps casting up new problems, and the cities have been, and certainly will continue to be, the places where they can be solved" (Jacobs et al., 2016:171). This is significant. The myth is that the success of a city depends on its capacity to innovate to be economically successful. Cities that are not innovative enough will perish, and the creative cities, the smart cities as they have been called recently, can problem-solve anything. Do cities still contain the seeds of their own success and destruction? The problem today is that the super-wealthy (see Chap. 2) are often not so much innovators as imitators. The idea of a national delivery system existed long before Jeff Bezos established Amazon. Electric cars predate the work of Elon Musk by over a century. People are becoming very rich by selling what we could already do, which is not the same as innovation.

At the same time, wealth can obscure some very serious problems. Greenhouse gas emissions and their impact are scientific facts. However, they are invisible to some people and very real to others. Emissions can be invisible for a fairly rich person from a developed country living in a house close to the city centre with its own outdoor private space and possibly with a sea view. The sea is a beautiful background that decorates and increases the house's sale price. For a family living in Tuvalu or Kiribati, the emissions caused by the development and wealth of some nations will mean the loss of their countries from too much sea as levels rise (Donner & Webber, 2014). Acknowledging these impacts could lead to unravelling the problems so they can be dealt with, but only if the impact of wealth and wealth creation is acknowledged. If the single, and very desirable, SDG was to end poverty, and in a way every SDG supports this over-arching goal, since ending poverty could include a more equitable distribution of wealth, and better educational opportunities for all, along with gender equality and climate change mitigation, then an essential first step would be to examine wealth and wealth creation.

Mapping Sustainable Development Goals

The need to address global societal and environmental problems is urgent, hence the creation of the 17 SDGs by the United Nations. These goals have become a reference for institutions and businesses across countries. Even though they do not have a legal status, and thus cannot be enforced, they represent a summary of the main challenges to current human civilisation, and therefore cover what is considered politically correct for inclusion in the intentions of any policies initiated by those guiding their respective societies.

Unfortunately, the SDGs form a list of concerns to fix the world by dealing with sustainability matters without losing interest in development. This seems like putting the cart before the horse. The first step, as noted above, would seem to be to find out what the environment can provide. The effort to amalgamate sustainability and development intentions overlooks the situation that these two concepts are not always complementary, especially when the ideal of development is linked to social and economic progress, as happens in the developed world. To be more precise, the SDGs still have faith in economic growth as the means of fixing all global concerns, including climate change, presumably in the hope that the environment will provide what is necessary (Table 1.1).

A closer look at the structure of the SDGs and the nature of their goals can be done by mapping them against their economic, political, environmental or technological concerns. Every SDG has more than one target corresponding with indicators that are supposed to facilitate measuring progress to their achievement. In the version of the 17 SDGs analysed for this book, there were 240 indicators, which is a lot. Hundreds of indicators might be handled by institutions, policymakers, and indicator lovers but hardly by ordinary people needing to make practical decisions in everyday life. This creates a gap between the SDGs and people, small businesses and even institutions. The gap is even more prominent in places lacking internet access to the World Bank database, a key source for the SDGs' indicators. The accessibility to data for measuring the indicators can be problematic, even at the country scale. The United Nations Economic Commission for Europe (UNECE, 2023) has an organised website with information and a database for every indicator. However, this only covers 56 countries, leaving a population of 6.6 billion between Asia, Africa, and Latin America unregistered. The data collection at the city scale or even smaller scales, like the scale at which architects, planners and urban designers work, will be hard to collect, especially for developing countries. The data collection and its processing through hundreds of indicators make the SDGs impractical for any layperson. Considering the amount of work needed before using the SDGs, it is evident that they were designed to be used by institutions. However, whether there is a genuine interest or a necessity to build a sustainable image, SDGs cannot be used to promise a future but only to measure the past.

Since there seems to be no clear hierarchy within the SDGs, ending poverty (SDG 1) or making partnerships (SDG 17) counts as achieving a goal, even though one can change the world and the other could simply be politics. Even though there is no hierarchy, there is a strong presence of economic factors in almost all the SDG indicators which creates a type of hierarchy defined by a domain of interest.

Figure 1.2 sets out the extent to which the measurable indicators for all the SDGs are reliant on political or economic change. The bar chart shows that, across all the targets and their indicators for each SDG, there is a consistent presence of economic and political factors in the definition of every indicator. Indicators related to economic variables are in every SDG, being the most relevant in 14 of the 17 SDGs. Indicators related to the environment are scarce and not even most prominent in SDG 13, Climate Action, where the mix between political and economic change indicators is still more important.

Table 1.1 The SDGs and development

SDG	Statement[a]	Link to development
1. No Poverty	Economic growth must be inclusive to provide sustainable jobs and promote equality.	Depends on economic growth
2. Zero Hunger	Economic growth must be inclusive to provide sustainable jobs and promote equality.	Depends on economic growth
3. Good Health and Well-being	Ensuring healthy lives and promoting the well-being for all at all ages is essential to sustainable development.	Resources needed to provide health and well-being to those currently without them
4. Quality Education	Obtaining a quality education is the foundation to improving people's lives and sustainable development.	Resources to supply education to those currently without it
5. Gender Equality	Gender equality is not only a fundamental human right, but a necessary foundation for a peaceful, prosperous and sustainable world.	Depends on social and cultural changes and the "prosperity" of economic growth
6. Clean Water and Sanitation	Clean, accessible water for all is an essential part of the world we want to live in.	Resources to supply such water
7. Affordable and Clean Energy	Energy is central to nearly every major challenge and opportunity.	Resources needed to supply clean energy and switch from fossil fuel energy
8. Decent Work and Economic Growth	Sustainable economic growth will require societies to create the conditions that allow people to have quality jobs.	How can economic growth for all be sustained within a finite planet?
9. Industry, Innovation, and Infrastructure	Investments in infrastructure are crucial to achieving sustainable development.	Investments require resources
10. Reduced Inequalities	To reduce inequalities, policies should be universal in principle, paying attention to the needs of disadvantaged and marginalized populations.	Current inequalities have resulted from development
11. Sustainable Cities and Communities	There needs to be a future in which cities provide opportunities for all, with access to basic services, energy, housing, transportation and more.	Resources needed to change existing cities to meet the goal
12. Responsible Production and consumption	Responsible Production and Consumption	If enforced would suggest less economic activity
13. Climate Action	Climate change is a global challenge that affects everyone, everywhere.	Requires huge changes in energy use, transport and agriculture
14. Life Below Water	Careful management of this essential global resource is a key feature of a sustainable future.	Suggests reduction in fishing activities
15. Life on Land	Sustainably manage forests, combat desertification, halt and reverse land degradation, halt biodiversity loss	Halting biodiversity loss suggests halting human use of unexploited land
16. Peace, Justice and Strong Institutions	Access to justice for all, and building effective, accountable institutions at all levels.	?
17. Partnerships	Revitalise the global partnership for sustainable development.	?

[a] United Nations (n.d.-b)

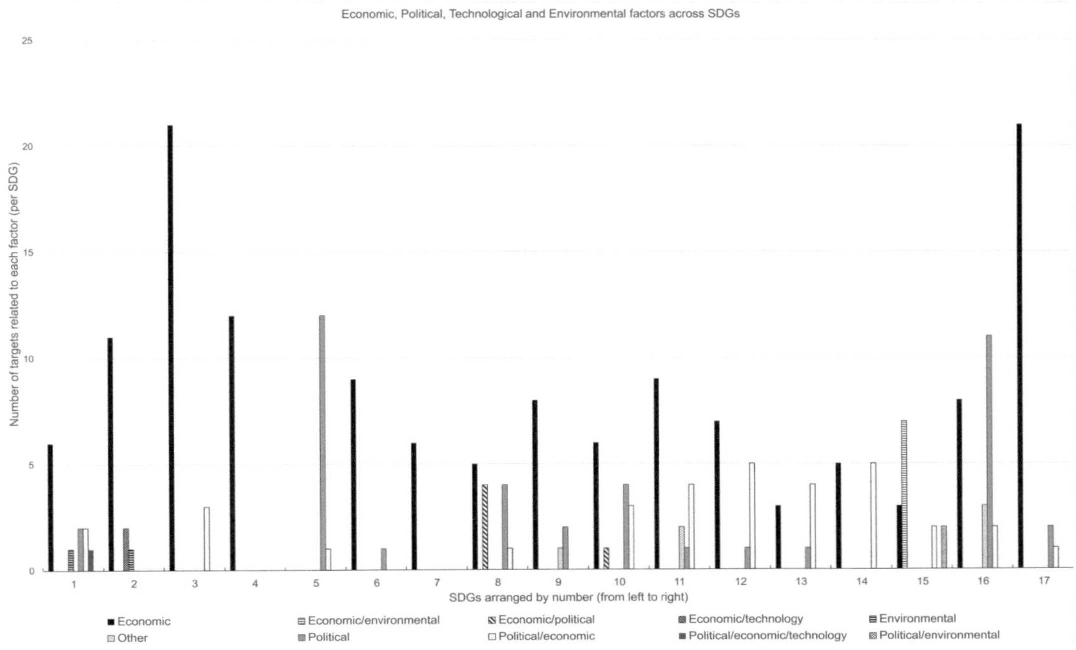

Fig. 1.2 Economic, political, technological and environmental factors across SDGs. (Data source United Nations (n.d.-b))

The analysis made for Fig. 1.2 shows that more than half of the indicators (59%) across all SDGs depend purely on change in economic variables. The proportion increases (66%) when economic variables are mixed with other interests, like political, environmental or technological (Fig. 1.3). This highlights the dependency of SDG indicators on economic variables, political variables being the second most relevant. Using this analysis the 17 goals and 240 indicators could be reduced to only two: dealing with money and politics.

The assumption behind the SDGs is very clear: Economic growth and political decisions to encourage economic growth will somehow change current development to sustainable development. However, what is not questioned is whether it is current economic growth for wealth accumulation for the few that has led to many of the unequal development and environmental problems. If unfair growth and wealth accumulation have been part of the problem, how can they be part of the solution?

Garcia et al. (2021:302–304) have argued that there is easily enough money in the world to fix the climate change problem through paying for clean energy systems to replace those using fossil fuels. Writing this towards the end of 2023, it is evident that there is as yet hardly any sign of the necessary sea change in moving to clean energy systems. What is stopping the spending of money on affordable clean energy, the seventh SDG? The 2023 COP28 started a relief fund for countries devastated by the impacts of climate change, such as flooding and drought. This fund "is the culmination of a 30 year effort by low- and middle-income countries to be compensated by high-income countries for the harm caused by the effects of climate change" (Sanderson, 2023). In that 30-year period global CO_2 emissions have continued to rise, increasing by more than 60% between 1990 and 2022 (Tiseo, 2023). Why have the developed countries not reined back their emissions and why have they taken 30 years to begin to compensate for the effect of them? Table 1.2 sets out the pledges made by a selection

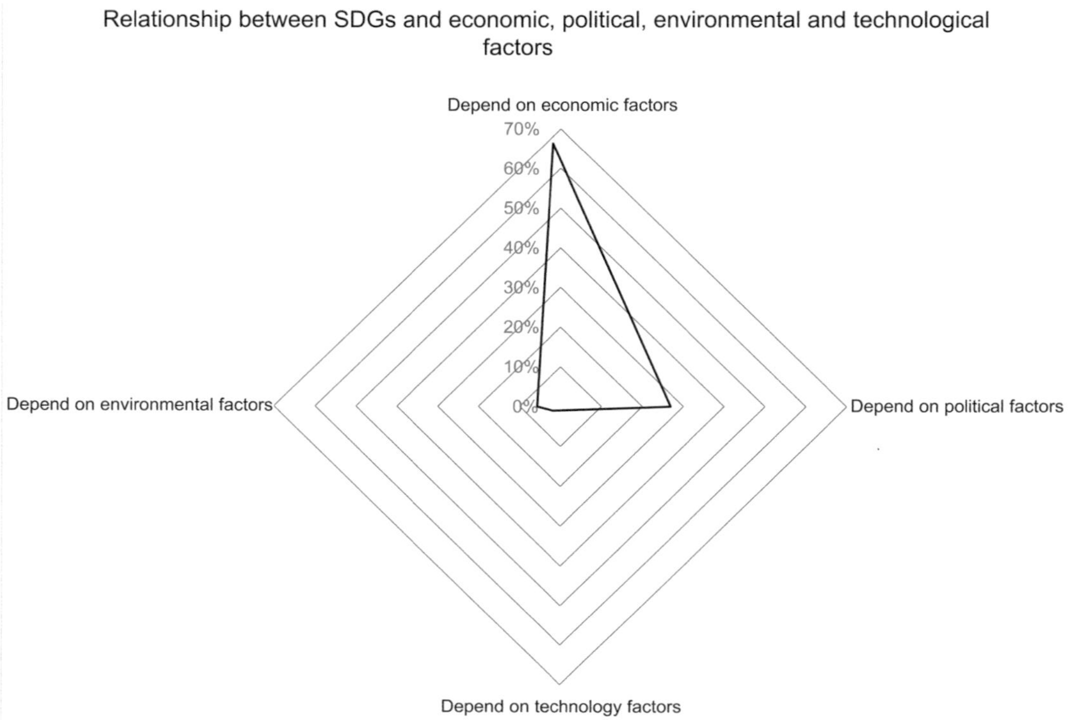

Fig. 1.3 Relationship between SDGs and economic, political, environmental and technological factors. (Data source United Nations (n.d.-b))

of developed countries, the percentage they contribute to 2022 global CO_2 emissions, 2022 emissions per head and GDP per head, together with rankings for percentage contribution to global CO_2 emissions and for GDP per capita.

What Table 1.2 reveals is that the polluter is not paying, at least not at present. With the exceptions of the UK and Germany, where the money pledged reflects their relative percentage contributions to global CO_2 emissions, the pledges from Japan and especially the USA do not reflect their contribution to climate change. Why are wealthy nations not choosing to compensate for the damage they have caused?

Climate change is not the only global problem. COVID-19 has reversed the decline in extreme poverty, defined as living on less than US$1.90 per day, and ending poverty is the first SDG. Whereas pre COVID-19, 8.6% of all people lived in extreme poverty this figure has risen to over 9%, and the recent increase in food prices and other impacts of the wars in Ukraine and the middle east could leave "the world even further from meeting the target of ending extreme poverty by 2030" (United Nations Statistics Division, 2023). Extreme poverty also reveals inequalities. Looking at those who are employed but fail to earn enough to take them out of extreme poverty, in 2021 most of these workers were located in Africa (37.9%), followed by 18.8% in Oceania (excluding Australia and New Zealand) and 6.6% in Central and Southern Asia (United Nations Statistics Division, 2023). Areas of the world like Europe and North and South America do not appear since those in work earn enough to take them out of extreme poverty. The tenth SDG is reducing inequalities. Extreme poverty and inequalities in the world are happening despite modern civilisation being the wealthiest in history (Carden, 2020). Carden goes on to point out that this wealth has been created because the market allows for innovation (innovation forms part of SDG 9—Industry, Innovation and Infrastructure), pointing out that the Soviet

Table 1.2 Countries that pledged money at COP28

Country[a]	Pledge US$[a]	% contribution to global CO_2 emissions[b]	CO_2 ranking[b]	CO_2 per capita (tonnes) (2021)[c]	GDP per capita US$[d]	GDP ranking[e]
UAE	100 million	N/A	N/A	21.8	53,757.9	N/A
Germany	100 million	1.8	7	8.1	48,432.5	19
UK	50 million	0.9	17	5.2	45,850.4	22
USA	17.5 million	13.5	2	14.9	76,398.6	7
Japan	10 million	2.9	5	8.6	33,815.3	29

[a] Sanderson (2023)
[b] Lu (2023)
[c] Our World in Data (n.d.)
[d] World Bank Group (2023)
[e] Knoema (2022)

Communist regime did not flourish because it stifled the market and with it the chance for innovation:

> In our commercial society, people have largely gotten a lot richer by buying low, selling high, and innovating—or as in the case of retail innovators such as Sam Walton, Sol Price, and Jeff Bezos, by innovating so as to create new ways for those who wish to buy low to cooperate with those who wish to sell high (Carden, 2020).

It should be noted that in the communist USSR there were jobs for everyone (a type of equity in a society), something that a market economy based on competition cannot promise even in the developed world. "Full employment was a promise honoured for more than a half century by successive leaderships in the Soviet Union" (Moskoff, 1992). The problem is that wealth creation can lead to unemployment and hence poverty for those who cannot earn. Britton (1994) in discussing rising unemployment in the UK states the following:

> [I do] not suggest that trade, technology, competition and so on are bad for the economy as a whole. On the contrary, the point is that they do raise average living standards and average pay, but this means that the minimum level of pay which employers will offer or workers accept also goes up—and that leaves a significant minority with no work at all.

The problem with the SDGs is that the fulfilment of many of them seems to rely on wealth creation without the fundamental concomitant of wealth redistribution. It seems there is one common driver of change affecting both sustainability and development that has not been included in the SDGs: wealth. Development means creating money so as to be able to alleviate poverty but development also has to be sustainable, meaning it has to be able to be sustained within the limits of the available resources, whether these are local or planetary. At the same time development also has to proceed so as to fulfil other SDGs such as clean water and quality education for all people.

This means that development cannot just rely on people consuming things newly invented in order to create wealth for someone, because within a finite system as some have more, someone else will be having less, whether that less is loss of private property through flooding from more severe weather events or loss of life in freak high temperatures. The other paradox is SDG 11—Sustainable Cities and Communities. Cities are the places where wealth is generated but are also the places with a higher environmental impact than small settlements and rural areas. It is the wealth generated in cities that leads to higher consumption of resources and higher emissions (Chicca et al., 2023:257–258).

At the same time as achieving the SDGs there is also the issue of conserving resources for future generations. Much of the existing built environment is a legacy from past generations. To what extent has it been formed with the idea of preserving resources for the future? Is preserving resources still the goal of those creating a built environment now that will be used by generations to come? Conversely, has its creation been dominated by ideas of wealth accumulation to be enjoyed in the present rather than thinking about the future? Will the type of built environment being created now be useful for future generations? Are those creating this built environment in tune with the goals of sustainable development?

The Built Environment Within the SDGs

This book focuses on the built environment and particularly those currently responsible for its design. The built environment includes cities because cities both generate and depend on wealth, while the same wealth accumulation is also responsible for social, economic and ecological problems. However, the built environment is wider than this, since cities are only one part of the extensive human transformation of the earth's surface. All the cultural landscape, every part of the planet transformed by humans, whether they inhabit, circulate through, or produce food from it, is the built environment. Urban and rural landscapes are part of the built environment since cities could not survive without their hinterlands and ecosystem services. This can be seen in the relationship between the built environment and the SDGs. Even though there is only one SDG directly related to cities, SDG 11, the number of indicators linked to the management and design of the built environment is spread across 12 of the 17 SDGs (Fig. 1.4). One explanation is that the built environment is more than its buildings; it is the source and the arena for everyday life, it is land, buildings and people.

Nine of the twelve SDGs related to the built environment have a predominance of economic variables, and 20 of their 33 indicators depend on economic variables (Fig. 1.5). The message is simple: The way to achieve a built environment that is more sustainable is through getting wealthier. However, previous generations have contributed to the present crisis by trying to do exactly this.

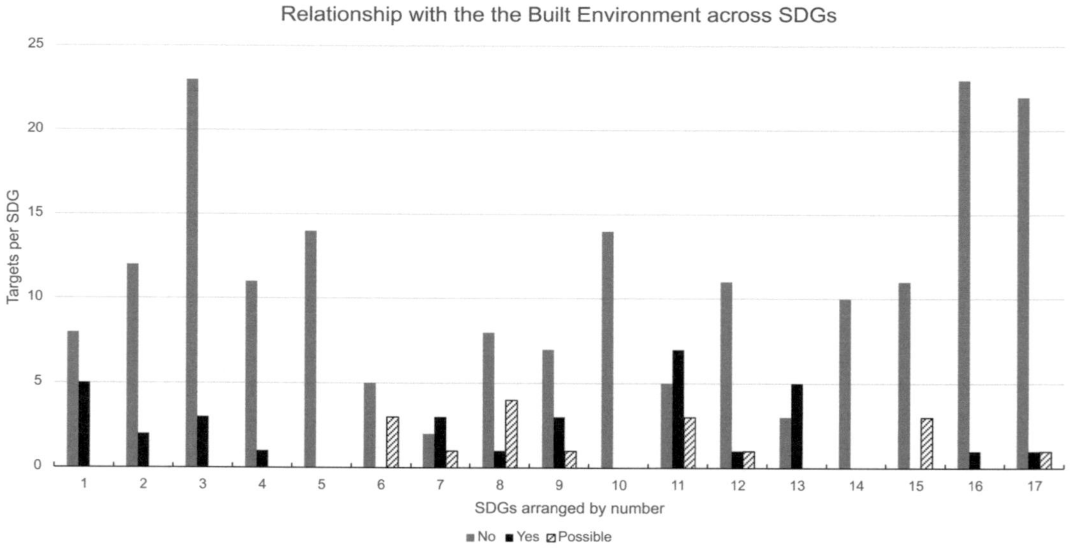

Fig. 1.4 The built environment across the SDGs

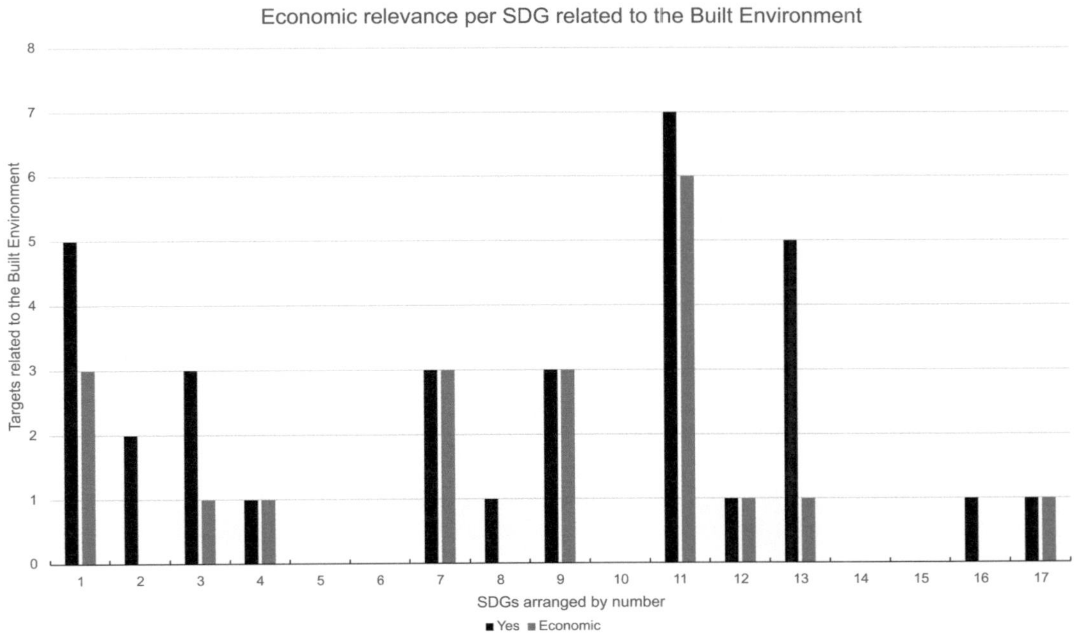

Fig. 1.5 Economic relevance in SDGs related to the built environment

What this book sets out to do is look first at the issue of wealth, and then at the relationship of wealth to the built environment. It will argue that the creation of a sustainable built environment means thinking beyond the iconic buildings created by wealth that often only serve the few, to what can benefit everyone. This may also mean a big change in the education of those trained to design the built environment.

References

Brecht, B. (1934). *Ein Kinderbuch*. das Kinderbuch verlag.

Britton, A. (1994). Full employment in a market economy. *National Institute Economic Review, 150*, 62–72.

Burton, R. (2021). In Angus Gowland (Ed.), *Anatomy of melancholy*. Penguin Random House.

Carden, A. (2020). Wealth. *Journal of Markets & Morality, 23*(1), 139.

Chicca, F., Vale, B., & Vale, R. (2023). *The environmental impact of living in cities*. Routledge.

Donner, S. D., & Webber, S. (2014). Obstacles to climate change adaptation decisions: As case study of sea-level rise and coastal protections measures in Kiribati. *Sustainability Science, 9*(3), 331–345.

Florida, R. L. (2002). *The rise of the creative class: And how it's transforming work, leisure, community and everyday life*. Basic Books.

Florida, R. L. (2017). *The new urban crisis: How our cities are increasing inequality, deepening segregation, and failing the middle class-- and what we can do about it*. Oneworld Publications.

Friedman, L. (2023). *Climate Summit leader tries to calm uproar over a remark on fossil fuels*. Retrieved from Sultan Al Jaber Tries to Calm Uproar Over a Remark on Fossil Fuels - The New York Times (nytimes.com), 7 December 2023.

Garcia, E., Vale, B., & Vale, R. (2021). *Collapsing gracefully: Making a built environment that is fit for the future*. Springer.

Jacobs, J. (1986). *Cities and the wealth of nations*. Penguin.

Jacobs, J., Zipp, S., & Storring, N. (2016). *Vital little plans: The short works of Jane Jacobs*. Random House.

Kelly, J. (2023). *The Great Wealth Transfer from Baby Boomers to Millennials will impact the job market and economy.* Retrieved from https://www.forbes.com/sites/jackkelly/2023/08/09/the-great-wealth-transfer-from-baby-boomers-to-millennials-will-impact-the-job-market-and-economy/, 23 September 2024.

Knoema. (2022). *World GDP per Capita Ranking 2022*. Retrieved from https://knoema.com/sijweyg/world-gdp-per-capita-ranking-2022-data-and-charts, 3 December 2023.

Lu, M. (2023). *Visualizing all the world's carbon emissions by country.* Retrieved from https://www.visualcapitalist.com/carbon-emissions-by-country-2022/#:~:text=Ranked%3A%20Global%20CO%E2%82%82%20Emissions%20by,biggest%20in%20terms%20of%20population. 3 December 2023.

Moskoff, W. (1992). *Unemployment in the Soviet Union.* National Council for Soviet and East European Research. Retrieved from https://www.ucis.pitt.edu/nceeer/pre1998/1992-806-36-2-Moskoff.pdf, 3 December 2023.

Our World in Data. (n.d.). *Per capita CO_2 emissions.* Retrieved from https://ourworldindata.org/grapher/co-emissions-per-capita, 3 December 2023.

Piketty, T., Yang, L., & Zucman, G. (2019). Capital accumulation, private property, and rising inequality in China, 1978–2015. *The American Economic Review, 109*(7), 2469–2496.

Sanderson, K. (2023). *First cash pledged for countries devastated by climate change: COP28 starts with historic decision.* Retrieved from https://www.nature.com/articles/d41586-023-03814-9#:~:text=Of%20197%20countries%20represented%20at,million%20and%20Japan%20%2410%20million, 3 December 2023.

Tiseo, I. (2023). *Annual carbon dioxide (CO_2) emissions worldwide from 1940 to 2022.* Retrieved from https://www.statista.com/statistics/276629/global-co2-emissions/, 3 December 2023.

Toth, G., & Szigeti, C. (2016). The historical ecological footprint: From over-population to over-consumption. *Ecological Indicators, 60*, 283–291.

United Nations. (2020). *Inequality—Bridging the divide.* Retrieved from https://www.un.org/sites/un2.un.org/files/2020/02/un75_inequality.pdf, 23 September 2024.

United Nations. (n.d.-a). *The 17 goals.* Retrieved from https://sdgs.un.org/goals, 12 August 2024.

United Nations. (n.d.-b). *Sustainable development goals.* Retrieved from https://www.un.org/sustainabledevelopment/sustainable-development-goals/, 21 February 2024.

United Nations Economic Commission for Europe (UNECE). (2023). *Progress in the UNECE Region.* Retrieved from https://w3.unece.org/sdg2023/progress-assessment.html, 5 May 2024.

United Nations Statistics Division. (2023). *End poverty in all its forms everywhere.* Retrieved from https://unstats.un.org/sdgs/report/2022/goal-01/#:~:text=Forecasts%20for%202022%20estimate%20that,be%20living%20in%20extreme%20poverty. 29 November 2023.

WHO (World Health Organisation). (2024). *Sustainable development goals.* Retrieved from https://www.who.int/europe/about-us/our-work/sustainable-development-goals#:~:text=The%20Sustainable%20Development%20Goals%20(SDGs,no%20one%20is%20left%20behind., 23 September 2024.

World Bank Group. (2023). *GDP per capita (current US$).* Retrieved from https://data.worldbank.org/indicator/NY.GDP.PCAP.CD, 3 December 2023.

Zitelmann, R. (2019). *No, the rich don't get rich at the expense of the poor.* Retrieved from https://reaction.life/no-rich-dont-get-rich-expense-poor/, 3 December 2023.

2 Wealth and the Rich

Why Might Wealth Be a Problem?

Looking at the SDG targets in Chap. 1 showed how much their achievement is tied up with economics and the idea of increased prosperity through economic activity. SDG 1 is No Poverty. To address poverty means one of two things—either everyone is rich enough to be above the poverty line, which would mean generating more money than there currently is in the world, or the money that already exists must be redistributed. To achieve SDG 10 (Reduced Inequalities) suggests redistribution must be the way forward. The normal way of redistributing money from the wealthier to the poor is via taxes or charitable handouts. Of these, taxes are the more certain way of achieving the goal. However, what becomes of taxes depends on the government that controls them, and there is no guarantee that the money raised by taxes will go to the poorest in society and raise their living standard.

Additionally, as Higgins and Lustig (2016) demonstrate, policies around taxes and benefits apparently designed to help the poor can make them poorer, "we show that taxes and transfers can be globally progressive, unambiguously equalizing, and unambiguously poverty-reducing and *still* make many poor worse off". It seems the problem lies less in how to redistribute wealth but rather in the desire for wealth and the idea that global problems can only be answered by creating more wealth. This is seen in an analysis undertaken by Parada-Contzen and Parada-Daza (2023), who stated that "for all the measures of wellbeing used, richer countries put higher weight on the homo economicus side of behaviour", or, in other words, people make decisions so that they will tend to have more not less. It seems that the SDGs are following this pattern, in a world where other measures like the Ecological Footprint show that there are limits to how much economic growth is possible when it comes to the resources available to fuel such growth (Wackernagel & Rees, 1996).

Perhaps wealth matters so much to people because it is the fabulously wealthy individuals in the past who feature in history books, such as Nero in Rome, who was known for his extravagant spending (Britannica, 2023) and the Romanovs in Russia with their jewellery and Fabergé eggs (Treble, 2014), while the workers that kept society going remain lost, un-named individuals. The wealth of emperors could be handed down in dynasties, unless squandered by some hapless member. Thomas Jefferson, the third president of the USA and part author of the *Declaration of Independence*, despite inheriting plantations lived beyond his means and died in debt (Gorman, 2018). Rather than inheritance, enormous wealth can also be acquired by having a good idea or developing an existing one and selling it to the masses, such as Henry Ford and the car (see Chap. 4).

E. J. García, B. Vale, *Wealth and the Built Environment*, Sustainable Development Goals Series,
https://doi.org/10.1007/978-3-031-81954-4_2

Why should personal wealth, in terms of all assets owned, be viewed as a good thing? It seems that the rich in these terms are admired: "Thus, it seems that individuals believe that the rich are in many ways different from the general population…rich people have a generally positive, competent and enviable image" (Rinn et al., 2023). Other investigations in psychology have found that the "rich are different from the poor" in having less empathy for others and exhibiting more self-interest (Kraus et al., 2012). This observation is important for climate change, since changing habits to reduce emissions may be harder for the wealthier if these work against such self-interests as having overseas holidays, a large house, several cars and an expansive wardrobe.

Poverty and wealth are normally discussed in the abstract, as if they are not part of everyday life. On the one hand, the poor are alluded to as the product of an economic system, a vulnerable sector of the community, its weakest link, but rarely as real people. On the other hand, the wealthy are not usually mentioned unless someone needs to be found responsible for the negative impacts of wealth, such as the alleged bribing of FIFA officials when it came to voting for who was to hold the next soccer world cup (Dunsheath, 2022). As suggested above, dealing with climate change cannot avoid dealing with wealth and the wealthy. It may also mean dealing with the superrich, noting that "the richest 1 percent grabbed nearly two-thirds of all new wealth worth $42 trillion created since 2020, almost twice as much money as the bottom 99 percent of the world's population" (Oxfam International, 2023).

Wealth and the Built Environment

Wealth is a sensitive subject for the built environment as "real estate and infrastructure represent two-thirds of global wealth and are the largest global asset class" (World Business Council for Sustainable Development, 2023:2). Wallach (2015:x) claims that the modern built environment is just a reflection of a world created to make money and that: "There's something else wrong with a world made for money: it's unbelievably, never-endingly, stupefyingly ugly and boring". This is a sad reflection on those who have designed this world as well as those who have paid for their services.

The building industry that creates this world made for money is also a big employer with 12% globally employed in construction activities (World Business Council for Sustainable Development, 2023:6), compared to the 25% globally employed in agriculture, which includes many from developing countries where the majority are engaged in farming as their only means of making a living (Roser, 2023). The good news for the building industry is that wealth appears here to stay. The number of millionaires and billionaires keeps rising globally along with their total wealth (Fleck, 2023). The number of architectural graduates as potential designers of the built environment also keeps growing, at least in the UK and the USA (Green, 2019; Parker, 2023). An Australian report also found that numbers of graduate planners had increased making planning a "growth profession" (Fielding and Williams, 2023:9). So, more money put into the built environment means more jobs for built environment professionals, so everyone should be happy. However, the clustering effect of wealth is also visible in the number and size of architectural firms as discussed in Chap. 3. Most are small, and only a few are huge. Then there is the detrimental effect of the built environment on climate change (Popescu & Luca, 2017). At the very least all architectural and planning efforts should be going into the design and layout of buildings and infrastructure that minimize these adverse effects on global temperature rise.

Wealth and Income

Wealth is commonly understood to mean financial or economic capital. A wealthy person usually has a lot of property, investments, and expensive tangible assets like paintings or money. Wealth and capital are often used interchangeably, although capital is not always linked with cash or properties. Human capital refers to education and skills, social capital is associated with relationships, and cultural capital represents the value of cultural legacies, like traditions and languages. A person may not necessarily be economically wealthy to have many friends always keen to help (social capital), or to be part of a religious community that provides support (cultural capital). Social capital could also be a commodity. A growing network of people who follow someone on Twitter, YouTube and Instagram can be monetized through social media. Income and wealth are also related terms in economics, but they differ. The winners of the Nobel Prize in economics Banerjee and Duflo (2019:238) explain this difference in simple words: "income is what people earn every year; wealth is their accumulated fortune".

The definition of capital and wealth is useful in economics because it helps to determine the wealth of a business, a household, a city, or a country. The wealth of a country, or national wealth, has been described by Piketty (2014:48) as being:

> …[the] total market value of everything owned by the residents and government of a given country at a given point in time…It consists of the sum total of nonfinancial assets (land, dwellings, commercial inventory, other buildings, machinery, infrastructure…) and financial assets (bank account, mutual funds, bonds, stocks, financial investments, insurance pension), less the total amount of financial liabilities (debt).

Piketty (2014:46) uses concepts of capital and wealth interchangeably. "Capital includes all forms of wealth that individuals (or groups of individuals) can own and can be transferred or traded through the market on a permanent basis". The American economist Suresh Naidu suggests that this understanding of wealth accepts many forms "which runs from real estates to financial assets, such as corporate shares and loans, and even slaves" (Boushey et al., 2017:100). What people and nations transfer, or trade, continues to change depending on politics, economic policies and social perceptions. However, what wealth is has not changed much in a couple of millennia. Aristotle defined wealth as "all those things whose value can be measured in money" (Curzer, 2012:84). In essence, this is very similar to the definition given by Piketty of capital. However, changes in the perception of wealth, its ethical and moral values, and its social and environmental impacts have consequences that need to be considered (see Chap. 9 for further discussion of ethics).

Having wealth does not come without environmental impact. According to the International Energy Agency (IEA) (Cozzi et al., 2023) "Wealth, energy use, and the consumption of goods and services are unevenly distributed across the world. Carbon dioxide (CO_2) emissions are no exception. Emissions vary across countries and across generations, but even more so across income groups". Inequality has been a part of most human societies since agriculture led to settlements and nomadic ways were lost. Alacevich and Soci (2018:13–14) note the consequences of this were the "evolution of much more complex economies and social structures—hence social stratification, clientage networks, and the emergence of ruling elites that were able to use violence freely to their own ends". The wealthy ruling elite were early on typified by Roman society, where under Augustus the senate members had to be wealthy and come from an appropriate family (Talbert, 1984). Although the power wielded by the senate changed much over the years of Roman dominance, somehow those in power were normally associated with being wealthy, suggesting that having wealth is a precursor for having power over those without it. Today this translates unreasonably to those with wealth having the power to change the climate for those without wealth.

Who Are the Modern Super-Rich?

In the modern era, especially from the late nineteenth century to the beginning of the twentieth century, being incredibly rich meant being a millionaire. In 1913, J. D. Rockefeller's (1839–1937) wealth, as shown in the Guinness World Records (2023a), was estimated at US$900 million. This amount is equivalent to US$189 billion today, but the point is that the largest unit of measurement for wealth was in the millions. This was going to change 3 years later, in 1916, when Rockefeller became the first billionaire. The fortune of Rockefeller was made through the Standard Oil company that, at its peak, controlled 90% of the oil consumption of USA. His only son, John D. Rockefeller Jr. (1874–1960), also involved in Standard Oil, commissioned one of the largest private built complexes in New York, the Rockefeller Center, which was constructed during the depression (American Experience, n.d.). One of his sons, David Rockefeller (1915–2017), whose wealth was still linked to shares in successor companies of Standard Oil and real estate, was behind the development of one of the most iconic buildings in New York, the Twin Towers in the World Trade Center. Minoru Yamasaki, the architect commissioned with the design of the Towers, was so impressed by the size of the budget initially estimated for the project that he was convinced it was a typo (Tognini, n.d.). The Rockefeller family were also philanthropists. They donated money to Harvard and the University of Chicago, among many others. They also founded their own university, Rockefeller University, and played a significant role in science and art. However, all this philanthropy had a dark side as "Rockefeller money funded eugenic scientists decades before Hitler put eugenic theories into practice" (Messall, 2004).

More recently the Rockefeller name has become directly associated with the built environment through the "100 resilient cities" project, although arguably the Rockefeller millions from oil have had far more effect on the modern built environment (see discussion of roads in Chap. 4). The resilient cities project was set up to provide financial help so cities could deal with the "physical, social and economic challenges" faced in the twenty-first century. From more than 1000 applications, 100 cities representing over 20% of global population were selected (The Rockefeller Foundation, 2023). Although the programme closed in 2019 it spawned a book *The Resilience Dividend*, which served as the theoretical background for the program (Rodin, 2014). The author, Judith Rodin, was chair of the Rockefeller Foundation and her book was praised by Bill Clinton, whose foundation had previously worked with Rodin, and by Michael Bloomberg, another very rich man who made money in finance, IT and the media. Rodin's book vaguely discusses the theory of ecological resilience and suggests examples of its application to cities but with no measurements (Garcia & Vale, 2017). Climate change and its impact are constantly present across all the chapters. However, the book never uses the words oil and carbon dioxide emissions, perhaps because this might lead to a conflict of interest. An analysis of other rankings of resilient cities (Garcia, 2022) shows that resilience is being ranked only as it affects the wealthy population in any city. Unfortunately, it is the wealthy in the city that tend to be the least sustainable and make the greatest contribution to climate change. Garcia (2022) summarises the situation:

> Cities compete to grow their economies to be more prepared and more resilient. In doing so they increase their impact on the environment and the exposure to threats, which in turn demands bigger investments and more economic growth. Getting richer and wealthier increases the chance for cities to be ranked as resilient.

Given this analysis, it is unsurprising that a book about resilient cities stemming from the Rockefeller Foundation failed to mention carbon emissions.

The Rockefeller family represents a dynasty of wealthy people across generations making money. The fortunes of both the Rockefeller family and Henry Ford were linked to energy and manufacturing since Ford's cars ran on oil, and the Rockefeller empire was built through investments in refineries. The more recent very rich, like Michael Bloomberg, have made their money differently, Bloomberg's coming from dealing in financial information and the media (Alcorn, 2019). Among the very rich Warren Buffet, and Rupert Murdoch come from privileged backgrounds. Buffet enlarged his fortune through finance and investments. At the same time, Murdoch made his fortune by investing in media, a business he knew well since his father was already a prominent figure in the Australian media. The next generation of the wealthy, mostly born after World War II, like Richard Branson, found more diverse means of making money, from music record labels to airlines. The following generation, with people like Elon Musk and Larry Page, define the first big change in the way of becoming rich (or richer), with wealth coming from investment in technology and manufacturing. A curious detail is that with the increase in investment and technology in part aimed at humanity being more sustainable through developments like electric vehicles, mining has returned to the portfolio of the richest in the world (Vendittii, 2021).

Billionaires

In 2022, there were more billionaires than ever before. Having an "estimated net worth of $1 billion per person is needed to make the cut" (Dolan & Peterson-Withorn, 2022) and be included in the list of 2668 richest in the world as defined by Forbes. This selected group represents 0.00005% of the adult population globally, mostly from developed countries. Almost two-thirds, 74 people, of the top 100 richest in the 2022 list were from developed countries, half (39) in USA. If Elon Musk, the richest person in the 2022 Forbes list of billionaires, had kept growing his fortune at the then current rate of 129% per year, he could have been the first trillionaire in 2024 (Batchelor, 2022), setting new boundaries for considering how much money is too much.

The minimum fortune to be considered among the richest keeps changing as well as their wealth. The number of listed people varies yearly. Their wealth also changes and produces shifts in position in the ranking (although the first positions have not changed that much in the last 3 years) or, to be more precise, every minute, since fortunes are updated every 5 minutes on the Forbes website. In 2022, the accumulated wealth of the richest 2668 people was worth US$12.7 trillion. As a reference, the GDP of China in 2022, which is the second largest in the world after the USA, was almost US$18 trillion (World Bank, 2023). The next highest were Japan and Germany at just over US$4 trillion. Most countries had a GDP below US$1 trillion, and just a few between US$1 trillion and US$5 trillion. This means that the fortune clustered in less than 3000 people is more significant than the GDP of any country except China and the USA. The 2022 wealthiest man in the world, Elon Musk, owned US$219 billion (Dolan & Peterson-Withorn, 2022), the same as the 2022 GDP of Greece. This is somewhat ironic given that the Ancient Greek view of wealth was having something that could be useful to you, not being wealthy for its own sake. This is discussed further in Chap. 9.

Batchelor (2022) notes that by 2033, there could be about 10 trillionaires, mostly males from the technology industries. One person having more money than a country's gross domestic product should be sufficient to suggest how much money is too much. The effect of having a growing group of individuals that can be wealthier than countries remains to be seen. What has been investigated is the environmental impact of billionaires (Oxfam, 2022). The fact that billionaires with their private jets and super yachts emit more carbon emissions is hardly unexpected. The Oxfam report investigated 125 billionaires and found annual emissions of an "average of 3.1 million tonnes per billionaire, which is over one million times higher than 2.76 tonnes—the average for someone in the bottom 90% of humanity". Perhaps more surprising is that a proportion of the emissions of the wealthy are generated by their investments in activities like mining and cement manufacture. The result is "that while the personal consumption emissions of billionaires can be a thousand times higher than those of ordinary citizens, emissions from billionaire investments can be a million times higher than normal people" (Oxfam, 2022:5). There is irony in the fact that the super-rich are also philanthropically helping the super poor to overcome some of the issues of climate change (Kumar, 2019) since the simplest action would be to stop being the cause of the problem in the first place.

Wealthy Built Environments and Expensive Buildings

When the shopping cart is billionaire in size, it is possible to buy or have made more and bigger things. Large objects seem like particles, and real estate products are like dolls' houses for this degree of wealth. Usually, a house is considered the most expensive asset for a family. Hypothetically, the most expensive home is Buckingham Palace in London, the home of the English monarchy, an old building but still an enormous palace of 77,000 m^2. Even though it is not on the market, it has been valued at US$1.6 billion (Yu, 2022). The top ten richest men worldwide are wealthy enough to buy around 100 Buckingham Palaces each. Buckingham Palace is probably not big or expensive enough to reflect that wealth, since the history of houses can also be read as a history of wealth (see Chap. 5). The bigger the wealth, the bigger, grander, and more ornate the house.

The built environment is everything constructed and altered by people for people, but it is the big and luxurious buildings that tend to catch the world's attention, not least because of competition to be the tallest or most expensive. Such record breaking is probably useful in promoting and attracting international tourists. This could explain why wealth is associated with buildings like the Burj Khalifa in Dubai, which cost US$1.5 billion (Emporis, 2022) and is the tallest building in the world, with a structure that reaches 828 m. The building also ranked 18th among the top 20 most expensive buildings (Emporis, 2022). It was completed in 2010 and achieved other records: tallest elevator (504 m), the most floors in a building and the highest restaurant (441.3 m) (Guinness World Records, 2023b). Skidmore, Owings & Merrill (SOM) was the firm in charge of the design. On their website, they have called themselves "Advocates of the Environment" and have their own "Climate Action Group", which clusters unknown people from different fields to do things not detailed on their website. As usually seen in big firms, they have very big statements about the environment and commitments to every green building council, organisation and campaign on earth. In the website section called "50 years of environmental activism through design", there is no picture or mention of the tallest building they made, the Burj Khalifa. Still, there is an inspiring quote from Nathaniel Owings, one of the founders of the firm: "Civilisations leave marks on the Earth by which they are known and judged. In large measure, the nature of their immortality is gauged by how well their builders made peace with the environment" (SOM, 2021a).

The quote links with other less well-known records and facts about Burj Khalifa like its record for the vertical pumping of concrete, previously held by Garda Hydroelectric Power Plant, a building that produces energy from renewable sources. In that case, at least, the emissions associated with the concrete are for generating renewable power rather than just for a tall building for the sake of being tall. Other construction facts include a "total weight of aluminium equivalent to five A380 aircraft, and 45,000 m^3 of concrete used in the foundation" (Burjkhalifa, 2022). The interior design includes expensive details like "silver travertine flooring, Venetian stucco walls, handmade rugs, stone flooring and dark, intricate Brazilian Santos rosewood [hopefully old stock or recycled as it is endangered (Kaiser, 2011)] to reflect shelter, comfort and above all, restrained luxury" (SOM, 2021b). The finishings illustrate the intention of making a building that would exceed the standard idea of a skyscraper (SOM, 2021b), even though excess could be a more precise term. Considering that the building is in a desert area, where water is scarce, the lake that forms part of its landscaping, with a spectacular fountain, must also be classified as a luxurious feature. All these features, facts and "restrained luxury" might not perfectly illustrate Owings's idea of a building "making peace with the environment".

A characteristic often forgotten about the Burj Khalifa is that it is mainly a residential building. The building has many attractions, including tennis courts, basketball courts, two swimming pools and 3000 parking spaces (Souha, 2021). The cheapest 2023 room at the Armani Hotel, one of the smaller buildings attached to the base of the Burj Khalifa, was over US$1000 a night (KAYAK, 2023). Buying an apartment in the building is very expensive as prices range from over half a million US dollars to nearly two million (Binayah Properties, n.d.).

Expensive buildings also involve structurally complicated ideas such as the Marina Bay Sands (Resort) in Singapore that opened in 2010, with a total construction cost of US$5.7 billion (Goh. 2023). This building features three towers connected high above the ground by a gigantic roof terrace that contains a swimming pool with the best views in the city (see also Chap. 8).

The built environment is so comprehensive that it covers many structures, from tunnels and bridges between countries to skyscrapers. One of the biggest and most expensive buildings is Masjid al-Haram, also known as the Great Mosque of Mecca, in the Mecca province of Saudi Arabia. The complex has been through many expansions since it became a monotheist building in 630 AD (Britannica, 2022). It is now undergoing one of its largest expansions estimated to cost US$100 billion (McCarthy, 2020). The website of the Great Mosque of Mecca states that the building complex is on the most expensive land in the world at US$100,000/m^2 (Mecca, 2022a, b). However, the building is not on the market, which makes numbers vary depending on how much religious and cultural value is attached to the building. There is no agreement on the total area of the Great Mosque of Mecca probably due to the constant renovations to accommodate a growing number of pilgrims. It ranges from 133,780 m^2 (Mecca, 2022a, b) to 356,800 m^2 (Britannica, 2022) for more than a million pilgrims. The building is also an international landmark for religious tourism and an important source of revenue. In 2018, Mecca ranked second in the top 20 destinations, with international overnight visitors spending US$20 billion (Mastercard, 2019). "Religious tourism in Saudi Arabia may not have the current revenue-generating capacity of the oil and gas sector, but the religious significance of Mecca and Medina will never run dry" (Al Lawati, 2022). However, aeroplanes still need oil-based fuels to fly international visitors to Mecca, just as energy is needed to make the materials that go into these mega structures. Billionaires or trillionaires will never be able to buy the Great Mosque of Mecca, the most expensive building on the planet in some rankings, because it is a state-owned building with a religious value. However, they can buy or build another type of expensive but perhaps less luxurious and appealing building that is still linked to their financial interests.

Nuclear Cathedrals

Nuclear power plants are one of those buildings that, until recently, rarely appeared on architecture websites or the covers of design magazines. However, they are one of the most expensive buildings in the world, although this has as much to do with the technology they contain as with the structure and envelope that contain this technology. In fact, about a third of the cost of a nuclear power plant involves paying the total design and administrative team behind its building (Potter, 2023). As a cost reference, the only plant under construction in the USA, Plant Vogtle (units 3 and 4) in Georgia, will cost US$34 billion if delays are controlled (Associated Press, 2022). Another nuclear power plant, the Hinkley Point C nuclear power station in the UK, in February 2023 was estimated to cost £32 billion (US$38.5 billion) (Nuclear Engineering International, 2023). These are costs greater than buying 20 Burj Khalifas. Despite the cost of nuclear power for many countries in the developed world, increasing its use is part of strategies for reducing the emissions from burning fossil fuels, thus aligning with SDG 7 Affordable and Clean Energy. How much nuclear power can be considered a renewable energy source is debatable since uranium has to be dug out of the ground and is finite. Given the decline in global use of nuclear energy, its ability to contribute significantly to reducing greenhouse gas emissions has also been questioned (Froggat & Scheinder, 2015). Nevertheless, this has not stopped the building of new nuclear power plants.

EDF, Électricité de France, is the company in charge of building two nuclear reactors in the UK at Hinkley Point and Sizewell. The French government mostly owns EDF. "Our purpose is to help Britain to achieve Net Zero" states the French company (EDF, 2022a:3). The road to zero-carbon proposed by EDF is based on promoting a combined strategy: developing "wind+nuclear+solar" (EDF, 2022b:6). The concept is that nuclear energy would provide a "firm low-carbon power option" (Committee on Climate Change, 2019:145) that would complement weather-dependent wind and solar and thus be the first step in transitioning to Net Zero electricity production. In this context, Net Zero means that electricity generation must not produce carbon dioxide emissions, so it does not account for the emissions that went into making the plant.

The energy dependency of the UK is around 38%, meaning that the country needs to import 38% of its energy from other countries to supply domestic demand. The world average per country is 10–15%, which makes the current situation of the UK concerning, even though the energy dependency has decreased from 47.6% in 2013 (Statista, 2021) and is much lower than the European average of 55.5% in 2021 (Eurostat, 2023). In 2022, since the boycott of Russian gas imports, the UK imported gas from Norway, and from the USA and Qatar as Liquified Natural Gas (LNG), although the UK also exported gas (Department for Energy Security and Net Zero, 2023). In 2017, 83% of UK households used gas for home heating (Foster, 2021). At that time, it seemed most UK citizens were doing a good job at consuming less gas and electricity, with median consumption decreasing from 2010. However, this achievement could not be attributed to all socio-economic levels of the population since gas was not evenly consumed across all households. Many people struggled to pay the bills and were under-consuming energy, especially young households and single parents with children, while high-income households owning and living in well-insulated big houses were over-consuming (Foster, 2021). This background is important in understanding who is responsible for the situation and who will benefit most from the energy produced by nuclear power plants in the UK.

As noted above, the EDF proposal to help Britain achieve Net Zero is to build more nuclear reactors and solar and wind farms with additional battery storage (EDF Renewables UK, 2023). The EDF Hinkley Point C nuclear power station has a planned capacity of 3.2 GW, as does the proposed plant at Sizewell. However, the operating date of Hinkley Point C has been rescheduled twice and now has a planned start of June 2027. The delays have led to an extra cost of £3 billion (Reuters, 2022). This is not surprising since most of the cost of supplying nuclear electricity is the upfront cost of building the power plant. Some 60–80% of the cost of electricity from a nuclear plant is due to the plant cost, whereas for electricity from natural gas, 70% comes from the fuel (Potter, 2023). Delays in construction occur in response to tightening regulations (Portugal-Pereira et al., 2018), and since money is borrowed to build a reactor, costs inevitably rise as interest on the loan has to be paid for longer.

EDF "recognises electricity from wind, solar and nuclear fuels produces [sic] zero carbon emission at the point of generation" (EDF, 2022b:4b). The misleading information is in the last part of the sentence "at the point of generation". What does it mean? It means EDF is not considering all the emissions generated by mining and acquiring the raw material, moving it, building the plant, decommissioning it, and treating, storing and disposing of nuclear waste. This is the limited-cycle-assessment version of a proper life cycle assessment. Lenzen (2008), from the University of Sydney, found that the most popular reactors, heavy water reactors (HWR) and light water reactors (LWR), with a 35 working year life, have an average greenhouse gas intensity of 65 g CO_2-e/kWh$_{el}$.

> While these greenhouse gases are expectedly lower than those of fossil technologies (typically 600–1200 g CO2-e/kWh$_{el}$), they are higher than reported figures for wind turbines and hydroelectricity (around 15–25 g CO2-e/kWh$_{el}$) and in the order of, or slightly lower than, solar photovoltaic or solar thermal power (around 90 g CO2-e/kWh$_{el}$). (Lenzen, 2008)

Lenzen's study is quite old and, for example, emissions from solar photovoltaic systems have dropped (Pincelli et al. 2024). However, the fact that nuclear plants are costly and that nuclear waste is always a problem should not obscure that emissions from nuclear electricity are less than those of electricity from fossil fuels and that governments feel they have a role to play in decarbonising developed societies. Consequently, most people will probably be happiest knowing that the plant is as accident proof as possible, whatever it looks like. Despite this, architects have been involved in nuclear power plant design, although Norman Foster and partners declined to design one for EDF (Rose, 2008). Commenced in 1958, Basil Spence and Partners were architectural consultants for the fourth nuclear power station to be built in the UK at Trawsfynydd in the Snowdonia National Park in North Wales (Fig. 2.1). The station began operations in 1965 and decommissioning started in 1991 and is expected to last 100 years. "The reactor buildings, which remain on the site, will be clad to become safestore buildings" (Historic Environment Scotland, n.d.), which will negate the idea put forward by Spence that he designed them to look good as a ruin (Rose, 2008).

The architectural consultants, if any, at Hinkley Point C remain unknown: "Like Chartres Cathedral, Hinkley Point C is being made by anonymous humans whose work is too rarely hymned. But unlike the former, it is not dedicated to the glory of God but to meeting British energy consumption, which is expected to double by 2050" (Jeffries, 2022). As a thought, ancient cathedrals used to have cemeteries and nuclear plants, as modern cathedrals dedicated to energy, are no exception. Nuclear power plants also require a place to bury their radioactive waste, but this is unlikely to become a tourist attraction like a cathedral cemetery. A tomb for nuclear waste is a complicated and extremely costly piece of infrastructure to build. However, if, as the UK intends, at least 25% of its electricity is to be nuclear, the amount of nuclear waste will increase and, with this, the need for storage space. The Nuclear Waste Services of the UK is promoting a new nuclear waste storage facility with the acronym GDF, which stands for Geological Disposal Facility. Essentially, this is a bunker for nuclear waste. Whereas the smallest village in the UK might hardly welcome a nuclear waste bunker, it might be keener to accept a Geological Disposal Facility: "Investment will flow into a community that hosts a GDF. There will be hundreds of well-paid jobs every year for over a century" (Nuclear

Fig. 2.1 Trawsfynydd nuclear power station. Source: Creative Commons (n.d.-a)

Waste Services, n.d.:15). A GDF will require digging a very deep underground facility made of vaults and tunnels up to 1 km below the earth's surface. The project is in its early stages, but it could cost as much as £53 billion (Vaughan, 2022), namely, the cost of building one or two nuclear power plants like Hinkley Point C. The GDF is planned to last 100 years from siting to closing, but what might happen if people find in 2080, 60 years after it was built, that it does not comply with the future standards anymore? In the meantime, the Nuclear Decommissioning Authority (NDA) estimates that the cost of decommissioning existing UK nuclear plants is around £132 billion (US$170 billion) and it will take 120 years (House of Commons Public Accounts Committee, 2020:5). However, these huge costs do not seem to have stopped the very rich turning to nuclear power, presumably as a way to become even richer.

Nuclear Plants and the Very Rich

There is an uncomfortable link between how the very rich make money and nuclear power. In theory, the top 10 billionaires could afford to buy more than one nuclear power plant just as an attraction for the backyard, but why would they want to own one? While some billionaires are already selling electric cars, others are thinking about the next business of producing more electricity to power these very cars. Bill Gates (ranked 4th on the Forbes list 2022) and Warren Buffett (ranked 6th on the Forbes list 2022) have joined forces to build a new type of small Natrium nuclear reactor that "will be a game-changer for the energy industry" in Gates's own words (Reuters, 2021). Bill Gates was the founder and founding investor of TerraPower, the company behind these new reactors, although TerraPower have since received millions from other investors. The US government are also part of funding the new Natrium reactor in Wyoming that is coupled with a storage facility (Freebairn, 2022).

> The Natrium sodium-cooled fast neutron reactor, co-developed with GE-Hitachi Nuclear Energy Inc., is backed by the U.S. Energy Department, which awarded an initial $80 million to the effort in 2020 and has committed to providing up to $2 billion for the project. TerraPower said the first unit is to begin operating in 2028 with total project costs, including design and a fuel fabrication facility investment, of around $4 billion.

The first reactor in Wyoming has a planned output of 345 MW of electricity though this can be lifted to 500 MW for a short time to meet peak demands (TerraPower, 2023). Hinckley Point C Light Water Reactor is rated at 3200 MW with an early 2023 estimated final build cost of US$40 billion (Cossins-Smith, 2023). Without the final built costs to go on, this would suggest the cost of investing in a MW of electricity from either plant will be approximately the same. Still, TerraPower (2023) only claims they will produce "carbon-free electricity at a competitive cost". How carbon-free it is if all the emissions of creating, running and disposing of the plant are accounted for remains to be seen. The one advantage of the small reactor is that there could be more of them to complement renewable wind and solar, and there has even been a proposal to have them on boats so they are off-shore, mobile and plugged into girds as required (Potter, 2023).

There have been plans for an even smaller 1.5 MW reactor designed in collaboration with the architectural firm of Gensler (see Chap. 3) to look like a ski lodge (Stamp, 2020) (Fig. 2.2). However, permission to build was declined by the US Nuclear Regulatory Committee because of insufficient evidence in the application to allow consideration of "key safety and design aspects" (USNRC, 2020).

Other Effects of the Super-Rich on the Built Environment

In medieval times, a church was built with the idea of housing God and for centuries this church would be one of the most prominent buildings in any town or city. In the twentieth century, churches were surpassed by skyscrapers as the tallest landmarks in the landscape. Owning an

Fig. 2.2 1.5 MW Aurora Nuclear Reactor. Based on (Stamp, 2020)

entire building that dominated the skyline became attractive to the wealthy, with examples including the Rockefeller Centre and Trump Tower. However, that is twentieth-century history. In the twenty-first century, what are billionaires and soon-to-be trillionaires building or planning to build?

Billionaire Urban Design

With the emergence of the super-rich, billionaires can buy things large enough to break any shopping cart. Among the most valuable real estate markets in the USA, are, for 2020 values, New York (US$2.8 trillion), Los Angeles (US$2.3 trillion) and San Francisco (US$1.3 trillion) (Real Estate Weekly, 2020). The top 15 richest men could buy San Francisco for themselves. The 2600 people in the Forbes list can buy all of them and still have money left over.

The inconvenience for the super-rich is that many big things like cities, towns and neighbourhoods usually belong to other people, institutions or governments. Even places with no houses, like rivers and mountains, cannot be bought if they belong to everyone (at least in theory). Famous old buildings also cannot be bought because they have an emotional or historical value, like the Pyramids of Egypt and the Eiffel Tower, or simply because they can make you unpopular on social media. However, it has not stopped people in the past from trying to buy them. Victor Lustig, recognised as America's greatest con man, sold the Eiffel Tower not once but twice in the 1920s (Howard, 1992).

Bigger Than a Building: Let's Buy a Country!

In James Bond movie, villains own big houses, but legendary villains such as Dr. Julius No in the first Bond movie of the same name, own islands. If you are interested and have the money, you can enter the website Private Island Inc. (2023), where several "private islands" are for sale. You can find good offers under US$500,000. However, it is not very clear how some low-lying examples will deal with sea-level rise.

Countries have bought land before. The USA bought Alaska from Russia in 1867 for US$7.2 million (Library of Congress, 2022), approximately US$150 million in 2022, so well within reach of a current billionaire. The USA tried buying land in recent times. In 2019, Donald Trump was interested in buying Greenland (Pilkington, 2019), the biggest island in the world and home to 56,000 people (Denmark.dk, n.d.). Greenland is believed to have considerable offshore oil and gas reserves (Associated Press, 2021). At first, the Danish government disbelieved Trump's suggestion thinking it was an April Fool's joke even though it was not April, but then Trump's level of intelligence is well known. The foreign minister of Greenland needed to remind the then-president of the USA: "Greenland is obviously not for sale" (Kennedy, 2019). When someone can buy bigger things than anything available on the market, none seems big enough, so the obvious next step is to create something for yourself.

New Cities

If buying New York is impossible, some designers can draw a city for you. This is the case of BIG, Bjarke Ingels Group (see Chap. 3), the international architectural firm partnering with real estate developers, engineers and other consultants, all led by e-commerce billionaire Marc Lore. This consortium developed a concept to build cities for five million people in desert locations somewhere in Utah, Nevada, Idaho, Arizona, Texas and the Appalachians. The project is named Telosa and intends to be a model of making cities that are better socially, economically and environmentally (Telosa, 2023) (Fig. 2.3). The idea is committed to everything politically correct and thus sounds positive in the media, from not allowing fossil fuel vehicles, minimising the heat-island effect, having zero-emissions buildings and reducing water consumption, to providing good jobs and economic growth that will be equitable. Each city is supposed to be built progressively. The first phase would need 1500 acres for 50,000 people at a cost of US$25 billion, and they promise that it will be ready in 2030. The total budget is estimated to be more than US$400 billion to house 5 million people in 40 years (by around 2070) (Telosa, 2023).

The concept of Telosa has already received a fair amount of criticism in the press (Crispin, 2021).

> Telosa certainly is a city of the future, but not in like a great way. Yes, it probably will have a very shiny public transportation system, but it seems futuristic more in the sense that, as the world deteriorates, the ultra-rich seem increasingly interested in telling the rest of us how to live. No longer content to just sneer down at us from their private jets, they take over our homes, our towns, our society.

Apart from not yet having a site, it is unclear who will own the city, why people would be attracted to live there, what the available jobs in the equitable economy might be, how expensive it would be to own or rent a house there, and how the new city will deal with state policies. On the website, the group has stated that the city will be for everyone and that they are "absolutely not attempting to create a utopia", which they define as trying to create a "perfect, idealistic state" but instead they are committed to new ideas for solving the difficult problems found in other cities (Telosa, 2023). Nevertheless, "In Utah, there's long been a preoccupation with building the promised land, or what some call Utopia" (Alsever, 2022).

The choice of Bjarke Ingels Group, the architectural firm in charge of the master plan, might not be a coincidence. Bjarke Ingels, in an interview with Time magazine, shared his idea of designing a master plan for the world to fix climate change (Nugent, 2020). The idea was called Masterplanet, and the intention was to "prove that a sustainable human presence on planet Earth is attainable with existing technologies" (Block, 2020). One of these ideas, presumably based on existing technology, was to ship stone flour made from ground-up rocks in four containers on a container ship and "inject some when you cross a marine desert" to enrich the soil, stimulate the development of flora and fauna and, in this way, reduce the greenhouse effect (Nugent, 2020). Maybe this and other enlightened ideas seduced the developer of Telosa and potential investors.

Fig. 2.3 The approach to one of the new cities in the desert. (Based on Telosa (2023))

Building cities in the desert is not a new idea, others have tried before and failed. Why these desert zones are not usually inhabited should be obvious: insufficient water and sometimes excessive temperature swings. Irregular and infrequent rain is why such areas are denominated as extreme environments. People can survive for some days without food but only a few days without water. Communities have traditionally found ways of living in the desert. Still, they were nomadic, and when they were in settlements, these were generally small, modest and compact, with no self-driving cars and flying taxis, as found in the Telosa illustrations.

Whether Telosa cities ever appear, the idea that the very rich can control human destiny like governments did in the past by creating new towns and cities is somewhat disturbing. This issue can be illustrated on a smaller scale. Two wealthy philanthropists set a foundation to raise money to support the ailing town of Kalamazoo in Michigan. They also controlled how the foundation money was to be spent. Semuels (2021) points out the potential problems of this.

> There's a long history in the US of the rich stepping in to fund cultural amenities like museums, but lately, they've started stepping in to fund projects—in Kalamazoo and elsewhere—that have long been perceived as the government's responsibility. It's a scenario that critics say sets the stage for the super wealthy to control more and more aspects of public life.

It is interesting to observe that it is more feasible for an individual to buy a city in the desert, given that money is no object, than for a country to extend its territory by having to go to war, the cost of which falls on the taxpayer (see Chap. 5). The super-rich are economically bigger than countries and can acquire land the size of towns with more autonomy. Perhaps they should be added to the list of countries, as a Republic of themselves.

Lines, Lies, and Dreams

The desert has proved to be an incredibly attractive scenario for wealthy investors making big promises as found in Telosa, but Saudi Arabia Crown Prince Mohammed bin Salman (MBS) has taken this approach to the next level. In 2017, MBS announced plans to build a city in the desert, or to be more precise, in the province of Tabuk, in the northwest of Saudi Arabia, between Egypt and the south of Jordan, and with access to the Red Sea. The new cluster of four cities, Neom, was planned to be a little bit of everything: smart, sustainable, touristic and a business centre, among a long list of goals. In the first illustrations of Neom, presented in 2020, one city looked Manhattan-inspired with an assembly of dense skyscrapers on reclaimed land, connected to the shore by many bridges, and surrounded by lower buildings following semi-circular street patterns (Saudi Gazette, 2018). However, this was not the project that caught the attention of the media worldwide. In 2021, The Line that is also part of the Neom project was unveiled (and later exhibited in August 2022) as a linear city. The city was proposed as a linear continuous building originally of 170 km that was to have an inner artificial canyon sandwiched by two 500 m-high mirrored walls, with a total area the size of Belgium, and with an estimated cost of $1 trillion (Al Jazeera, 2022) (Fig. 2.4). The city would only be 200 m wide which questions the quality of the spaces that can be potentially created in the interior. Currently under construction the idea behind the now reduced 112 km line is that it will be divided into 140 modules each 800 m long and housing 80,000 people (Ravenscroft, 2023).

The various websites describing the project are full of sentences that try to define what Neom is or does without really saying how such a city can flourish in economic terms (Neom, 2023a). How will the people earn the money to live? What sort of jobs will they do? There do not seem to be any factories in The Line. Can a city exist without making things or engaging in some form of trade? The Neom project is promising everything and more and has been deeply criticised for this for good reasons. However, all projects of this kind are making the same promises, including redefining sustainability by using Artificial Intelligence to make meeting the Sustainable Development Goals easier (International Institute for Sustainable Development, 2023). When MBS first announced the project, it was said that the city would have no cars, no streets, and zero carbon emissions. It was to be run on 100% renewable energy, providing a pollution-free and healthy environment for residents displaced from their original homes by sea-level rise (Stuff, 2021) (Fig. 2.5)

The Neom project links to the Saudi Vision 2030, which expresses the country's aims to reach net zero by 2060 (Vision 2030, 2023) despite the fact the economy of Saudi Arabia is heavily dependent on oil exports. One of the multiple intentions of Neom is to diversify the economy and encourage investments and the development of the private sector. The vision and all its projects are supported by the Saudi Public Investment Fund (PIF), one of the largest sovereign funds in the world with estimated assets of around US$600 billion (Elimam and Al Sayegh, 2023). A sovereign fund is a way of investing state assets, oil and gas revenues in the Saudi case, in global business to maximize long-term investment returns. Since 2015, the PIF of Saudi Arabia has been controlled by Crown Prince MBS, not without international concern as The Wall Street Journal has called it the "world's least transparent" wealth fund (Clark, 2016). Controversies about the funding and the people involved do not stop there. In 2022, "The Biden administration asked a US judge for a 60-day extension before it formally weighed in on whether Mohammed bin Salman, the Saudi crown prince, ought to be granted sovereign immunity in a case involving the murder of Washington Post columnist Jamal Khashoggi" (Kirchgaessner, 2022). This call was made after Biden met MBS in Saudi Arabia. What is surprising is that Biden, during his campaign, promised

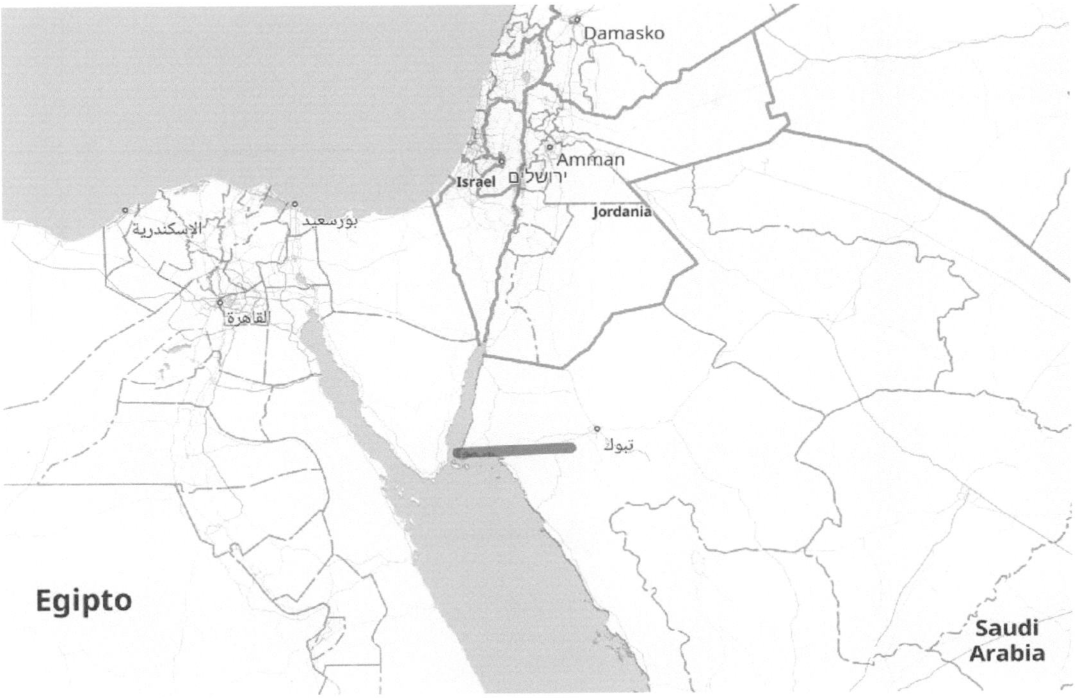

Fig. 2.4 The Line. (Source: Creative Commons (n.d.-b))

to treat Saudi Arabia as a pariah but then campaign promises are not perhaps the same as real ones.

The Neom website is plagued with many sentences and video clips that increase the feeling of fragmentation. There are no plans, measurements or analysis of how zero carbon will be achieved. “You cannot build a 500-metre-tall building out of low-carbon materials”, noted Philip Oldfield, head of the School of the Built Environment at the University of New South Wales. He further states, “This would require a phenomenal quantity of steel, glass and concrete” and estimates that emissions from building The Line would equate to more than the entire UK emissions for 4 years and that “this vast embodied carbon will overwhelm any environmental benefits that a small footprint provides” (Barker, 2022).

These super ambitious projects that appear to be created to build the brand of a business rather than a city, have something in common in that they use the same logic to build ghost businesses in the name of solving environmental problems. They promise people a profit from something that they do not have. The energy invested in creating and using new technologies to transition toward reduced environmental impacts is never counted in the final emissions of the project, only the potential benefits of something not yet built. Turning desert land into valuable real estate, however, can only increase the wealth of those involved, so the building of Neom has begun and at the start of 2023 a video of the works underway was made available, with an opening date of 2024 for the first Neom city. This will be the luxury resort city of Sindalah on an island in the Red Sea, presumably as a way of making money from tourist dollars. Elsewhere the video shows diggers in the desert but not much more (Neom, 2023b).

It is not a coincidence that utopian projects for cities in the desert and on the ocean are becoming attractive to billionaires and other powerful characters. In the financial capitals of the world, land is scarce and extremely expensive. Every inch has been sold and bought countless times. Many

Fig. 2.5 Impression of living between the mirrored walls. (Based on Ravenscroft (2023))

people want to live in, and many firms want to have a presence in, major cities like New York, San Francisco, London and Paris. However, deserts and oceans represent potential new real estate. From this point of view, climate change and responses to it, like creating and selling new zero-carbon cities, could benefit this powerful minority. Although Neom was presented as a place where no one lived at the launch of the project, there are Saudi communities, like the Huwaitat tribe, that have been living in the area for centuries. They have been progressively evicted, and armed confrontations have happened in which one of the activists was killed. "For the Huwaitat tribe, Neom is being built on our blood, on our bones", says Alia Hayel Aboutiyah al-Huwaiti, an outspoken activist and member of the tribe living in London. "It's definitely not for the people already living there! It's for tourists, people with money. But not for the original people living there" (Michaelson, 2020).

Utilitarian Buildings

The very rich are not only involved in building new cities. You cannot build electric cars to mitigate climate change without making a factory in which to build them. Elon Musk has five what he terms gigafactories in "Fremont, California; Sparks, Nevada; Berlin, Germany; Shanghai, China; Austin, Texas; and Buffalo, New York" for making cars and batteries and with plans to build a sixth in New Mexico (Rapier, 2023). Such large factories do not come without an environmental impact. For the German factory, expansion problems were delayed because of fear there would be insufficient water for both the factory and the surrounding settlements. Locals have also complained "there was not sufficient community input before the sprawling site was approved" (Allyn, 2022). Musk through Tesla is keen on the expansion into Germany in order to meet the company's European growth targets and presumably make money in doing this. The gigafactory in Austin has also run into criticism, being placed in a community east of the I-35 highway that was deprived and already home to industries not wanted in the more prosperous areas of the city (Chen, 2022) (see Chap. 4 for a similar discussion of the factory development of Henry Ford). There are worries that the presence of the factory has raised real estate values which could lead to gentrification and the loss of affordable housing for the residents who would have nowhere else to go. Tesla also organized a clean water supply for the factory whereas some residents in the area are paying for private supplies which are all but clean. However, the biggest fear from the new gigafactory is flooding (Chen, 2022):

> The massive Gigafactory spreads over 2000 acres of cleared swampland. The concrete that has replaced formerly permeable surfaces has effectively removed flood buffers in an area that is next to the Colorado River. And with climate change increasing the odds of heavy rain, that makes the risks of runoff and flooding more serious than ever before.

A similar story has been told for the gigafactory in Shanghai. This was built on a reclaimed swamp in a former agricultural area and has also seen homelessness as a result of the rise in real estate prices since the coming for the plant. Overall, the "Gigafactory and surrounding buildings contribute to the UHI [urban heat island] effect, decrease natural vegetation, increase warming temperatures, and contribute to pollution" (Kenig-Ziesler et al., 2021).

A similar problem of unlooked for impact has been created by Amazon warehouses, noting that in 2023 Amazon chairman Jeff Bezos was the third wealthiest person in the world with a fortune of over US$160 billion (Forbes, 2023). In a Consumer Report study of the Red Hook neighbourhood of Brooklyn, which currently has two large Amazon warehouses and a third planned, problems with air quality and increased traffic were revealed. The report stated: "...facilities like Amazon's Red Hook operations are very different. They're massive. So far, they make up nearly half a million square feet of warehousing and parking space, which if laid flat, would fill six and a half football fields. And they generate a steady stream of delivery trucks and vans" (Jaff, 2023). The report also notes "These are multi-million-dollar companies—even trillion-dollar companies—that are building these facilities, and they need to monitor and mitigate their impact on their neighbors". Similar objections have been raised to Amazon warehouses in California, especially as they are being built on prime farmland (Asher-Schapiro, 2022).

The issue seems to be that the very wealthy easily lose sight of the effect they are having on the poor and lowly when making money is involved. It is very easy to be philanthropic and give money to have a college or art gallery named after you for posterity, but harder to put yourself in the position of someone struggling to make ends meet who suddenly finds their home perpetually shaken by a steady stream of delivery trucks ensuring same or next day delivery of something that is probably not a human necessity like food, fresh water or unpolluted air.

Who Are the Wealthy?

The environmental impact of the super-rich was discussed at the start of this chapter. What should not be forgotten is what it means to be not super-rich but just rich. You are rich if you own a boat, even a small one. If you own a car, you are rich. If you own, or are even in the throes of buying, a house, you are rich, at least compared to those who live on or just above the poverty line. In the USA, home of the first and third richest men in the world, "the poverty rate has remained between 11 and 15 percent for decades. Millions more live just above the poverty line" (Howard et al., 2017). For each country, the poverty line is defined as an income below which people find it difficult to provide basic requirements such as food and housing. This is what the SDGs are attempting to counter but unless the issue of who has wealth is tackled it is hard to see how they can achieve this within the resource limits of a finite planet. Additional is the fact that globally human society is becoming richer as "wealth per adult had grown from US$31,380 to 87,489 over the period 2000-21, an average growth rate of 5.0% per annum" (Shorrocks et al., 2022:18–19). The same document also notes that in the same period in the world 2.8 billion adults had annual wealth of only US$10,000 (Shorrocks et al., 2022:2). Achieving the SDGs will mean a curb on how rich a person can become. There is something wrong in human society where one person can accumulate so much wealth whilst others live in poverty.

The other issue with being rich is that the current climate crisis is caused by the wealthy, not the poor. The poor cannot afford foods with a high carbon footprint, nor can they afford to drive far, or take holidays, or fly. Many cannot afford adequate heating or cooling. Yet the poor will suffer the effects of climate change more than the rich. According to the OECD (2022), fragile environments are those exposed to risks that the state or community system do not have the resources to manage. In 2022, fragile environments housed 24% of the global population but 73% of those living in extreme poverty, and the latter figure is predicted to increase to 86% by 2030. It is these same fragile environments that will be most affected by climate change. The implication is that anyone with the resources to do so should not live beyond their carbon footprint, so as to mitigate global warming as quickly and effectively as possible. However, what has been observed is the fact the wealthy in both developed and developing countries are not making efforts to curb their carbon footprints. Those who can afford it in developing countries are adopting westernised lifestyles (Zheng et al., 2023), which have become aspirational. Such lifestyles demand energy to support them and very expensive nuclear power plants are seen as a way of supplying this energy but with a lower carbon footprint. Electric cars and nuclear power plants will not, together, solve climate change. How people live in the developed world has to change so that walking, cycling and public transport become the norm. The problem is that no-one has yet become rich from promoting walking and cycling. The proposed cities in the desert do not seem to be for those living at or below the poverty line, and yet, as shown above, these are the people living in fragile environments who will be most affected by climate change. Will the new cities in the desert really be places for those whose fragile environments are destroyed by climate change?

The big problem with the super-rich is that they set aspirational goals that make people dissatisfied with what they have and leave them wanting more. This also affects the design of buildings. Luxury liners, large cars and aeroplanes were all admired by Le Corbusier in *Towards a New Architecture* as perfections in machines that he failed to see in most buildings, both old and contemporary. Through the eyes of climate change, these same standards of machine perfection that have influenced architecture are things that the people must stem the use of to avoid a very changed world in the future. Does not this suggest that the design of buildings and the built environment also needs to change?

References

Al Jazeera. (2022). *Saudi Arabia plans $1 trillion mirrored skyscraper in Neom.* Retrieved from https://www.aljazeera.com/news/2022/7/25/saudi-arabia-to-build-1tr-mirrored-skyscraper-in-neom, 12 September 2023.

Al Lawati, A. (2022). *The Hajj is back and Saudi Arabia is hoping to cash in.* Retrieved from https://edition.cnn.com/2022/07/06/business/saudi-hajj-economy-mime-intl/index.html, 14 September 2022.

Alacevich, M., & Soci, A. (2018). *Inequality: A short history.* Brookings Institution Press.

Alcorn, C. (2019). *How Michael Bloomberg became the richest man in media.* Retrieved from https://edition.cnn.com/2019/11/08/business/bloomberg-wealth/index.html, 25 March 2024.

Allyn, B. (2022). *Tesla's first European factory needs more water to expand. Drought stands in its way.* Retrieved from https://www.npr.org/2022/11/03/1131695382/

tesla-ev-electric-vehicles-europe-germany-drought-climate-change-factory, 13 September 2023.

Alsever, J. (2022). *The futuristic city of Telosa might be coming to Utah*. Retrieved from https://www.utahbusiness.com/telosa-utah-might-be-a-thing/ , 4 July 2022.

American Experience. (n.d.). *The Rockefellers*. Retrieved from https://www.pbs.org/wgbh/americanexperience/features/rockefellers-timeline/, 1 November 2023.

Asher-Schapiro, A. (2022). *World's biggest Amazon warehouse raises fear over toxic air.* Retrieved from https://www.context.news/big-tech/worlds-biggest-amazon-warehouse-raises-fears-over-toxic-air, 13 September 2023.

Associated Press. (2021). *Greenland suspends oil exploration because of climate change*. Retrieved from https://www.pbs.org/newshour/science/greenland-suspends-oil-exploration-because-of-climate-change#:~:text=The%20U.S.%20Geological%20Survey%20estimates,harsh%20weather%20have%2-0limited%20exploration, 11 September 2023.

Associated Press. (2022). *Georgia nuclear plant's cost now forecast to top $30 billion*. Retrieved from https://www.usnews.com/news/best-states/florida/articles/2022-05-08/georgia-nuclear-plants-cost-now-forecast-to-top-30-billion, 14 September 2022.

Bachelor, M. (2022). *Who will become the world's first trillionaire?* Retrieved from https://www.theceomagazine.com/business/finance/worlds-first-trillionaire/, 6 September 2023.

Banerjee, A. V., & Duflo, E. (2019). *Good economics for hard times* (1st ed.). Penguin Books.

Barker, N. (2022). *Sustainability and liveability claims of Saudi 170-kilometre city are "naive" say experts*. Retrieved from https://www.dezeen.com/2022/08/08/sustainability-liveability-the-line-saudi-170km-city-naive/?li_source=LI&li_medium=bottom_block_1, 13 September 2023.

Binayah Properties. (n.d.). *Get a luxury apartment in world famous building Burj Khalifa*. Retrieved from https://www.binayah.com/get-a-luxury-apartment-in-world-famous-spot-burj-khalifa/#:~:text=Question,2%20million%20(USD%20544%2C000), 1 November 2023.

Block, I. (2020). *Masterplanet is Bjarke Ingels' plan to redesign Earth and stop climate change*. Retrieved from https://www.dezeen.com/2020/10/27/bjarke-ingels-big-masterplanet-climate-change-architecture-news/, July 20, 2022.

Boushey, H., De Long, J. B., & Steinbaum, M. (2017). *After Piketty: The agenda for economics and inequality*. Harvard University Press.

Britannica, The Editors of Encyclopaedia. (2022). *Great Mosque of Mecca. Encyclopedia Britannica*. Retrieved from https://www.britannica.com/topic/Great-Mosque-of-Mecca, 14 September 2022.

Britannica, The Editors of Encyclopaedia. (2023). *Nero*. Retrieved from https://www.britannica.com/biography/Nero-Roman-emperor, 26 October 2023.

Burjkhalifa (2022). Construction. Retrieved from https://www.burjkhalifa.ae/the-tower/architecture-design/, 4 January 2025.

Chen, A.X. (2022). *Tesla's Austin Gigafactory could be an environmental disaster waiting to happen*. Retrieved from https://www.protocol.com/climate/tesla-austin-gigafactory-water-flooding, 13 September 2023.

Clark, S. (2016). *Saudi Wealth Fund may be world's least transparent*. Retrieved from https://www.wsj.com/articles/saudi-wealth-fund-may-be-worlds-least-transparent-1477997912, 13 September 2023.

Committee on Climate Change. (2019). *Net Zero: The UK's contribution to stopping global warming*. Committee on Climate Change.

Cossins-Smith, A. (2023). *Cost of EDF's Hinkley Point C nuclear project rises to $40bn*. Retrieved from https://www.power-technology.com/news/hinkley-point-c-project-costs-rise-again/?cf-view, 1 November 2023.

Cozzi, L., Chen, O., & Kim, H. (2023). *CO_2 emissions are highly unequal across income groups and regions*. Retrieved from https://www.iea.org/commentaries/the-world-s-top-1-of-emitters-produce-over-1000-times-more-co2-than-the-bottom-1, 4 September 2023.

Creative Commons. (n.d.-a). "Trawsfynydd Nuclear Power Plant" by Original uploader was William M. Connolley at en.wikipedia is licensed under CC BY-SA 3.0.

Creative Commons. (n.d.-b). "The Line context" by Luistxo is licensed under CC BY-SA 4.0.

Crispin, J. (2021). *A billionaire wants to build a utopia in the US desert. Seems like this could go wrong*. Retrieved from https://www.theguardian.com/commentisfree/2021/sep/20/a-billionaire-wants-to-build-a-utopia-in-the-us-desert-seems-like-this-could-go-wrong, 4 July 2022.

Curzer, H. J. (2012). *Aristotle and the virtues*. Oxford University Press.

Denmark.dk. (n.d.). *Greenland: The world's largest island*. Retrieved from https://denmark.dk/people-and-culture/greenland, 11 September 2023.

Department for Energy Security and Net Zero. (2023). *Digest of UK Energy Statistics Annual data for UK, 2022*. Retrieved from https://assets.publishing.service.gov.uk/government/uploads/system/uploads/attachment_data/file/1182229/DUKES_2023_Chapters_1-7.pdf, 1 November 2023.

Dolan, K. A., & Peterson-Withorn, C. (2022). *Forbes world's billionaires list: The richest in 2022*. Retrieved from https://www.forbes.com/billionaires-2022/, 6 September 2023.

Dunsheath, F. (2022). *Qatar FIFA World Cup controversy: What you need to know.* Retrieved from https://www.rnz.co.nz/news/what-you-need-to-know/480588/qatar-fifa-world-cup-controversy-what-you-need-to-know, 20 October 2023.

EDF. (2022a). *Sustainable business update. What we're doing today, tomorrow and in the coming years to help Britain Achieve Net Zero*. Retrieved from https://www.

edfenergy.com/sites/default/files/edf-sustainable-business-update-160922.pdf, 20 September 2022.

EDF. (2022b). *About Hinkley Point C. What is Hinkley Point C?* Retrieved from https://www.edfenergy.com/energy/nuclear-new-build-projects/hinkley-point-c/about, 14 September 2022.

EDF Renewables UK. (2023). *EDF Renewables UK and Ireland.* Retrieved from https://www.edf-re.uk/, 24 October 2023.

Elimam, A., & Al Sayegh, H. (2023). *Saudi Arabian Public Investment Fund says assets surpasses 2.23 trillion riyals in 2022.* Retrieved from https://www.reuters.com/world/middle-east/saudi-public-investment-fund-managed-assets-surpasses-223-trillion-saudi-riyals-2023-08-06/#:~:text=DUBAI%2C%20Aug%206%20(Reuters),1.98%20trillion%20riyals%20for%202021. 13 September 2023.

Emporis. (2022). *The most expensive buildings in the world.* Retrieved from https://www.emporis.com/statistics/most-expensive-buildings?utm_medium=website&utm_source=archdaily.com, 14 September 2022.

Eurostat. (2023). *Energy statistics—An overview.* Retrieved from https://ec.europa.eu/eurostat/statistics-explained/index.php?title=Energy_statistics_-_an_overview#:~:text=EU%20energy%20import%20dependency%20rate%20stood%20at%2055.5%25%20in%202021.&text=Gross%20available%20energy%20in%20the,by%206.0%25%20compared%20with%202020, 7 September 2023.

Fielding, J., & Williams, D. (2023). *Planners in Australia: State of the profession report.* Planning Institue of Australia.

Fleck, A. (2023). *The number of millionaires has skyrocketed.* Retrieved from https://www.statista.com/chart/30671/number-of-millionaires-and-share-of-the-population/, 1 November 2023.

Forbes. (2023). *The world's realtime billionaires.* Retrieved from https://www.forbes.com/real-time-billionaires/#4eae2f843d78, 13 September 2023.

Foster, H. (2021). *Energy follow up survey: Household energy consumption & affordability.* Retrieved from https://assets.publishing.service.gov.uk/government/uploads/system/uploads/attachment_data/file/1018725/efus-Household-Energy-Consumption-Affordability.pdf , 22 September 2022.

Freebairn, B. (2022). *PacificCorp, Terrapower study deployment of five more Natrium nuclear reactors.* Retrieved from https://go.openathens.net/redirector/wgtn.ac.nz?url=https://www.proquest.com/trade-journals/pacificorp-terrapower-study-deployment-five-more/docview/2730882484/se-2, 11 September 2023.

Froggart, A., & Schneider, M. (2015). Nuclear power versus renewable energy—A trend analysis. *Proceedings of the IEEE, 103*(4), 487–490.

Garcia, E. Vale, B. (2017). *Unravelling sustainability and resilience in the built environment.* London; New York: Routledge, Taylor & Francis Group. 2017.

Garcia, E. (2022). How cities can be resilient. In R. C. Brears (Ed.), *The Palgrave encyclopedia of urban and regional futures* (pp. 859–880). Palgrave Macmillan.

Goh, A. (2023). *When was Marina Bay Sands built?* Retrieved from https://bestofsingapore.co/when-was-marina-bay-sands-built/#, 6 September 2023.

Gorman, M. (2018). *Thomas Jefferson might have been our most financially challenged founding father.* Retrieved from https://www.forbes.com/sites/megangorman/2018/07/03/thomas-jefferson-might-have-been-our-most-financially-challenged-founding-father/?sh=3db5ac642212, 26 October 2023.

Green, B. (2019). *Rising number of local enrolments should keep UK profession well supplied.* Retrieved from https://www.ribaj.com/intelligence/market-analysis-architecture-students-international-gender-grade-inflation-regional-balance-brian-green. 1 November 2023.

Guinness World Records. (2023a). *Richest person (ever) - Adjusted for inflation.* Retrieved from https://www.guinnessworldrecords.com/world-records/74539-richest-person-ever-%E2%80%93-adjusted-for-inflation, 4 September 2023.

Guinness World Records. (2023b). *Burj Khalifa: The tallest building in the world.* Retrieved from https://www.guinnessworldrecords.com/records/hall-of-fame/burj-khalifa-tallest-building-in-the-world, 1 November 2023.

Higgins, S., & Lustig, N. (2016). Can a poverty-reducing and progressive tax and transfer system hurt the poor. *Journal of Development Economics, 122*, 63–75.

Historic Environment Scotland. (n.d.). *Trawsfynydd Nuclear Power Station.* Retrieved from https://canmore.org.uk/site/286966/trawsfynydd-nuclear-power-station, 8 September 2023.

House of Commons Public Accounts Committee. (2020). *The Nuclear Decommissioning Authority's management of the Magnox contract.* Retrieved from https://publications.parliament.uk/pa/cm5801/cmselect/cmpubacc/653/653.pdf, 21 September 2023.

Howard, N. (1992). King Con. *Dun and Bradstreet, Inc.D & B Reports, 40*(3), 44.

Howard, C., Freeman, A., Wilson, A., & Brown, E. (2017). Poverty. *Public Opinion Quarterly, 81*(3), 679–689.

International Institute for Sustainable Development (IISD). (2023). *Event highlights potential of Artificial Intelligence in enabling SDGs.* Retrieved from dg.iisd.org/news/event-highlights-potential-of-artificial-intelligence-in-enabling-sdgs/#:~:text=Vinuesa%20highlighted%20that%20AI%20was,potential%20positive%20effects%20of%20AI. 12 September 2023.

Jaff, D. (2023). *New investigation shines a light on massive amazon warehouses and the strain on one small neighborhood.* Retrieved from https://advocacy.consumerreports.org/press_release/new-investigation-shines-a-light-on-massive-amazon-warehouses-and-the-strain-on-one-small-neighborhood/, 13 September 2023.

Jeffries, S. (2022) *Building Britain's Biggest Nuclear Power Station review—So boring it's a masterpiece!*

Retrieved from https://www.theguardian.com/tv-and-radio/2021/jun/02/building-britains-biggest-nuclear-power-station-review-so-boring-its-a-masterpiece, 14 September 2022.

Kaiser, J-A. (2011). *Santos Rosewood.* Retrieved from https://www.woodworkingnetwork.com/wood/wood-explorer/santos-rosewood, 6 September 2023.

KAYAK. (2023). *Armani Hotel Dubai.* Retrieved from https://www.nz.kayak.com/Dubai-Hotels-Armani-Hotel-Dubai.333333.ksp, 1 November 2023.

Kenig-Ziesler, L., Dixon, K., Xi, Y., & Labruna, N. (2021). *A comparison of Tesla Gigafactories.* Retrieved from https://storymaps.arcgis.com/stories/f40a028932d747c69a0312b71d6edde3, 13 September 2023.

Kennedy, R. (2019). *Trump ridiculed over plan to buy Greenland - 'Final proof that he has gone mad'.* Retrieved from https://news.sky.com/story/trump-ridiculed-over-plan-to-buy-greenland-final-proof-that-he-has-gone-mad-11786395, 11 September 2023.

Kirchgaessner, S. (2022). *White House seeks to delay decision on Prince Mohammed immunity over Khashoggi murder.* Retrieved from https://www.theguardian.com/us-news/2022/jul/18/prince-mohammed-jamal-khashoggi-immunity-white-house, 13 September 2023.

Kraus, M. W., Piff, P. K., Mendoza-Denton, R., Rheinschmidt, M. L., & Keltner, D. (2012). Social class, solipsism, and contextualism: How the rich are different from the poor. *Psychological Review, 119*(3), 546–572.

Kumar, R. (2019). *The business of changing the world: How billionaires, tech disrupters, and social entrepreneurs are transforming the global aid industry.* Beacon Press.

Lenzen, M. (2008). Life cycle energy and greenhouse gas emissions of nuclear energy: A review. *Energy Conversion and Management, 49*(8), 2178–2199.

Library of Congress. (2022). *Treaty with Russia for the purchase of Alaska.* Retrieved from https://www.loc.gov/rr/program//bib/ourdocs/alaska.html#:~:text=On%20March%2030%2C%201867%2C%20the,United%20States%20Edouard%20de%20Stoeckl, 11 September 2023.

Mastercard. (2019). *Mastercard Global Destination Cities Index. Bangkok tops Mastercard's Global Destination Cities Index for the fourth consecutive year.* Retrieved from https://www.mastercard.com/news/insights/2019/global-destination-cities-index-2019/, 14 September 2022.

McCarthy, N. (2020). *The world's most expensive buildings.* Retrieved from https://www.statista.com/chart/19852/estimated-cost-of-the-most-expensive-buildings-on-the-planet/, 14 September 2022.

Mecca. (2022a). *Mecca facts.* Retrieved from https://mecca.net/mecca-facts/, 14 September 2022.

Mecca. (2022b). *Home.* Retrieved from https://mecca.net,14 September 2022.

Messall, R. (2004). The long road of eugenics: From Rockefeller to Roe v. Wade. *The Human Life Review, 30*(4), 33–74.

Michaelson, R. (2020). *'It's being built on our blood': The true cost of Saudi Arabia's $500bn megacity.* Retrieved from https://www.theguardian.com/global-development/2020/may/04/its-being-built-on-our-blood-the-true-cost-of-saudi-arabia-5bn-mega-city-neom, 13 September 2023.

Neom. (2023a). *What is Neom.* Retrieved from https://www.neom.com/en-us/about, 1 November 2023.

Neom. (2023b). *Neom releases first progress film, showing construction so far—up to January 2023.* Retrieved from https://www.neom.com/en-us/newsroom/neom-progress-film, 13 September 2023.

Nuclear Engineering International. (2023). *EDF increases cost estimates for Hinkley Point C.* Retrieved from https://www.neimagazine.com/news/newsedf-increases-cost-estimates-for-hinkley-point-c-10612738, 1 November 2023.

Nuclear Waste Services. (n.d.). *Introduction to geological disposal.* Retrieved from https://assets.publishing.service.gov.uk/government/uploads/system/uploads/attachment_data/file/1057183/Introduction_to_Geological_Disposal.pdf, 21 September 2023.

Nugent, C. (2020). *The climate is breaking down. Architect Bjarke Ingels has a masterplan for that.* Retrieved from https://time.com/collection/great-reset/5900743/bjarke-ingels-climate-change-architecture/, 20 July 2022.

OECD. (2022). *States of fragility 2022.* OECD Publishing.

Oxfam. (2022). *Carbon Billionaires.* OxfamGB.

Oxfam International. (2023). *Richest 1% bag nearly twice as much wealth as the rest of the world put together over the past two years.* Retrieved from https://www.oxfam.org/en/press-releases/richest-1-bag-nearly-twice-much-wealth-rest-world-put-together-over-past-two-years#:~:text=Billionaire%20fortunes%20have%20increased%20by,rising%20food%20and%20energy%20profits. 1 November 2023.

Parada-Contzen, P., & Parada-Daza, J. R. (2023). On the weighting of homo economicus and homo virtus in human behaviour. *Humanities and Social Sciences Communications, 10*(1)., 13pp.

Parker, J. (2023). *How many architecture students graduate each year?* Retrieved from https://www.architecturemaker.com/how-many-architecture-students-graduate-each-year/. 1 November 2023.

Piketty, T. (2014). *Capital in the twenty-first century* [Capital au XXIe siècle.]. The Belknap Press of Harvard University Press.

Pilkington, E. (2019). *Donald Trump reportedly wants to purchase Greenland from Denmark.* Retrieved from https://www.theguardian.com/us-news/2019/aug/15/donald-trump-greenland-purchase-denmark, 11 September 2023.

Pincelli, I.P., Hinkley, J. and Brent, A.C. (2024). Life cycle assessment of a virtual power plant: Evaluating the environmental performance of a system utilising solar photovoltaic generation and batteries. *Renewable Energies* 2. 1–17.

Popescu, A.-L., & Luca, O. (2017). Built environment and climate change. *Theoretical and Empirical Research in Urban Management, 12*(4), 52–66.

Portugal-Pereira, J., Ferreira, P., Cunha, J., Szklo, A., Schaeffer, R., & Araújo, M. (2018). Better late than never, but never late is better: Risk assessment of nuclear power construction projects. *Energy Policy, 120*, 158–166.

Potter, B. (2023). *What does it cost to build nuclear power?* Retrieved from https://progress.institute/nuclear-power-plant-construction-costs/, 7 September 2023.

Private Island Inc. (2023). *Find your private Island.* Retrieved from https://www.privateislandsonline.com/, 11 September 2023.

Rapier, G. (2023). *Tesla Gigafactories: A look at the manufacturing hubs and their future.* Retrieved from https://www.businessinsider.com/tesla-gigafactory, 13 September 2023.

Ravenscroft, T. (2023). *The Line architects explain Saudi mega city in Discovery Channel Documentary.* Retrieved from https://www.dezeen.com/2023/07/18/the-line-architects-discovery-channel-documentary/, 12 September 2023.

Real Estate Weekly. (2020). *Report finds NYC most valuable city in America.* Retrieved from Report finds NYC most valuable city in America | Real Estate Weekly | Property Value Appeal · NYC311 (4744000.com), 24 October 2023.

Reuters. (2021). *Bill Gates and Warren Buffett to build new kind of nuclear reactor in Wyoming.* Retrieved from https://www.theguardian.com/us-news/2021/jun/03/bill-gates-warren-buffett-new-nuclear-reactor-wyoming-natrium, 11 September 2023.

Reuters. (2022). *Britain's new nuclear plant faces more delays and costs.* Retrieved from https://www.reuters.com/business/energy/edf-says-hinkley-point-start-now-june-2027-costs-rise-2022-05-19/, 7 September 2023.

Rinn, R., Ludwig, J., Fassler, P., & Deutsch, R. (2023). Cues of wealth and the subjective perception of rich people. *Current Psychology, 42*, 27442–27457.

Rodin, J. (2014). *The resilience dividend: Being strong in a world where things go wrong.* Public Affairs.

Rose, S. (2008). *Sun, sand—and a nuclear reactor.* Retrieved from https://www.theguardian.com/artanddesign/2008/oct/14/architecture-nuclearpower, 8 September 2023.

Roser, M. (2023). *Employment in agriculture.* Retrieved from https://ourworldindata.org/employment-in-agriculture, 25 March 2024.

Saudi Gazette. (2018). *NEOM aims to drive transformation of Saudi Arabia.* Retrieved from https://saudigazette.com.sa/article/543977, 12 September 2023.

Semuels, A. (2021). *Two rich men decided to fund a Failing City. Some people say they made it worse.* Retrieved from https://time.com/6110450/kalamazoo-foundation-for-excellence/, 4 July 2022.

Shorrocks, A., Davies, J., & Lluberas, R. (2022). *Global Wealth Databook 2022.* Credit Suisse Research Institute. Retrieved from https://www.google.co.nz/url?sa=t&source=web&rct=j&opi=89978449&url=https://www.credit-suisse.com/media/assets/corporate/docs/about-us/research/publications/global-wealth-databook-2022.pdf&ved=2ahUKEwjomez85vqIAxW-SGcHHaO2ASQQFnoECBkQAQ&usg=AOvVaw0jGQuYRCBsOd5_MxnfBYS-, 10 October 2024.

SOM. (2021a). *50 years of environmental activism through design.* Retrieved from https://www.som.com/story/50-years-of-environmental-activism-through-design/, 15 September 2022.

SOM. (2021b). *Burj Khalifa.* Retrieved from https://www.som.com/projects/burj-khalifa/, 6 September 2023.

Souha, A. (2021). *Owning an apartment in Burj Khalifa.* Retrieved from https://properties.emaar.com/en/blog/property-advice/apartment-in-burj-khalifa/?gclid=Cj0KCQjw94WZBhDtARIsAKxWG-8f3lrgjKgmKq0dJ8A1-5ZmVnbuUqMQwq1puETzVz0uTjpT8tVAkpwaAuVuEALw_wcB, 15 September 2022.

Stamp, E. (2020). *This sleek building is actually a nuclear reactor.* Retrieved from https://www.architecturaldigest.com/story/sleek-building-actually-nuclear-reactor, 11 September 2023.

Statista. (2021). *Dependency rate on energy imports in the United Kingdom (UK) from 2008 to 2018.* Retrieved from https://www.statista.com/statistics/267591/dependency-on-energy-imports-in-the-united-kingdom/, 22 September 2022.

Stuff. (2021). *Welcome to 'The Line': The zero-carbon city with no cars and no streets.* Retrieved from Welcome to 'The Line': The zero-carbon city with no cars and no streets | Stuff.co.nz, 13 September 2023.

Talbert, R. J. A. (1984). Augustus and the senate. *Greece & Rome, 31*(1), 55–63.

Telosa. (2023). *Telosa: Creating and more equitable and sustainable future.* Retrieved from https://cityoftelosa.com/, 11 September 2023.

TerraPower. (2023). *The Natrium Technology.* Retrieved from https://natriumpower.com/reactor-technology/#:~:text=The%20Natrium%20technology%20is%20a,hours%20to%20serve%20peak%20demand, 11 September 2023.

The Rockefeller Foundation. (2023). *100 resilient cities.* Retrieved from https://www.rockefellerfoundation.org/100-resilient-cities/, 6 September 2023.

Tognini, G. (n.d.). *20 years and $20 billion after 9/11, The World Trade Center is still a work in progress.* Retrieved from https://www.forbes.com/sites/giacomotognini/2021/09/10/larry-silverstein-durst-oculus-20-years-and-20-billion-after-911-the-world-trade-center-is-still-a-work-in-progress/?sh=4b9bd70461d2, 4 September 2023.

Treble, P. (2014). The clan that laid the golden eggs: Absurdly rich, painfully private, the Romanovs intrigue us now as much as then. *Maclean's (Toronto)*, 127(22), 62.

United States Nuclear Regulatory Committee (UNSRC). (2020). *Aurora—Oklo Application.* Retrieved from https://www.nrc.gov/reactors/new-reactors/large-lwr/col/aurora-oklo.html, 11 September 2023.

Vaughan, A. (2022). *Cost of new UK underground nuclear waste facility jumps to £53 billion*. Retrieved from https://www.newscientist.com/article/2309444-cost-of-new-uk-underground-nuclear-waste-facility-jumps-to-53-billion/, 21 September 2023.

Venditti, B. (2021). *Ranked: The top 20 metals and mining billionaires*. Retrieved from https://elements.visualcapitalist.com/top-20-metals-and-mining-billionaires/, 20 October 2023.

Vision 2030. (2023). *Leadership Message*. Retrieved from https://www.vision2030.gov.sa, 13 September 2023.

Wackernagel, M., & Rees, W. (1996). *Our ecological footprint*. New Society Publishers.

Wallach, B. (2015). *A world made for money*. University of Nebraska Press.

World Bank. (2023). *GDP (current US$)*. Retrieved from https://data.worldbank.org/indicator/Ny.Gdp.Mktp.Cd?most_recent_value_desc=true, 6 September 2023.

World Business Council for Sustainable Development. (2023). *Finance and built environment primer.* Retrieved from https://www.wbcsd.org/contentwbc/download/17485/245100/1, 25 March 2024.

Yu, E. (2022). *How much for Buckingham Palace? £1.3 billion in heated UK market*. Retrieved from https://www.bloomberg.com/news/articles/2022-05-25/buckingham-palace-valuation-hits-1-3-billion-on-hot-ukmarket#xj4y7vzkg, 30 June 2022.

Zheng, H., Wood, R., Moran, D., Kuishuang, F., Tisserant, A., Meng, J., & Hertwich, E. G. (2023). Rising carbon inequality and its driving factors from 2005 to 2015. *Global Environmental Change, 82*. 10pp, 102704.

3 Why Is Wealth an Issue for the Built Environment?

Introduction: Sustainability and Business Goals

Chapter 1 demonstrated that achieving the SDGs was largely reliant on increasing economic development, whilst Chap. 2 showed how much of the wealth generated by current development is going to the few rather than the many. This chapter sets out to probe the connections between the built environment, those responsible for its design and wealth. The first step in doing this is to examine what is meant by being sustainable and then look at how this has been applied to the built environment.

Sustainability first appeared in eighteenth-century Germany in connection with sustaining the yield from old forests (Garcia & Vale, 2017:18–19). It thus implies sustaining resources such that an output, in this case, one of timber, can be maintained. Applying this to a commercial enterprise would mean the business was not interested in growing, with the chance of making more profit, but rather would aim to sustain sufficient turn-over to maintain the workforce and create a constant profit measured over many years. Some years might be more profitable, but the business would be sustained reasonably constant in the long term. Under the demand of responsible consumers, businessmen who can maintain beneficial outputs for everyone in the long term would match intentions in SDG 12 (Responsible Consumption and Production). Sadly, this is not how modern businesses which call themselves sustainable seem to behave. At one end of the scale, a modern sustainable business might set out to make money by selling products that, for various reasons, can carry a sustainable label. In contrast, other enterprises might aim to consider sustainability values in their organisations and their outputs. A more comprehensive description of sustainable business models is provided by Geissdoefer et al. (2018).

> The definitions in the literature have in common that they see sustainable business models as a modification of the conventional business model concept, with certain characteristics and goals added to it; and, they either (1) incorporate concepts, principles, or goals that aim at sustainability; or (2) integrate sustainability into their value proposition, value creation and delivery activities, and/or value capture mechanisms.

Although issues of a circular economy within an enterprise are raised by Geissdoefer et al. (2018), the potential conflict between an enterprise, whether sustainable or not, that aims to grow and hence increase profit irresponsibly and one that wishes to sustain output responsibly is not explored by them. The nearest to this in their work is a description of a possible sustainable business strategy as being one that: "Aims at providing information and incentives that encourage less consumption" but whether that is necessary or unnecessary consumption nor what makes the

E. J. García, B. Vale, *Wealth and the Built Environment*, Sustainable Development Goals Series,
https://doi.org/10.1007/978-3-031-81954-4_3

level of consumption so environmentally detrimental that less consumption is desirable is not discussed. Fundamentally, a business enterprise is there to sell something, whether a product or a service. Perhaps the real issue for sustainable business is the level of such a product or service that can be sustained long-term, even into future generations. This long-term view should be the goal of any business that aims to contribute to the SDGs and the key variable to define what "responsible" means. These points are not clearly stated in the SDGs or measured. What are the implications of shortsighted goals and business outputs for the built environment and their professionals?

The Business of Modern Architectural Practice

Those responsible for the design of the modern built environment are professionals—architects, planners and urban designers—and the developers and clients who provide the money. At the same time, all these parties are also part of businesses that need to make sufficient money to pay for staff and overheads. As noted in Chap. 10, looking at the history of architectural education, although developers have always been with us, planning and urban design are relatively new occupations. Before their appearance, the layout of infrastructure and the design of buildings were in the hands of the architect or builder, and in medieval times, these were often the same person. At the same time, whoever is responsible for creating the built environment would still be running a business unless they were working for the government in some form. As most architects in the developed world are in some form of private practice, these form the next topic of discussion.

Modern architectural practice is a business. The business might aim to be sustainable by not seeking to grow, as discussed above, or it might be engaged in designing buildings intended to add to a more sustainable built environment, or both. However, practices involved with producing more "sustainable" buildings do not yet seem to be having a big effect on the built environment, not least because there is no clear definition of what makes a sustainable building nor any collection of data to demonstrate that a building performs as a sustainable building (Williams & Lindsay, 2007), and even less interest in making sure that the building is run in a sustainable way over its life. Rather than asking the question, "how should this building be designed so that those who use it can behave sustainably?", sustainable buildings seem more to do with adding bits of technology to conventional designs, such as adding photovoltaic panels to the roof or permeable pavements to the car park. This fails to question whether conventional designs are the route to a sustainable built environment. The building of houses in the developed world offers a good example.

From Building a Lot to Housing More

The emphasis on the family home when it comes to housing fails to observe that single-person households are rising in numbers, which must move against sustainability since a single-person home will need similar resources—in terms of creating a bathroom and kitchen with their fittings and appliances—as a multi-person family home. In 2021 in the USA, single person households formed 28% of all households, a rise from only 13% in 1960 (US Census Bureau, 2021). In 2005, Williams noted, "the growth in one-person households is potentially a resource time bomb. The rapid growth in household numbers, resource inefficiencies and increasing affluence amongst one-person households make the group potentially a major resource consumer and waste producer". Despite this, in the developed world, the family home is still the norm when it comes to building houses. The purpose-designed retirement accommodation in countries like the UK and New Zealand could be seen as diverging from the family house model, but of the two main types of retirement accommodation—independent living in villas or apartments and rest home or aged care accommodation—the former are just smaller family homes with kitchen and bathroom and many rooms in rest homes will have ensuite bathrooms. In the 2023 ANZ Report on retirement villages in New Zealand, 89% had no target date for becoming carbon neutral (Hinchliffe, 2023:11). The issue for sustainability is questioning what sort of retirement housing would be optimal in terms of the resources needed to build and maintain it, such as co-housing and more options with shared facilities (Yavari, 2019), something perhaps equally applicable to family housing.

Given that a house built in 1960 is probably still being lived in in 2023, when this chapter was written, producing an environment now that conserves resource use by building appropriate housing will need to be looking forward to changing demographics rather than reproducing the status quo with sustainability add-ons. The common misunderstanding of sustainable architecture is seeing it as a synonym for using the latest technology and gadgets, such as ground source heat pumps and nanogel windows (Buratti et al., 2016). In turn, this sends the message that sustainable buildings are more expensive than conventional buildings because of the need to incorporate new technologies, and hence, sustainable architecture and buildings are only for the rich. This does not have to be the case, as sustainable buildings can have comparable costs to conventional buildings, (van Hal, 2006; Cohen & Snell, 2018), but often it seems to be that those with the money to have a sustainable building will not build modestly to minimise their use of resources but want the large house with multiple garaging for electric vehicles and with the latest sustainable technology add-ons: This is not sustainable building. A New Zealand example is given by the design for a complex consisting of a lodge for 24 guests, various administration buildings and an "owner's pod" designed for a local millionaire, where the owner's "pod" alone is 565m^2 (Cook, 2022), although planning permission to build was refused. Despite its claimed sustainability credentials through having a green roof throughout the complex (Jamieson, 2022), the size of the "pod", a term which implies a small space for living, such as the 34m^2 living pod available off the shelf in NZ (Podlife, 2022), seems a lot of space for one person and their family. This built floor area would encompass the houses of a hamlet in the UK or a complete rural Indonesian village (Vale et al., 2018:210–211), apart from which anyone who cannot design a decent three-bedroom house in 100m^2 should probably resign from the profession. It is hard to see why one person needs so much space, even allowing for additional bedrooms.

Building larger than necessary will generate a profit for someone, something equally true of sustainable component add-ons. This does not accord with the fundamental principle of sustainable design, which is to do more with less to lessen the impact of humanity, especially those living in the developed world, on the planet. Wealthy people can buy and customise the built environment as they want, and their wealth allows them to do it. The sense of entitlement and freedom given by wealth gets in the way of aligning individual and collective interests to be able to contribute to the SDGs. Most building codes deal with minimums that cannot be exceeded but not with the maximums that might be deemed acceptable. This raises the issue of whether wealth is standing in the way of moving not only to a sustainable built environment but to a sustainable society, issues which this book aims to explore.

The Appeal of Wealth: Large Buildings as Great Architecture

Being small enough for purpose is a vital step in making sustainable buildings. Although tiny houses have recently enjoyed more publicity, the building achievements generally admired are large rather than small. As long ago as the first or second centuries BC, the seven wonders of the ancient world were compiled as a guide for sightseers (Adhikari, 2022). Three were in Greece (the Statue of Zeus at Olympia, the Colossus of Rhodes and the temple of Artemis at Ephesus), one in Turkey (the Mausoleum of Halicarnassus), one in Iraq (the Hanging Gardens of Babylon) and two in Egypt (the Lighthouse at Alexandria and the Great Pyramid of Giza), of which only the latter remains. In 1994, the American Society of Civil Engineers produced a list of the "seven wonders of the modern world", and by rank these were the Channel Tunnel, CN Tower, Empire State Building, Golden Gate Bridge, Itaipu Dam, Delta Works/Zuiderzee Works and Panama Canal (WorldAtlas, 2023). One characteristic both sets have in common is their size compared to the ordinary structures and buildings of the time, and to create buildings and engineering works of great size takes a great deal of money. The Gloholiday Community (Gloholiday, 2023) have further created a list of the seven architectural wonders of the world, again presumably in the interest of directing travellers to what they should see (Eiffel Tower, Guggenheim Museum at Bilbao, Burj Khalifa, Golden Gate Bridge, Taj Mahal, Chrysler Building, Gateway Arch in St Louis).

While some wonders on this list seem more engineering than architectural icons, the buildings are either large (Guggenheim Museum, Chrysler Building), or both large and expensive (Taj Mahal, Burj Khalifa—see Chap. 2). The point is that humanity has always admired artefacts that take both a superhuman effort and a superhuman budget to produce. Perhaps this stems from childhood when the objects in the world appeared much bigger to the child than to an adult, given that brain development for understanding of the true size of objects does not occur until after the age of about five (Turesky et al., 2021). This leads to considering whether such admiration of the large affects how buildings are viewed. Setting aside the issue of whether people visit cities only to see specific buildings or to experience the city as a place to walk around and through, guidebook in hand, in the form of "self-directed learning" (Roberson, 2015), a 2017 book has focussed on contemporary world architecture worth the effort to visit. The preface (Anon, 2017) states, "*Destination Architecture* presents 1,000 of the most compelling works of architecture completed in the last 30 years". This echoes the idea of the seven wonders of the ancient world as a guide for the visitor. However, unlike common tourist guidebooks, only addresses are given, and there are no maps or additional information, such as which bus to catch to get there or where to eat cheaply near a famous example of architecture. The buildings are arranged by country within continents, and Europe has by far the largest number of entries, followed by Asia and North America. In 2022, the top ten countries by GDP fell within these three continents (Dimitropoulou, 2022). It would be reasonable to expect more money to be spent on architecture in wealthier economies.

Of the countries represented in this guide in South America, Brazil ranks 11th for GDP, Argentina 31st, Chile 43rd, Colombia 44th, Peru 51st, and Uruguay 83rd, so this economic mix suggests the guidebook might include some smaller and possibly less expensive but architecturally worthy buildings in these countries. To explore this further, there are 42 entries for South America. These consist of 10 museums or galleries, 6 cultural/library/exhibition buildings, 5 buildings associated with sports, 3 each of hotels, performance spaces and buildings related to transport (two airport buildings and a bus station), followed by 2 each of civic centres, offices and wineries. None of these buildings are small nor linked to the major problems of these countries, which are a lack of infrastructure and mass housing (Habitat for Humanity, 2019, 2023). Of the remaining building types selected, there is one civic centre, one innovation centre, one high-rise apartment building and a Ministry of Education, and again, all are large buildings.

The two remaining architectural places to visit are a somewhat smaller. One is a 2008 three-storey photographic studio, designed by Studio MK27 mainly for taking pictures of food. Located in São Paulo the building appears to be made of stacked concrete boxes, with an area of 1160m^2 (Studio MK27, n.d.-a) (Fig. 3.1). This is also about the floor area of 12 of the 38 houses featured on the Studio MK27 website, the largest of which is the 2018 Panorama House in Los Angeles with a floor area of 3100m^2 and the smallest the 2007 Osler House in Brasilia with an area of only 270m^2 (Studio MK27, n.d.-b), still about three times the size of the 2015 recommended 93m^2 for a three-bedroom home in the UK (RIBA, 2015). On the ground floor of the Osler House are the main bedroom suite, another bedroom and bathroom, and a utilities area,

Fig. 3.1 Photographic Studio. (Based on Studio MK27, n.d.-a)

whilst above are the living room, kitchen and a "small office" (ArchDaily, 2023a, b). To design a house with only two bedrooms in an area of 270m^2 takes some skill. In 2017, only 4 of the 29 European countries had an average house floor area greater than 100m^2, the largest being 130m^2 in Cyprus (Laakso & Heiskanen, 2017).

In the South American entries of the architectural guidebook, the other smaller building as a chosen destination is a 2014 chapel in Rio de Janeiro known as the Joa Chapel. Designed by Bernardes Arquitetura, this is only 43m^2 but it is a chapel in a private garden for small family ceremonies (Bernardes Arquitectura, 2012) (Fig. 3.2). Private chapels have a long history and were for those with money. Thomas (2018:89) states: "The principal function of a private chapel was to provide a space where a select group of people could hear Mass". Originally only found in the houses of the aristocracy, by the late Middle Ages they were also associated with the homes of the gentry (Thomas, 2018:90). If you could not afford a private chapel, you worshipped in the parish church along with everyone else, thus sharing facilities. Although modern, the Joa Chapel is still an architectural product of relative wealth and exclusivity.

The point of examining what passes for great modern architecture, at least in the eyes of the Phaidon editors who compiled the list, is to show the lack of the small and ordinary in the choice, somehow suggesting that the small and ordinary can never be great architecture. This has not always been the case. At the start of the twentieth century, architects were obsessed with affordable housing design, leading to competitions for the best low-priced cottage (Weaver, 1913). Nor have buildings that have attempted to reduce environmental impact significantly been considered worthy of inclusion in a list of great architecture.

The 22 examples of architecture listed as worth visiting in Central America were also

Fig. 3.2 Joa Chapel. (Based on Bernardes Arquitectura, 2012)

examined to ensure South America is not an aberration. Of these, four were museums, four were office buildings, three were higher education buildings, and two were apartment buildings. Of the remaining buildings, a performance space and an embassy are both large. The fire station is also large but is at least a community structure. The pavilion has no function (for further discussion of pavilions, see Chap. 6), and the entry that consists of six villas for artists in residence, although not large, seems elitist, since living in a special residence is not a condition of being a great artist. The other recommended buildings are a winery, a parish church, a chapel and a private guest house with a pool.

Perhaps these are the only types of buildings designed by architects, which leaves the issue of who designs the houses and buildings that are too small and ordinary to be included in *Destination Architecture*. Only the wealthy could afford a large house from Studio MK27, but this does not mean that architects are not involved in small buildings. As the saying goes, involving architects in the design of small buildings could be the equivalent of turning bread and water into tea and toast, or making the ordinary and basic something special. Such work goes on, but it seems it is seldom recognised as architecture worthy of comment or not even as architecture. Large buildings are evidence of wealth accumulation in the built environment. If great architecture can only be large and made by architects, can large architectural firms design sustainable buildings that contribute to achieving to the SDGs?

The Sustainability of Large and Wealthy Architectural Firms

Every year *Building Design Online* publish a list of the largest architectural firms ranked by the number of "fee-earning architects they employ" (Building Design, 2022). Table 3.1 sets out the top ten for 2022.

Looking just at the top two practices in Table 3.1, Gensler grew from a small practice established in 1965 in San Francisco that Art Gensler set up with his wife as office manager and accountant and one draftsman. To keep solvent Art Gensler worked mornings for another architect (Bernstein, 2021). Beginning by taking jobs no-one else wanted, the firm specialised in interior design but described as "an astute business man" (Gensler, 2023a), Gensler grew the firm to be the huge enterprise it is today with 52 offices across 22 countries (Genlser, 2023b). This was done by avoiding the idea of the star architect and by recognising that everyone working in the firm brings talent to the whole enterprise. Apart from a focus on the client and the people working in the practice, Gensler also have a sustainability objective to become carbon neutral within the decade. Given they are the largest firm in terms of fee income in the world this seems very important for the future of the built environment. The firm states: "Achieving carbon neutrality entails eliminating or offsetting all CO_2 emissions from the built environment" (Gensler, 2023c). The strategies for achieving the goal are "right-sizing, selecting low- and zero-impact materials, reusing buildings instead of engaging in new construction, and offsetting impact through net-positive production and/or increases in carbon sequestration (captured through absorption)". However, this is still adding sustainability to normal architectural practice and does not question how and where humanity should live in order to keep within the resource limits of the planet. If the money is there, then it is possible to make buildings and pay for any required carbon off-sets, although the idea of persuading people to build small rather than big through "right-sizing" is a good step into a more sustainable future.

A building that appears as an example of sustainability on the Gensler website is the Harvey B. Milk terminal building in San Francisco airport, where a Gensler team were brought in as sustainability experts. The question is whether, despite a 79% reduction in total carbon emissions, an airport can ever be a sustainable building in that it encourages those who can afford it to travel. Although aviation only makes up some 3.5% of total global CO_2 emissions (Chicca et al., 2023:233), flying also affects ozone, methane and nitrous oxide at high levels, and the overall effect of flying is to increase the global warming effect (Baer, 2018). Perhaps more importantly Baer goes on to state that: "As in many other areas of a stratified world system, the affluent participate in [flying] for a variety of cultural, status, and income reasons, and contribute much

Table 3.1 2022 top ten largest architectural firms

Firm	Country	Architects employed	Fee income US$	Rank 2022	Rank 2021
Gensler	USA	2692	1bn–1.5bn	1	1
Zhubo Design	China	1320	110 m–119 m	2	-
HDR	USA	1312	500 m–599 m	3	3
Nikken Sekkei	Japan	1300	600 m–699 m	4	2
IBI Group	Canada	977	290 m–299 m	6	5
Perkins Eastman	USA	793	270 m–279 m	7	8
Aedas	China	750	230 m–239 m	8	6
HKS	USA	700	240 m–249 m	9	9
Haeahn Architecture	South Korea	697	120 m–129 m	10	13
Heerim Architects and Planners	South Korea	693	180 m–189 m	11	10

more overall to greenhouse gas emissions from flights than do working-class people and particularly the poor around the globe". Gensler are hence involved in the design of a "sustainable" building that will only benefit the relatively wealthy, whereas the emissions from flying affect everyone, but then when you have 2692 employees to support, you are going to have to keep designing buildings whether these seem part of a truly sustainable future or not.

The second largest firm, Zhubo Design, also claim a sustainability agenda: "With the market development and the company's accumulation of more than 20 years, the company has expanded the design of prefabricated buildings, BIM technology, green buildings, sponge cities, building intelligence etc. consultation service" (Zhubo Design Co Ltd., 2021a). Zhubo Design-Shenzhen received a Gold LEED award for the Qianhai China Metallurgical Technology Building (Zhubo Design Co Ltd., 2021b) (Fig. 3.3). A Gold award means that out of 110 points, the maximum possible, a building has scored 60–79 points (USGBC, 2022) (55–72% of the maximum possible). The highest LEED award is Platinum, where a building must score 80+ points (at least 73% of the maximum). However, 73% would only be a clear B grade under the NZ university grading system (New Zealand Education, 2023). Unlike Michelin Stars for a restaurant, which involves anonymous visits over the years to check that the culinary standard, including value for money, is maintained before a star is awarded (Pendrill, n.d.), LEED scores seem easy to obtain. Most are gained at the design stage, not when the building is occupied and in use, so they are a measure of intentions rather than performance. If you win a gold medal at the Olympic Games, you are the best in the world based on performance, not intention. Unless well versed in the LEED certification system, a Gold award sounds like the best possible, suggesting the building must have excellent environmental credentials, although this is not the case. Indeed, at a time when there is a particular urgency to achieve climate change mitigation, buildings that are claimed to be sustainable with a "Platinum" rating should be scoring higher than a B grade, but then large practices have to keep finding jobs for their numerous employees, so clients need to be attracted, and a LEED award sounds prestigious.

Claimed to be inspired by the shape of a carp, one of the sustainability features of the Qianhai China Metallurgical Technology Building is its use of sponge city principles. These involve permeable paving, a rainwater garden and rainwater

Fig. 3.3 Qianhai China Metallurgical Technology Building. (Based on Zhubo Design Co Ltd., 2021b)

reuse. The sponge city concept originated in China as a means of managing storm water and flooding, but in the past, we have had sponge cities and the problem of designing them has only arisen from increased urbanisation. Rather than thinking what human lifestyle changes are needed to combat climate change, one means of reducing emissions has been urban intensification in order to reduce transport emissions, with a dramatic increase in non-permeable surfaces. Alas, with increasing severe weather events due to climate change, urban flooding becomes a strong possibility. This is an example of added urban complexity, where the answer to one problem only brings on another problem that then has to be solved (Garcia et al., 2021:65).

To counter the idea that large practices are not wholly wedded to making sustainable buildings the UK practice of Feilden Clegg Bradley Studios were ranked 86th in 2022, down from 82nd in 2021, in terms of the number of architects employed, with a turnover of US$10–19 million (Building Design, 2022). From their beginnings as Feilden Clegg in 1978 with the design of low energy houses (Prasad, 2005), the firm has maintained a reputation for more sustainable buildings to the point of developing a free carbon calculator to help others reduce carbon emissions at the design stage (Feilden Clegg Bradley Studios, n.d.).

Wealthy Architects and Sustainable Buildings

Perhaps wealthy architects do better than large practices when it comes to making true sustainable buildings. Table 3.2 sets out the top ten wealthiest architects (Arch20, n.d.) together with their 2022 ranking, if any, for the size of the practice (see Table 3.1.).

Although the fee incomes in the final column should only be taken as a guide since they come from a variety of sources and the date of the information is not always given, two things emerge from comparing Tables 3.1 and 3.2. To be a wealthy architect you do not necessarily need a very large practice, but it does seem important that your name is prominent in what the practice is called. However, is making sustainable buildings a path to becoming a very rich architect?

Norman Foster is the wealthiest architect in Table 3.2. In 2022, Foster + Partners completed a 28 storey tower in Budapest as new headquarters for MOL Group, an oil and gas company (Fig. 3.4). Whatever the sustainability credentials of the building, can a building for an oil and gas company be considered a sustainable building even if it is "on track to achieve LEED Platinum" (ArchDaily, 2023a) —still a B grade as noted above. What MOL Group say about sustainability is even more telling: "Sustainable Development, for us, means a corporate commitment to the balanced integration of economic, environmental and social factors into our everyday business operations, to maximise long-term stakeholder value and to safeguard our licence to operate" (MOL Group, n.d.-a). This suggests "sustainability is something we do as long as it does not interfere with oil and gas exploration and extraction". One project of MOL is to inject CO_2 into gas fields so as to extract more gas and oil and, at the same time, dispose of CO_2, which sounds a bit like having your fossil fuel cake and eating it (MOL Group, n.d.-b).

The new headquarters has the usual sustainability credentials: "The building utilizes low and zero-carbon technologies, such as integrated rooftop photovoltaic panels, and ground source heat pumps, and features a greywater recycling and rainwater harvesting system" (ArchDaily, 2023a), and its heating and cooling system is also connected to the district heating. However, whether connection to district heating in Budapest is a sustainability credential is debateable, given at least part of its fuel comes from Russian natural gas (Balkan Green Energy News, n.d.).

Frank Gehry, now 93 and equal second in terms of wealth in Table 3.2, does not really enter the sustainable building debate. In 2010 he "slammed the LEED certification, saying it is awarded for 'bogus stuff' and that climate change and sustainable design are 'political' issues" rather than design ones (Singh, 2010). Despite this, for the new Business School for UTS Gehry said he would include all the usual sustainability

Table 3.2 Top ten wealthiest architects

Architect	Net worth US$	Practice	2022 Ranking	Fee income US$
Norman Foster	240 million	Foster + Partners	22	240-249 million
Santiago Calatrava	100 million	Santiago Calatrava	–	N/A
Frank Ghery	100 million	Gehry Partners	–	63.4 million
Moshe Safdie	100 million	Safdie Architects	–	13 million
Zaha Hadid	95 million	Zaha Hadid Architects	33	70-79 million
Renzo PPiano	25 million	Renzo Piano Building Workshop	–	1-10 million
Bjarke Ingels	17 million	BIG (Bjarke Ingels Group)	–	92 million (2020)
Kongjian Yu	15 million	Turenscape	–	N/A
Maya Lin	12 million	Maya Lin Studio	–	<1 million
David Adjaye	10 million	Adjaye Associates	–	77.1 million

[a] Datanyze (2023)
[b] zoominfo (n.d.-a)
[c] IncFact.com (2023)
[d] zoominfo (n.d.-b)
[e] Glassdoor (2023)
[f] Growjo (2022)

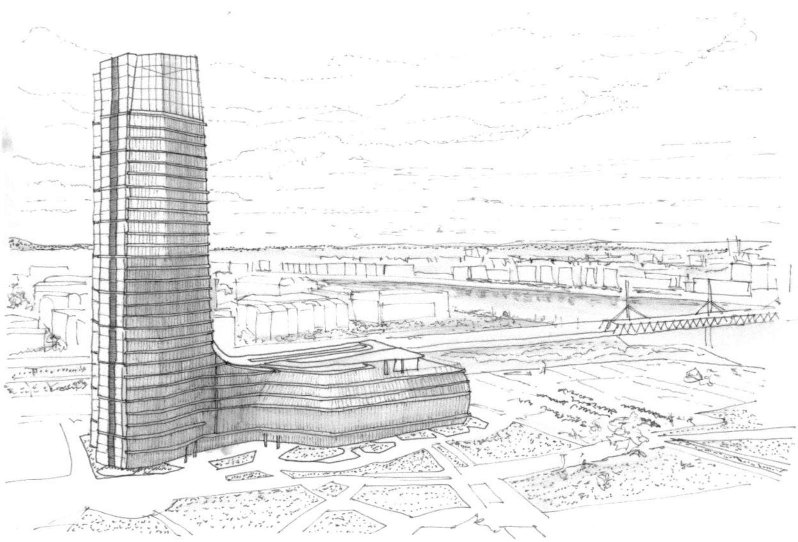

Fig. 3.4 MOL Headquarters. (Based on MOL Group, n.d.-c)

elements plus "co-generation for energy efficiency and the use of paperless design to enable a strong focus on the elimination of waste that typically added up to 30 per cent of a project's total resources" (Jewel, 2010) (Fig. 3.5). Described at its opening as a squashed paper bag, albeit a beautiful one, it was built using 320,000 custom-designed bricks, which hardly seems a way to reduce resource use in creating a building.

Seventh in Table 3.2 is Bjark Ingels founder of BIG (Bjark Ingels Group). CopenHill is an award-winning waste to energy plant in Copenhagen designed by this firm. They describe this building as an example of "hedonistic sustainability" but as Garcia et al. (2021:26–30) argue at length, a building that is "an over-sized, expensive and unnecessary project" could hardly be described as sustainable design.

This discussion may be considered unfair, as surely some, at least, of these wealthy and well-known architects, must have designed a building that does sit lightly on the earth by minimising its

Fig. 3.5 UTS Business School. (Source: Creative Commons (n.d.))

lifetime use of resources and maximising the function and pleasure it gives to its users and those passing by? The problem perhaps lies in the nature and size of the commissions that make architects both famous and rich whilst also providing work for the firm's employees. This suggests a wealth-centric approach to architecture where the aim is to become well-known for iconic buildings with a creative edge, which probably also means a large budget. Chicca et al. (2023:229) argue that there is a relationship "between the production of new architectural icons as the materialisation of current cultural values, and how these new buildings are shaping and promoting cities to attract businesses and people". Thus, even a building designed to be sustainable or have sustainable aspects may become a wealth generator or tourist attraction with all their accompanying environmental impacts. While it is hard to pin down the environmental impact of wealth generation, the impact of increasing tourism is clear. Tourism is the third largest global export earnings category after fuels and chemicals (Chicca et al., 2023:233). Gössling and Peeters (2015) have pointed out that the impact of a holiday trip near home by bicycle is vastly different from that of a long-distance flight followed by a cruise. They go on to state:

> …documented energy use per guest night may vary between 3.6 and more than 3700 MJ, and corresponding emissions between 0.1 and 260 kg CO…any development of tourism that focuses on below-average RUIs [resource use intensities] will constitute low-intensity consumption and contribute to a de-materialization of the global tourism system. Current trends, however, point in the opposite direction, as the tourism system becomes more energy, freshwater, land and food intense on a per trip per arrival per guest night basis, while the total number of tourists continues to increase, a result of a growing and wealthier world population engaging in more resource intense tourism.

Iconic buildings, which can be viewed as a product of wealth, and with many examples among the pages of *Destination Architecture,* have the potential to add to this tourism traffic, which itself is a product of wealth. The tragedy is that this is what most architecture students see as architecture, something that is probably a long way from their eventual experience of practice, as discussed next.

Everyday Architectural Practice

Hamil (2020), writing for the National Building Specification, notes: "The RIBA [Royal Institute of British Architects] describes firms that have fewer than ten employees as ‘small practices’, and estimates that they represent almost 80% of UK architecture practices”. Of more importance is the fact that Hamil describes small practice work as simple and using traditional methods where most, if not all, information will be carried on the drawings without the need for taking off quantities. Such work will have adequate budgets for the job in hand but that will probably not stretch to creating the types of building that will find their way into *Destination Architecture.* Marks (2018) goes further by stating that only 15% of architectural practices have a significant influence on the world of building, being built around a “named” architect. Are those educating the next generation of architects responding to the implications of these statistics, with their suggestion that most graduates will be working in small practices on relatively small projects with small budgets, in the hope of turning bread and water into tea and toast for their modest clients?

From decades spent by the authors in architectural education, it seems that this is not the case. Rather, potential architects are being educated with a view of being future “named” architects with the possibility of attracting budgets that will allow for the creation of the unusual and complex building as part of establishing a “name”. The moral dilemma of projects with big budgets is explored further in Chap. 9, but it would seem that wealth has impacted not only on architecture but also on architectural education. For a building to become a household name or a wonder of the world it has to be special, large and have an appropriate budget. The use of iconic buildings as part of wealth generation through tourism has already been mentioned. Building very high is a more expensive option than building low, through the need to embody more energy into the building’s structure and systems, but perhaps more importantly, building high is seen as a way of concentrating more people on less land in cities, while simultaneously the bringing together of people in cities is seen as a strategy for wealth generation. The generation of wealth, therefore, goes hand-in-hand with buildings with increased environmental impact through the resources that go into making them (Garcia et al., 2021:209–210). This is moving away from, rather than towards, more sustainable human settlements. The counterargument to this is that the expanding human population must go somewhere but nature is not kind to species that outgrow their resources and human ingenuity with its link to SDG 9—Industry, Innovation and Infrastructure—has not yet found a way through innovation to create a built environment that can overcome these problems (for attempts and their consequences see Chap. 2).

The other issue with the large and complex building, apart from its environmental impact, is that moving into unknown territory when it comes to design innovation can prove expensive. In a list of large-scale projects with significant cost overruns, most of which are infrastructure such as railways, tunnels and bridges, two buildings come in at second and third, these being the Scottish Parliament Building with a 1600% overrun (Fig. 3.6) and the Sydney Opera House with a 1400% overrun (Flyvbjerg, 2014). The only other building in the list of 33 projects is the Bank of Norway headquarters coming in at 11th with a cost overrun of only 440%.

The Scottish Parliament is one of the 1000 pieces of recent architecture that supposedly ought to be visited (preferably by bicycle?), though the comments on the entry are worth repeating.

> Some regard the Scottish Parliament building as a disaster—energy-inefficient and poorly constructed, despite coming in three years late and ten times over budget. Yet those who love it praise a tour de force in local granite, concrete and decorative wood detailing that reflects the drama of the surrounding landscape. (Anon, 2017:179)

The Sydney Opera House was completed in 1973, so is too old to appear in *Destination Architecture*, and the Bank of Norway headquarters was not selected for inclusion.

Before leaving the consideration of what are regarded as engineering and architectural wonders, the Empire State Building, which was part of the 1994 selection of the American Society of Civil Engineers, is worth examining. Built between 1930 and 1931 at the time of the great depression, this was an expensive building at a

Fig. 3.6 Scottish Parliament Building by Enric Miralles. (Source Emilio Garcia)

cost of US$24.7 million. This was far less than the suggested cost of US$43 million, and it took only 13 months to build rather than the 18 estimated (Jacobsson & Wilson, 2018). Jacobsson and Wilson have examined the many aspects that led to this remarkable achievement, among which was the design issue of repetition: "It is well documented that the architect Lamb designed the floors identically for expeditious construction. Thus, as construction proceeded, it was repetitious, in a manner similar to the line in Ford's plants and enabled the building to grow with an average of 4.5 stories per week". This is a long way from the design of the Sydney Opera House or the Scottish Parliament Building with their non-repeated curves and calls into question how we view architecture. To complete the story, 2010 saw the completion of an energy upgrade of the Empire State Building, which cut energy use by around 40% at a cost of US$31 million (de la Garza, 2021).

Rhythm, based on some form of repetition, is part of design, not least because it is fundamental to nature. In the past, the great cathedrals used repetition of form as a means of being able to prefabricate stones in the quarry and construct them on site, yet these were still the buildings that pushed the technology of the time. Now, while ignoring all the work done mostly on time and within budget by small practices with unknown names, it seems that for a building to be recognised as architecture, it cannot be a reworking of the same theme, as happened with the great European medieval cathedrals, but instead, has to look like nothing that has ever been seen before. This is a long way from the natural world, where a forest of beech trees is more like a bunch of medieval churches or cathedrals, recognisably the same species but each an individual because of where it stands. This idea calls into question the whole nature of what architecture is. As noted in the introduction to the 2023 Venice Biennale (La Biennale di Venezia, 2023): "In architecture particularly, the dominant voice has historically been a singular, exclusive voice, whose reach and power ignores huge swathes of humanity—financially, creatively, conceptually—as though we have been listening and speaking in one tongue only". The SDGs remind us that we have to act together, but we shall be using a very different "tongue" from that of the "exclusive voice" by changing to one globally inclusive with all that this implies.

References

Adhikari, S. (2022). *The 7 Wonders of the Ancient World.* Retrieved from https://www.ancienthistorylists.com/ancient-civilizations/7-wonders-of-the-ancient-world/#:~:text=The%20Seven%20Wonders%20of%20the%20Ancient%20World%20were%3A,Rhodes%2C%20Greece%207%20The%20Lighthouse%20at%20Alexandria%2C%20Egypt, 29 January 2023.

Anon. (2017). *Destination architecture*. Phaidon Press Ltd.

Arch20. (n.d.). *10 architects from a basic architect salary to the richest in the world.* Retrieved from https://www.arch2o.com/10-architects-from-a-basic-architect-salary-to-the-richest-in-the-world/, 14 February 2023.

ArchDaily. (2023a). *MOL Headquarters / Foster + Partners.* Retrieved from https://www.archdaily.com/993638/mol-headquarters-foster-plus-partners, 15 February 2023.

ArchDaily. (2023b). *Osler House / Studio MK27.* Retrieved from https://www.archdaily.com/806178/osler-house-studio-mk27-nil-marcio-kogan-plus-suzana-glogowski, 6 February 2023.

Baer, H. A. (2018). Grappling with flying as a driver to climate change: Strategies for critical scholars seeking to contribute to a socio-ecological revolution. *The Australian Journal of Anthropology, 29*(3), 298–315.

Balkan Green Energy News. (n.d.). *Budapest to install geothermal district heating system of up to 20 MW.* Retrieved from https://balkangreenenergynews.com/budapest-to-install-geothermal-district-heating-system-of-up-to-20-mw/, 15 February 2023.

Bernardes Arquitectura. (2012). *Joa Chapel.* Retrieved from https://www.bernardesarq.com.br/en/projeto/joa-2/, 29 January 2023.

Bernstein, F. A. (2021). *Obituary: Art Gensler, 1935-2021.* Retrieved from https://www.architecturalrecord.com/articles/15123-obituary-art-gensler-19352021, 10 February 2023.

Building Design. (2022). *WA100 2022: The big list.* Retrieved from https://www.bdonline.co.uk/wa100-2022-the-big-list/5115702.article, 10 February 2023.

Buratti, C., Moretti, E., & Belloni, E. (2016). Nanogel windows for energy building efficiency. In F. Pacheco Torgal, C. Buratti, S. Kalaiselvam, C. G. Granqvist, & V. Ivanov (Eds.), *Nano and Biotech based materials for energy building efficiency* (pp. 41–69). Springer.

Chicca, F., Vale, B., & Vale, R. (2023). *The environmental impact of cities: Death by democracy and capitalism.* Routledge.

Cohen, A., & Snell, C. (2018). Climate change and the bottom line: Delivering sustainable buildings at market rate. *Architectural Design, 88*(1), 110–115.

Cook, M. (2022). *Time set for Damper Bay consent hearing.* Retrieved from https://www.odt.co.nz/regions/wanaka/time-set-damper-bay-consent-hearing, 29 January 2023.

Creative Commons. (n.d.). UTS Business School, II | Openverse) Credit Newtown grafitti is licensed under CC BY 2.0.

Datanyze. (2023). *Gehry Partners.* Retrieved from https://www.datanyze.com/companies/gehry-partners/15041724, 14 February 2023.

de la Garza, A. (2021). *The Empire State Building's green retrofit was a success. Will other buildings follow suit?* Retrieved from https://time.com/6026610/empire-state-buildinggreen-retrofit/, 22 February 2023.

Dimitropoulou, A. (2022). *Largest countries by GDP, 2022.* Retrieved from https://ceoworld.biz/2022/03/31/economy-rankings-largest-countries-by-gdp-2022/, 1 February 2023.

Feilden Clegg Bradley Studios. (n.d.). *FCBS Carbon.* Retrieved from https://fcbstudios.com/fcbscarbon, 15 February 2023.

Flyvbjerg, B. (2014). What you should know about megaprojects and why: An overview. *Project Management Journal, 45*(2), 6–19.

Garcia, E., & Vale, B. (2017). *Unravelling sustainability and resilience in the built environment.* Routledge.

Garcia, E., Vale, B., & Vale, R. (2021). *Collapsing gracefully: Making a built environment that is fit for the future.* Springer.

Geissdoefer, M., Vladimirova, D., & Evans, S. (2018). Sustainable business model innovation: A review. *Journal of Cleaner Production, 198*, 401–416.

Gensler. (2023a). *Art Gensler 1935–1921.* Retrieved from https://www.gensler.com/art-gensler-legacy, 10 February 2023.

Gensler. (2023b). *Offices.* Retrieved from https://www.gensler.com/offices, 10 February 2023.

Gensler. (2023c). *The Gensler Cities Climate Challenge (GC3).* Retrieved from https://www.gensler.com/the-gensler-cities-climate-challenge, 13 February 2023.

Glassdoor. (2023). *Maya Lin Studio.* Retrieved from https://www.glassdoor.com/Overview/Working-at-Maya-Lin-Studio-EI_IE1019358.11,26.htm, 14 February 2023.

GloHoliday. (2023). *7 Architectural Wonders of the World.* Retrieved from 7 Architectural Wonders of the World | GloHoliday, 16 January 2023.

Gössling, S., & Peeters, P. (2015). Assessing tourism's global environmental impact 1900-2050. *Journal of Sustainable Tourism, 23*(5), 639–659.

Growjo. (2022). Adjaye Associates Revenue and Competitors. Retrieved from https://growjo.com/company/Adjaye_Associates, 14 February 2023.

Habitat for Humanity. (2019). *Argentina.* Retrieved from https://www.habitat.org/sites/default/files/documents/English_LAT_CountryProfiles_2019_Argentina.pdf, 6 February 2023.

Habitat for Humanity. (2023). *The housing need in Brazil.* Retrieved from https://www.habitat.org/where-we-build/brazil, 6 February 2023.

Hamil, S. (2020). *NBS for small practices.* Retrieved from https://www.thenbs.com/knowledge/nbs-for-small-practices#article-1, 1 February 2023.

Hinchliffe, R. (2023). *ANZ Retirement Villages Survey Report.* Retrieved from file:///C:/Users/Brenda/Downloads/ANZ-Retirement-Village-Sector-Insight%20(2).pdf, 3 April 2024.

IncFact.com. (2023). *Renzo Piano Building Workshop.* Retrieved from https://incfact.com/company/renzopianobuildingworkshop-newyork-ny/, 14 February 2023.

Jacobsson, M., & Wilson, T. L. (2018). Revisiting the construction of the Empire State Building: Have we forgotten something? *Business Horizons, 61*(1), 47–57.

Jamieson, D. (2022). *Peter Thiel's luxury Wānaka lodge idea rejected by council planner.* Retrieved from https://www.stuff.co.nz/business/property/128522034/peter-thiels-luxury-wnaka-lodge-idea-rejected-by-council-planner, 29 January 2023.

Jewell, C. (2010). *Frank Gehry on inspiration and sustainability for the UTS in Sydney.* Retrieved from https://thefifthestate.com.au/articles/franks-ghery-on-inspiration-and-sustainability-for-the-uts-in-sydney/, 15 February 2023.

La Biennale di Venezia. (2023). *Biennale Architettura 2023: The Laboratory of the Future.* Retrieved from https://www.labiennale.org/en/news/biennale-architettura-2023-laboratory-future-0, 1 March 2023.

Laakso, S., & Heiskanen, E. (2017). *Good practice report: Capturing cross-cultural interventions.* European Network for research, Good Practice and Innovation for Sustainable Energy. Retrieved from https://www.researchgate.net/publication/323175360_Good_practice_report_capturing_cross-cultural_interventions, 1 September 2023.

Marks, K. (2018). *Guide to architectural practices.* Retrieved from https://jobs.architecture.com/article/different-types-of-practices/, 1 February 2023.

MOL Group. (n.d.-a). *MOLGROUP.* Retrieved from https://molgroup.info/en, 15 February 2023.

MOL Group. (n.d.-b). *MOL Group: Sustainability Case Study.* Retrieved from https://molgroup.info/storage/documents/case_studies/climate_change/co2_eor_project_croatia_origin.pdf, 15 February 2023.

MOL Group. (n.d.-c). *Take a look around from the top of Budapest.* Retrieved from http://molcampus.hu/en/, 20 October 2024.

New Zealand Education. (2023). *Academic Grading in New Zealand.* Retrieved from https://www.newzealandeducation.info/education-system/academic-grading-in-new-zealand.html, 14 February 2023.

Pendrill, K. (n.d.). *How to get a Michelin star: 5 criteria for restaurateurs.* Retrieved from https://www.touchbistro.com/blog/how-to-get-a-michelin-star/#subheading=3, 22 February 2023.

Podlife. (2022). *Tiny Houses for NZ—Start Your Podlife Today.* Retrieved from https://podlife.co.nz/, 29 January 2023.

Prasad, S. (2005). *Richard Feilden (obit).* Retrieved from https://www.independent.co.uk/news/obituaries/richard-feilden-495665.html, 15 February 2023.

RIBA (Royal Institute of British Architects). (2015). *Space standards for homes.* Retrieved from file:///C:/Users/Brenda/Downloads/HomewiseReport2015pdf.pdf, 4 April 2024.

Roberson, D. N. (2015). The travel guidebook: Catalyst for self-directed travel. *Tourism Analysis, 20*(1), 123–128.

Singh, T. (2010). *Frank Gehry Slams LEED, Calls Sustainable Design "Political".* Retrieved from https://inhabitat.com/frank-gehry-calls-sustainable-design-political/, 15 February 2023.

Studio MK27. (n.d.-a). *SC.* Retrieved from https://studiomk27.com.br/sc/, 29 January 2023.

Studio MK27. (n.d.-b). *Houses.* Retrieved from https://studiomk27.com.br/category/houses/, 6 February 2023.

Thomas, S. E. (2018). *The Parish and the Chapel in medieval Britain and Norway.* The Boydell Press.

Turesky, K., Vanderauwera, J., & Gaab, N. (2021). Imaging the rapidly developing brain: Current challenges for MRI studies in the first five years of life. *Developmental Cognitive Neuroscience, 47*, 1–14.

US census Bureau. (2021). *Census Bureau Releases New Estimates on America's Families and Living Arrangements.* Retrieved from https://www.census.gov/newsroom/press-releases/2021/families-and-living-arrangements.html, 27 February 2023.

US Green Building Council (USGBC). (2022). *LEED rating system.* Retrieved from https://www.usgbc.org/leed#:~:text=Projects%20go%20through%20a%20verification,Platinum%20(80%2B%20points)., 14 February 2023.

Vale, B., Vale, R., & Chicca, F. (2018). Dwelling. In F. Chicca, B. Vale, & R. Vale (Eds.), *Everyday lifestyles and sustainability* (pp. 208–217). Routledge.

van Hal, A. (2006). Leveraging the economical benefits of sustainable building. *WIT Transaction on Ecology and the Environment, 98*, 35–41.

Weaver, L. (1913). *The "Country Life" book of cottages.* Country Life.

Williams, J. (2005). One-person households—A resource time bomb? *WIT Transactions on Ecology and the Environment, 81*, 409–417.

Williams, K., & Lindsay, M. (2007). The extent and nature of sustainable building in England: An analysis of progress. *Planning Theory and Practice, 8*(1), 31–49.

WorldAtlas. (2023). *7 Wonders of the Modern World.* Retrieved from https://www.worldatlas.com/articles/seven-wonders-of-the-modern-world.html, 29 January 2023.

Yavari, F. (2019). *House conversions and sharing for an ageing population in New Zealand.* PhD Thesis. Victoria University of Wellington.

Zhubo Design Co Ltd. (2021a). *Zhubo Design.* Retrieved from https://www-zhubo-com.translate.goog/about/enterprise?_x_tr_sl=zh-CN&_x_tr_tl=en&_x_tr_hl=en&_x_tr_pto=sc&_x_tr_sch=http, 13 February 2023.

Zhubo Design Co Ltd. (2021b). *Qianhai China Metallurgical Technology Building Won LEED-CS Gold Pre-Certification.* Retrieved from http://zhubo.com/en/news/jingjinzhubo/645.html, 13 February 2023.

zoominfo. (n.d.-a). *Safdie Architects.* Retrieved from https://www.zoominfo.com/c/safdie-architects-llc/68126132, 14 February 2023.

zoominfo (n.d.-b). *Bjarke Ingels Group (BIG).* Retrieved from https://www.zoominfo.com/c/bjarke-ingels-group/, 14 February 2023.

Wealth, Cars and the Built Environment

4

Billionaires, Trillionaires and Cars

At the time of first drafting Chap. 2, Elon Musk was the wealthiest man in the world with a net worth of US$219 billion (Doland & Peterson-Withorn, 2022) but by September 2023 he had fallen to second place because of the money lost through a fall in shares after acquiring Twitter and in August 2023 was only worth US$180 million (Doland & Peterson-Withorn, 2023). Out of the six companies related to him, Elon Musk rapidly increased his wealth as the maker of Tesla electric cars and as a rocket producer through Space X, a company dedicated to sending things into space, including satellites. Contrary to most non-wealthy people, the numbers related to the wealthy are almost impossible to check. We do not know how many things they own, how much debt they have or how much money they make. The information available can be very accurate, partially right or a fantasy story. What is highly likely is that the manufacturing and selling of Tesla cars probably contributes significantly to Elon Musk's wealth, made possible by his being the biggest shareholder and owning almost 13% of the company (Oberoi, 2023). Tesla and Space X have in common that they are companies linked with moving people and things via terrestrial or aerial means, with strong emphasis on manufacturing and selling vehicles, these being cars and rockets. They also rely on using technology in an appealing way for the market. A big part of the success of Tesla was producing an electric vehicle that could compete with the look, comfort and features of any other luxurious car in the market. The success of Elon Musk interrupts the reign of billionaires related to software and computers, like Bill Gates, who was the first person to reach US$100 billion in 1999 (Harmon, 1999). When the wealthiest person in the world pushes the boundary of wealth one digit further, it forms a trend in the industry, a trend that many others will try to follow.

The climate change crisis set the perfect scenario for revamping a very old idea coined in the nineteenth century of motor vehicles powered by electricity. Nowhere in the SDGs, however, is there any indication that having electric rather than petrol cars might be a good idea. Rather target 11.2.1 of SDG 11 (Sustainable Cities and Communities) concerns access to public transport, not private transport. Nevertheless, changing to electric vehicles has been viewed as a positive step in climate change mitigation (International Energy Agency., 2022:6), although some doubt has now been cast on their effectiveness since it depends on how the electricity is generated and where the lifecycle boundaries are drawn (Ren et al., 2023). The problem is it seems easier to make money selling luxurious electric cars than public transport systems, while it is the latter that are needed in a sustainable future that addresses inequalities and poverty reduction.

E. J. García, B. Vale, *Wealth and the Built Environment*, Sustainable Development Goals Series,
https://doi.org/10.1007/978-3-031-81954-4_4

Perhaps it is simply the private car that is the problem, however, it is powered.

During the nineteenth century, the development of steam powered cars that used either kerosene or petrol as fuel, petrol cars using an internal combustion engine (ICE), and electric cars was progressing side by side with no clear front runner. In 1900, in the USA, more than one-third (38%) of automobiles were powered by electricity, 40% by steam and only 22% by petrol (Richardson, 2018). The number of automobiles in 1900 was only 33,000, which looks insignificant against the 2000 total of 133.6 million (Federal Highway Administration, 2023), and horses and bicycles were still used for private transport, while trains, street-cars and buses were public transport options (Sovacool, 2009). The same author notes:

> The early market was divided into at least six distinct segments including those rooted in the world of the horse, those wanting to rely on bicycles, those in favor of trains and subways, those displaying enthusiasm for new steam-powered "horseless" carriages, those desiring gasoline automobiles, and those preferring electric-powered vehicles.

Nevertheless, the fact at that time that 38% of the total private vehicle fleet was electric was years ahead of the current situation. In 2020, all electric vehicles formed 2.2% of US vehicle sales and this had only climbed to 6.1% by 2022, making EVs around 1% of the total fleet (Kilgore, 2023).

At the turn of the twentieth century, electric cars were ideal for going about town and given the state of the non-urban roads, driving in towns accounted for much of all driving. (The development of the road system in the USA is discussed below). However, by 1912 an electric vehicle cost US$1750 against US$650 for an ICE petrol car, sounding the death knell for the electric car (US Department of Energy, 2014). To revive the electric car, Elon Musk was interested in the business of manufacturing and selling electric cars that looked luxurious and as attractive to buyers as petrol cars. In analysing the triumph of the petrol car over the alternatives, Kirsch (1997) noted: "The automobile came to symbolize a specific constellation of social objectives — speed, technological prowess, the experience and conquest of road-accessible nature". The new electric Telsa, rather than being the product of socially focussed thinking as how best to move humanity around in an era of climate change mitigation, just aimed to compete with these same attributes of petrol cars.

Oil Availability Plus Internal Combustion Engine Vehicles Equal Billionaires

Elon Musk is not the first billionaire to have made money out of private vehicles. Around 1925, Henry Ford became one of the first billionaires by manufacturing and selling ICE vehicles (Daugherty, 2016). The emergence of the first billionaire in the car industry would not have been possible without the wider accessibility of petrol. At the beginning of the twentieth century, another billionaire, Rockefeller (see Chap. 2), built an economic empire through the acquisition of refineries and the economically successful expansion of the Standard Oil Company. In the 1880s, this company already had control of 90% of the oil refineries in the USA (Burclaff, 2021). The expansion of refineries made petrol more available and affordable which created the environment for innovation in the construction and manufacturing processes of vehicles (Fig. 4.1).

Toys for Rich Folks

In 2022, the White House announced that President Biden believes that "the future of transportation is electric" and that the USA must "lead the EV future" to "create jobs, make more in America, and fight climate change while advancing environmental justice" (The White House, 2022). The goal of the White House is "for electric vehicles to make up 50% of all vehicles sold in the United States by 2030" in 8 years (The White House, 2022). In 2022, there were 290.8 million cars in the USA and 2.86 million new cars were sold (Carlier, 2023), making about

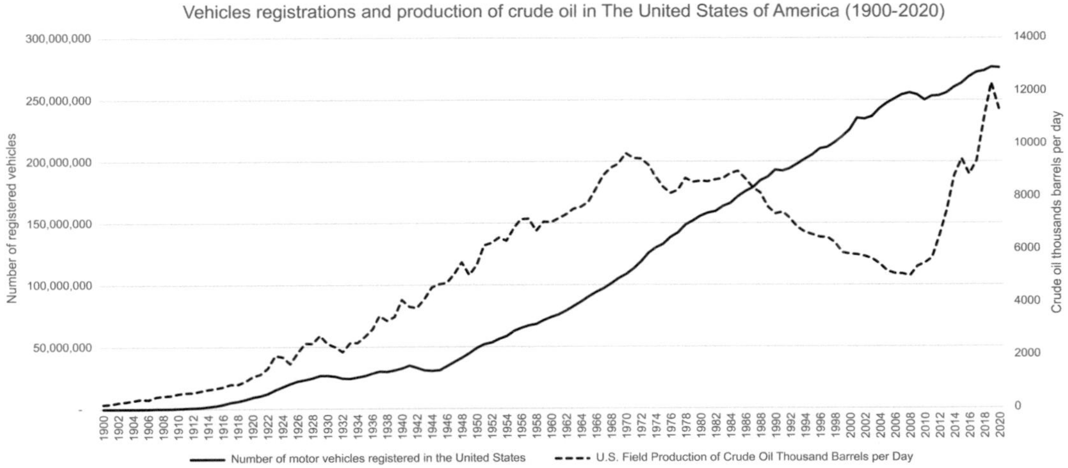

Fig. 4.1 Vehicle registrations and crude oil production in the USA (1900–2020). (Sources for number of registered vehicles (APTA, 2021) and for crude oil (EIA, 2023))

1% of all cars. This slow replacement rate means that increasing sales of electric vehicles will make little impression on the total car fleet. However, the goal of changing the vehicle fleet of a country from one source of energy is sensible. In terms of the environment, electric cars are more efficient than petrol cars in converting the fuel to motion: "EVs convert over 77% of the electrical energy from the grid to power at the wheels. Conventional gasoline vehicles only convert about 12–30% of the energy stored in gasoline to power at the wheels" (US Department of Energy, n.d.). Another advantage of electric vehicles is the lack of pollutants although this does depend on the source of electricity, as the pollutants may be generated elsewhere by the plant making electricity, especially if coal powered. Chicca et al. (2023:70) make the point that lower overall emissions may be had by a hybrid petrol car in a state like West Virginia where electricity is 88% from burning coal. Where electricity comes from renewable resources then electric vehicles always make sense environmentally, although the more carbon intensive production of electric vehicles over petrol powered ones also needs to be considered (Vitta, 2021). However, a study by the International Council on Clean Transportation "found that the life-cycle GHG emissions of average battery electric mid-size cars sold in the United States in 2021 were 57–68% lower than internal combustion engine (ICE) vehicles, depending on the average carbon intensity of the electricity grid" (O'Malley & Slowic, 2024).

The current interest in electric vehicles is, therefore, not new. The electric vehicle industry developed fast between the end of the nineteenth century and the beginning of the twentieth century. The most significant innovations involved creating an electric motor and experimenting with portable and rechargeable batteries. Even the inventor Thomas Edison was involved in developing the automotive industry. He created nickel-iron batteries (Abarro et al., 2023) that were used by the Baker company (Fig. 4.2). The company built a name for producing luxurious cars that in 1910 were priced from US$850 for the Imperial to US$3500 for both the Brougham and Landaulet, with most models coming in at above US$2000 (New York Public Library, 2023). This meant only rich people could afford them. A typical wage in 1910 would have been 30cents/hour for a 60-hour week (United States Bureau of Labor Statistics, 1913:68), equating to approximately US$900 per annum.

Even after the introduction of Henry Ford's Model T in 1908, electric cars were still prevalent in rural areas and among female consumers who found them safe, practical and easy to drive since they started instantly, did not need hand cranking

Fig. 4.2 Advert for Baker electric car. (Source: Creative Commons, n.d.-a)

and there was no need to shift gears (Sovacool, 2009). Ford's wife, Clara, drove a Detroit Electric car (The Henry Ford, 2022) because she found her husband's product "dirty and noisy" (Wilson, 2022). The female market was so crucial that Detroit Electric employed the daughter of the company's superintendent to further promote the brand among women (The Henry Ford, 2023). The Detroit Electric car company built 35,000 vehicles between 1908 and 1939 (Wilson, 2022), but even though they tried to make the cars more affordable, they still had a price of US$600, while ICE vehicles were much cheaper. In October 1922, Ford's Model T Touring was US$298 and the Model T Runabout was US$269 (MTFCA, 2023). Ford's cars experienced a massive drop price of almost 30% over 14 years since the first Model A Runabout was sold for US$850 in July 1903 (Transportation History, 2021). This was possible due to the innovation of the assembly line production process that transformed the way cars were produced (Esch, 2018:2).

Ford advertised his cars to both men and women (Flammang, 2022). Contrary to the success of electric vehicles (EVs) among women at the start of the twentieth century, they do not have the same appeal to women now. Modern EV buyers have been predominantly affluent white males (S & P Global Mobility, 2023). Tesla, the biggest EV seller in the USA, has also struggled to reach a female market that appears not to feel identified with the macho image projected by Elon Musk. In the first quarter of 2019, 69% of Tesla vehicles were registered by men, leaving 31% registered by women (Bomey, 2019). The situation has not changed much since. In 2022 "The industry average for car buyers is 41.2% women, but this percentage drops to 28% for vehicles with an electric powertrain", while Tesla stands at 33.1% female buyers (Pappas, 2023). Analysts have found that women tend to be more "practical than emotional" (Bomey, 2019). When buying a car, it seems men are seduced by the aspects of driving that might lead to driving faster, hence the attraction of the Tesla. The Insurance Institute for Highway Safety (2023) based on data from the US Department of Transportation's (2021) Fatality Analysis Reporting System states that men travelled more than women and accounted for 72% of passenger vehicle driver deaths, twice the number of female crash deaths. Key drivers of car crashes like speeding and alcohol were substantially higher and more frequent among males than females. It is clear that more female driving should be encouraged since women drive less, speed less, and wait to get home to have a drink. Because women are more responsible drivers, when buying EVs perhaps they want to avoid the hassle of getting stranded in the car and the impact that buying an expensive EV could have on family savings. These are only sales trends and do not describe all women's attitudes toward EVs. However, the cost of EVs is a more objective parameter since they are only affordable for the global population living in developed countries normally with a high GDP per capita. Since 2021, China has led the world in electric car sales (IEA, 2023a) despite in 2022 being ranked 72 in terms of GDP per capita (Worldometer, n.d.). However, an article published in 2019 showed that inequality in China was approaching the level of that in the USA (Piketty et al., 2019), which suggests the electric vehicles belong to the relatively wealthy in China.

In the USA, "two-thirds of households who own electric or plug-in hybrid vehicles earn more than $100,000" (Cage, 2022). The lack of affordability of EVs remains the same after more than a century, and this does not only apply to the USA. When writing this in September 2023 the cheapest EVs in New Zealand were the MG 415 Excite with a starting price without any rebate of NZ$46,990 and the GWM Ora with a starting price of NZ$47,900 (EVDB, 2023). Acknowledging that the median annual salary in New Zealand is NZ$56,160 (US$33,049) and a family of four in Auckland, the largest city, would need to earn NZ$162,600 (US$96,857) before tax to live comfortably (Swift, 2022), it is evident that EVs are entirely out of range for the middle class. Nevertheless, worldwide sales of electric cars have increased dramatically: "A total of 14% of all new cars sold were electric in 2022, up from around 9% in 2021 and less than 5% in 2020" (IEA, 2023b). Even though the research organisation BloombergNEF have suggested that EVs will be cheaper to produce (something that might not necessarily be reflected in their sales prices) than IEC vehicles by 2027 (Partridge, 2021), for now, they are still toys for wealthy people, especially rich white male folks.

Redefining Mass Transportation Patterns

The most important disruption caused by the diffusion of ICE vehicles in the first half of the twentieth century was their impact on modes of public transport, like streetcars and buses. However, this was not a sudden change. Ford released his first car in 1903, the Model A, but the model that made him move from millionaire to billionaire was the Model T which appeared in 1908. The car and company were so successful that between 1908 and 1919 they sold 3,487,455 vehicles, including trucks and ambulances. The biggest sellers were the Model T Touring and Model T Runabout. Together they sold three million cars (McCalley, 2007). The impact was so important that by 1922 47% of cars in the USA were Ford's Model T (Eli et al., 2023). A Forbes list of 1918 had Henry Ford within the top 10 millionaires with US$100 million and John D Rockefeller at the top of the list as the only billionaire with US$1.2 billion (Peterson-Withorn, 2017).

Contrary to the later habit of US car manufacturers of releasing a new model every year (see Chap. 6), Ford produced the same Model T for 19 consecutive years until 1927, at which time the Model T was "accounting for half of all the cars on the planet" (Esch, 2018:1). Famously, the only colour choice was black between 1914 and 1925, although at other times it could be grey, blue, red or green. By 1927 the production of the Model T ceased after selling more than 15 million units (Ford Motor Company, 2020a).

Regardless of the success of Ford's ICE vehicles and their marketed affordability, from 1918 to the mid-1920s, streetcars (trams) were still the most popular mode of public transportation. In 1923, streetcars accounted for almost all passenger trips, more than 13 million in total (APTA, 2021). In 1910, buses began to replace streetcars: "In a few cities, auto and auto-supply companies, including General Motors, Firestone Tire and Rubber Company and Standard Oil California, bought an interest in transit companies and encouraged the conversion from street car to bus" (National Museum of American History, n.d.). By 1940, passenger trips on streetcars (included in the light rail category in Fig. 4.3) decreased to six million, more than one-half in 17 years. In the same period, 1923 to 1940, the registration of ICE vehicles doubled from 15 to 32 million (See Fig. 4.3). Trolleybuses appeared in 1929 and were most popular in the 1950s.

Most car production ceased during WWII. Ford factories, for example, were used to produce B-24 bombers, tanks and armoured vehicles, trucks and ambulances among "military gear of every type, from aircraft engines to guns to helmets to tents" (The Henry Ford, 2017). Gasoline was also saved for war purposes and was not as readily available as before. The crisis and scarcity of resources favoured interest in revamping electric streetcars. They became more popular from 1941 to 1946 when passenger trips increased from 6 to 9 million (APTA, 2021); however, after WWII, ICE vehicle companies made a strong comeback supported by production of crude oil and availability of gasoline. The political decision was to favour the development of ICE buses instead of streetcars and trolleybuses, which were found very expensive to run and maintain. Bus passengers increased from 1946 to 1948, reaching more than ten million passenger trips, while streetcar and trolleybus passenger trips went down rapidly to four million in 1949, half the passenger trips made by bus that year (APTA, 2021). From 1956, the use of trolleybuses and streetcars remained very low, a situation that could have contributed to making the

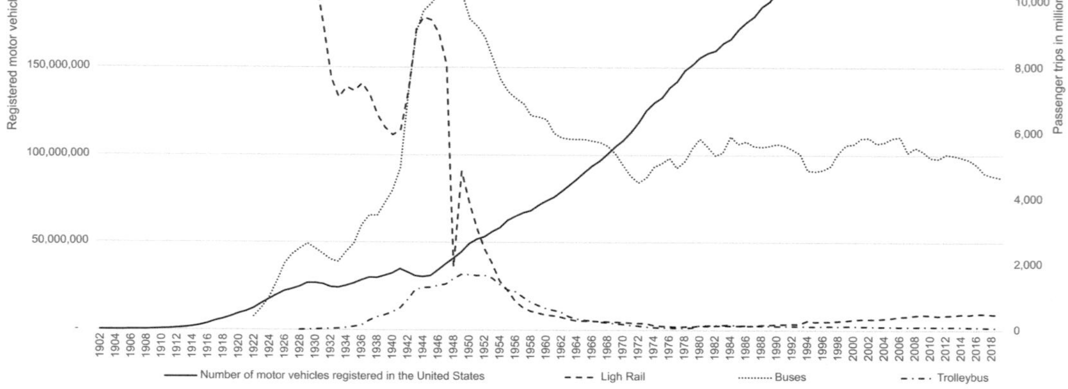

Fig. 4.3 Growth in registered vehicles and decrease in public transportation use. (Source for data APTA, 2021)

use of buses more popular but never did. The use of ICE buses went down from 1950 to 1973, even though the country's population kept growing and reached 215 million that year (O'Neill, 2022). The use of buses never surpassed five million passenger-trips, representing 40% of total trips done by streetcars in 1923 when the population was 115 million (O'Neill, 2022). However, the number of private vehicles registered in the USA rarely went down significantly, only dropping from 2008 to 2010 with the economic crisis (See Fig. 4.3). Since then, the number of registered vehicles (cars and trucks) reached 290.8 million in 2022 with a projection of there being 292.3 million in 2023 (Hedges and Company, 2023), placing the USA among the top countries for registered vehicles per 1000 people (Our World in Data, n.d.). This is working against the SDG 11 (Sustainable Cities and Communities) target 11.2.1 of access to public transport.

The Impact of ICE Vehicles on the Urban Landscape

Mass production and affordability of cars allowed people to cover greater distances and gave them more freedom in their trajectories. They could live further from the city centre because they did not have to depend on the public transportation network. However, many factories, including Ford's, were on the periphery of cities while most of their workers lived in city centres because it was cheaper (Barrow, 2018:20). The decision to build factories outside the city centre was deliberately taken to avoid paying more taxes. As a reporter noted in 1940: After all "beyond the city limits [there is] little traffic congestion, land is cheap, and taxes low". Eventually, many manufacturing jobs left the Detroit metropolis altogether, headed for the other Rustbelt cities, where they would be close to steel production..." (Barrow, 2018:149). This meant that some form of transport was required to bring the workers to the factories.

Public transport had already allowed the developments of suburbs in nineteenth century USA (National Museum of American History, n.d.):

> Suburbs were made possible by railroads, horsecars, cable cars, and electric streetcars. Some suburbanites left the city to get away from poor immigrants and migrants. Others believed that a quiet, less-congested area was better for health and family. In 19th century cities, people of different races and incomes lived in close proximity. With the rise of suburbs, communities became more sharply divided by race, wealth, and ethnicity.

This posed a problem for company owners building their factories on the cheaper land on the city periphery since workers on the factory floor could well be housed nearer the city centre. However, Ford had the answer. In 1914, he announced he would be paying his factory workers US$5/hour, double the normal rate for the job. It has been claimed that by doing this Ford was simply making sure that he had customers who could afford to buy his cars (Lazdowski, 2020:52), although other motives are discussed below. More critically for this part of the chapter, the business of selling cars and gasoline affected the growth and planning of cities by adjusting the shape of built environments so they could be both used by vehicles and accommodate them.

Roads

The success of the Model T was in part its power to weight ratio leading to its ability to deal with driving over the very poor roads in the USA at that time. Ford had used vanadium steel which made the car lighter, also meaning the engine could be smaller so the car was cheaper to run (Holland, 2019). All that was needed were roads for the Model T cars to drive on. A campaign by cyclists for better roads had begun as earlier as the 1880s (Wells, 2007), but roads for cycling could be both smaller and use less materials and hence be much cheaper than roads for the larger and heavier motor vehicles (Weigand et al., 2013). At the beginning of the twentieth century, better roads in the USA meant mobility so that people could look beyond their immediate environment to supply all their needs but it also changed the nature of who could use the roads (National Museum of American History, n.d.).

> In Washington, as in most cities in 1900, people usually walked or took public transportation. Some used bicycles. Wealthier residents owned their own carriages and usually stored carriages and stabled horses at commercial liveries. Improved streets allowed more traffic, but vehicles were still slow enough that pedestrians could walk in and cross the street at any point. Over the next 20 years this would change, as growing numbers of autos took over city streets.

Because of speed, roads became the place for cars and others like cyclists and pedestrians ventured on to them at their peril. Jacobs (1961:10) went further by stating: "Automobiles are often conveniently tagged as the villains responsible for the ills of cities and the disappointments and futilities of city planning. But the destructive effects of automobiles are much less a cause than a symptom of our incompetence at city building". Jacobs' comments were directed at highway designers and city planners that "do not know what to do with automobiles" because they do not know what makes and constitutes a liveable city in the first place "with or without automobiles" (Jacobs, 1961:10). When it comes to city design cars have limitations. They cannot scale steps as walkers can and so traverse steep gradients. They have to take a longer and less steep route, meaning more land is taken up. Rather than grapple with multiple means of transport going at different speeds and designing an environment in which this could be possible, most city planners seem to assume that motorised transport has priority, maybe because it is faster and therefore looks more efficient, and that a small sidewalk will cope with all remaining transport modes (pedestrians, pushchairs, scooters, cycles, skateboards, roller blades). The real problem is that it is easier to solve the needs of automobiles instead of facing the complexity of movement in cities since that would imply understanding who is moving where and when and designing streets to be take account of this, thus making them more liveable (Fig. 4.4).

Although decentralisation began with public transport and the development of suburbs connected to the city centre by tram or rail, the car gave greater freedom for those wishing to live away from the centre. Jacobs's (1961) criticism noted above highlighted the fact that well-renowned planners and designers contributed to destroying cities through ideas that promoted decentralisation and homogeneity, such as Garnier's 1917 Cité Industrielle where industry was clearly separated from housing (Wiebenson, 1969:11). For Jacobs, concepts of decentralisation came from Ebenezer Howard's Garden Cities and the work of Patrick Geddes. Their ideas were later welcomed by Lewis Mumford, Clarence Stein, Henry Wright and Catherine Bauer in the USA (Jacobs, 1961:27). This group saw overcrowded metropolitan cities as bad. The solution was to leave them behind and live in smaller new self-sufficient, scientifically planned communities. Providing decent housing was the primary concern and all other activities that would provide employment for the garden city citizens were to be self-contained and separated. In the UK, Letchworth, known as the first Garden City and with a layout designed by Parker and Unwin, was started with a temporary railway station in 1903 with the construction of houses starting in 1904 (Miller, 1983). In 1905, the new manager of the garden city company began work to turn Letchworth into an industrial town (Pastor et al., 2023). In 1910, the Spirella Company (a corset manufacturer) started production in the town in temporary premises, building the first factory there in 1912 (Historic England, 2023). The town attracted other factories, but the British Tabulating Machine Company, which moved from London, was its biggest source of employment, developing up to 30 installations in Letchworth before WWI (Martin, 1990:24). However, in the case of Ford's business plan, his second factory at Highland Park was established in 1910, after outgrowing the 1904 Piquette Avenue Plant in Detroit, and was in an enclave of Detroit away from the city centre (Ford Piquette Avenue Plant, 2023), thereby forcing his workforce to commute.

Jacobs has argued that the Garden Cities influenced planners and architects, impacted planning legislation, and, most importantly, killed the economies and liveability of cities. Although

Fig. 4.4 Separation of pedestrians, tram and road vehicles at Vauban, Germany. (Source: Creative Commons, n.d.-b)

Howard's original cities had been organised around the railway as the means of transportation, the increasing availability of the private car was to affect this. The homogenised landscape of the Garden City meant that the vast majority of the urban landscape was dedicated to a single function, whether residential, industrial or for leisure. As the population grows in such a Garden City, with land devoted to a single purpose such as residential or commercial, different activities end up far from each other. Jobs become far from home, along with banks, supermarkets, libraries, sports facilities and even schools. By decentralising and homogenising the urban area, people needed to use the car more and cover long distances to do more than one thing daily. The ultimate quality affected was the diversity of cities, which for Jacobs was the most crucial component of their complexity.

Put this way gives the impression that architects and planners changed the urban landscape of the USA by themselves. However, this would be underestimating the power of the main driver of urbanisation: wealth accumulation. As noted above, Ford moved his factory to Highland Park in 1910 where in 1913 an assembly line was established so cars could be mass-produced more efficiently (Ford Motor Company, 2020b). The expansion of Ford's business was ambitious, but it was based on the sudden and solid sales of the Model T. Automobiles were assembled at Highland Park, but the factory relied on a network of providers to produce the different car parts. With the continued success of the Model T, Ford needed to find a site with a massive land availability that was cheap and where taxes were lower than those of the city to maximise profits.

> He decided to scrap most of his existing multistory assembly plants by 1919. Very few manufacturers who had just invested millions of dollars in a network of assembly plants would have been willing to underwrite a second major expansion, but Ford was committed to making a car for the masses and he needed a new building type. (McMahon, 2016:57)

The site for the Ford River Rouge complex in Dearborn, Michigan, ticked all these boxes. Moreover, Ford had grown up in the area. The factory was designed by architect Albert Kahn and was built between 1917 and 1928. From the start, an area was set aside for parking as "an estimated 7500 workers parked their cars at Ford's Dearborn plant, and by 1929 fifty thousand workers were driving to the Rouge plant which had set aside 122 acres for parking" (McMahon, 2016:57–58). Ford had provided some housing at Dearborn for those working in his tractor factory there to avoid the long commute by tram from the city. Between 1919 and 1921 when the tractor factory had been transferred to River Rouge 250 houses had been built (Eckert, n.d.). However, the vast majority of Ford workers were commuters and as noted above, many came by car.

The question is who made the first move toward decentralising cities in America, industrialists like Ford with their factories on the edges of settlements or architects and planners? Jacobs proposed that decentralisation originated with the Garden City movement, and that this influenced planners and architects in the USA. The first US experiment with garden city principles came with the development of Sunnyside Gardens in 1924–28 in Queens, New York. "The Site was convenient for workers in the office centers of Manhattan. The rapid transit station at the corner of the property was 15 minutes travel time from the 42nd Street center. Land was cheap…" (Stein, 1949). Nevertheless, cars were present in Sunnyside Gardens, garaged together in a court. In the same article, Stein described the need for the next experimental settlement of Radburn because the car had become a "disrupting menace to city life in the USA—long before it was in Europe". Radburn began building in 1927 and has been described as a satellite "Town for the motor age" being sited 12 miles from New York City in fields but with a "planned highway and rail line that was projected to provide easy access to the midlevel employment of most of the residents" (McClelland et al., 2005). It was assumed that every family in Radburn would have a car (Buder, 1990:168). The great thing about land on the edge of the city was that it was cheap but the downside was locking residents into a pattern of commuting even though home was in harmonious surroundings. Radburn was never finished as planned because of the great depression but its trademark layout of pedestrian paths separated from roads has become legendary (Fig. 4.5). The settlement has also been linked to discriminatory policies since there was discussion during the planning as to whether people other than those of white race should be allowed to live there (Birch, 1983:126). There is evidence that this discrimination took place in a 1929 letter from City Housing Corporation, the institution behind the building of Sunnyside Gardens and Radburn. This letter was in response to an enquiry from the civil rights activist W.E.B. Du Bois. The City Housing Corporation letter states that: "We have in mind embarking shortly on a separate negro community" (University of Massachusetts Amherst, n.d.). However, this racial segregation in housing was common in the USA at that time and continued for many decades (Massey & Denton, 1993:19).

Despite the intentions of planners and designers to create garden cities in the USA, albeit only for what they saw as the right sort of people, the spread of suburban America must owe something to Henry Ford. Ford started the process of industrial decentralisation on a big scale when he bought the land to build his River Rouge complex in 1915. By 1930, Dearborn had 50,538 inhabitants, a growth of 1938% in population since the arrival of the development of the Ford River Rouge complex:

> In the Detroit area, suburbanization was led mostly by Henry Ford, who not only located an entire manufacturing complex over the city's border in Dearborn, but was first among the Big Three to make the automobile a mass-consumer item. The city's suburbanization, in other words, was spurred simultaneously by the migration of the automobile industry and the mobility of automobile users, and Ford was involved with both (Barrow, 2018:16)

If Howard's idea was for the Garden Cities to be "magnets", Ford pulled thousands of people out of the city with one building and no particular

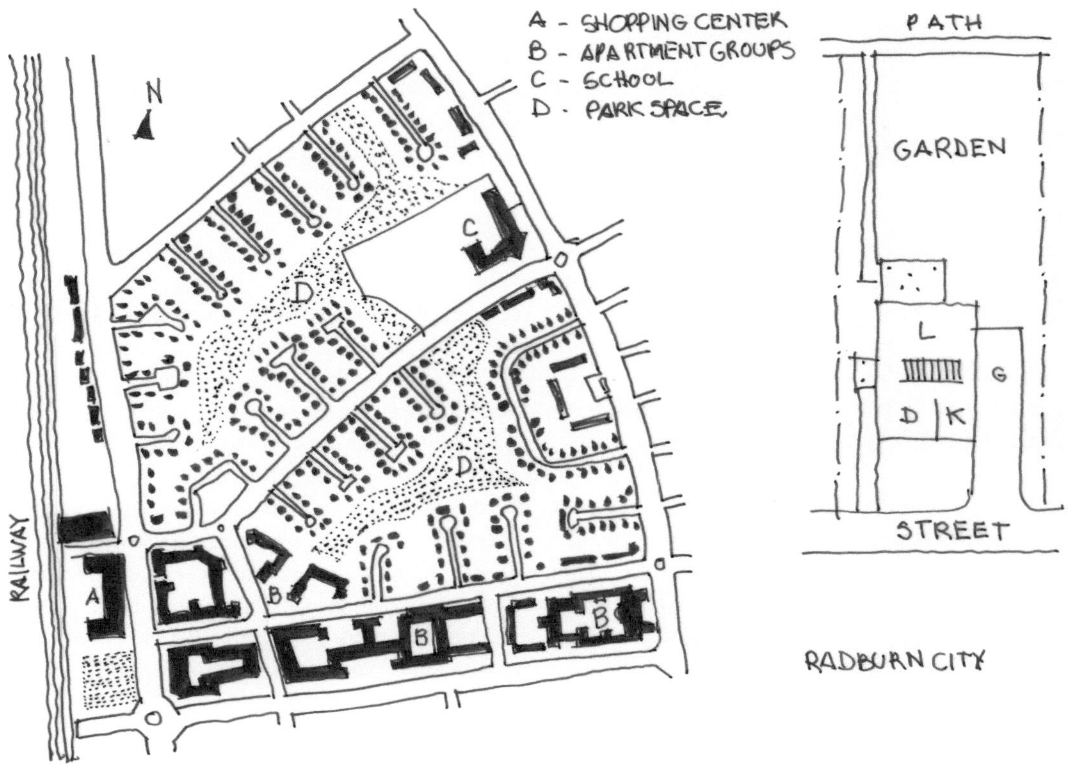

Fig. 4.5 The Radburn layout. (Based on Archi-Monarch, n.d.)

interest in planning. Orlando (2022:22) goes further:

> It's hard to overstate the effect of the automobile—and especially his [Ford's] affordable Model T—on the marketability of American land. It didn't just revolutionize transportation. By changing where people could go and when they would get there, the automobile revolutionized how they used the land, how they organized their lives across the great expanse of space. It opened vast new lots—newly valuable assets—for buying and selling.

The massive development of Dearborn was to the detriment of the city of Detroit, which had to cope with allocating and providing for all the blue-collar workers segregated and discriminated from the new suburbs predominantly designed for white-collar workers from the Ford factory. Barrow (2018:16) and Orlando (2022:22) both make the point that the growth in suburban America for those who could afford the car left the centre of the city for those who could not and who were more likely to live in overcrowded conditions in existing older dwellings, especially as the promise of work in the new factories attracted people to the cities. Meanwhile roads were being built and paid for not just by road users but by all tax payers (see below). "In essence, cities were redistributing wealth from the poorest residents, who couldn't afford a car, to the wealthier suburbanites" (Orlando, 2022:22).

The desire to accumulate wealth could be seen as the main driver of change in the built environment of American cities. It was not the ideas of idealistic planners that transformed the urban landscape of cities in the USA but wealth accumulation. The wealthiest sectors of the population made decentralisation possible by producing or buying cars. The built environment amplified the negative impact of automobiles on the cityscape and economy of cities. Responsibilities for dealing with cars were thrown on to the city via its taxpayers and distributed unequally.

The case of Ford has been used here as a sample of how billionaires adversely affect society and challenges the idea that when wealthy people do well economically, so does everyone else. In Ford's case, producing his wealth created unintended consequences that leave a clear footprint in the built environment. People profiting from producing cars created a problem for land use because cars need roads to drive on and land to occupy when they are not being driven.

Public and Private Debt: The Legacy of Cars and Billionaires

At the end of the 1920s, Ford successfully sold cars, and Rockefeller provided the oil they needed to function. There was one missing element in the equation. For people to move from point A to point B in the city through motorised means, they needed a car, gasoline and roads. The problem is that everybody needs roads, primarily for cars, but nobody makes money out of them.

In the USA, the highway system debt has been increasing steadily since the aftermath of WWII, especially after the Federal Highway Act of 1956. Until 1946, the total existing miles of the system were 3.3 million. By 2021, there were almost 4.2 million miles built (US Department of Transportation, 2021). However, the debt as a result of building the extra 30% of total mileage from 1946 to 2021 has gone from 0.07% of the real GDP of the country in 1946 to 1.2% of the real GDP in 2021, which means the debt has increased 17 times (See Fig. 4.6).

Ford was in the business of making money, not losing it, and he needed people to invest in roads. Since the Federal Highway Act of 1956 was signed by President Eisenhower, the highway system has been considered one of the "greatest public works projects" (US Department of Transportation, 2006). However, the "modern interstate highway system owes its existence to Ford's invention" (Anon, 2005). Without the mass production of cars, there would be no need for an interstate highway system. If Eisenhower was the father of the Federal Highways, Ford is the father of its debt, a legacy that has lasted for decades and that has been more difficult to deal with since the debt per existing mile has increased at least 6 times in less than 30 years (from 1996 to 2021) (see Fig. 4.7). The debt of building for cars will last many generations.

Many roads were needed as Ford's idea of paying a US$5-a-day rate has, as noted above, been claimed to allow workers to buy a Ford vehicle and contribute to creating a blue-collar class capable of consuming and therefore helping the company to sell more cars (Lazdowski, 2020:52). Stated in this way, it looked like a win-win situation; workers get better wages and Ford does better business. However, is it a win-win situation? Firstly, the impact on sales would have been marginal even if workers bought a new car every year. The real advantage for Ford from the higher wages was staff retention to avoid the previous high turnover of staff and the costs of training new workers (Worstall, 2012). Ford was a terrific salesman with marketing techniques and ideas ahead of his time. Elon Musk is also a skilful seller with innovative marketing strategies for the industry, like not investing in a marketing department or marketing strategy at all. Selling the Tesla is all word-of-mouth or, more precisely, word-on-social-media, thus selling a product and making the buyers into an unpaid advertising department.

The Impact of Roads

In New York City, almost 32% of land is devoted to transportation compared to 13.5% in Seoul (Huu & Ngoc, 2021) but this is total transportation not just roads. A better indication of the impact of the car on the built environment is shown by "the time-area requirements of various modes for a 20-minute commute with 8 hours of vehicle parking (no parking is required for walking or public transport)" (Victoria Transport Policy Institute, 2021) (Fig. 4.8).

With the information in Fig. 4.8, a city designer would surely plan for bus travel supplemented by cycling and walking and the car would be a non-starter given its huge land take, especially when just sitting idle. Free parking for the

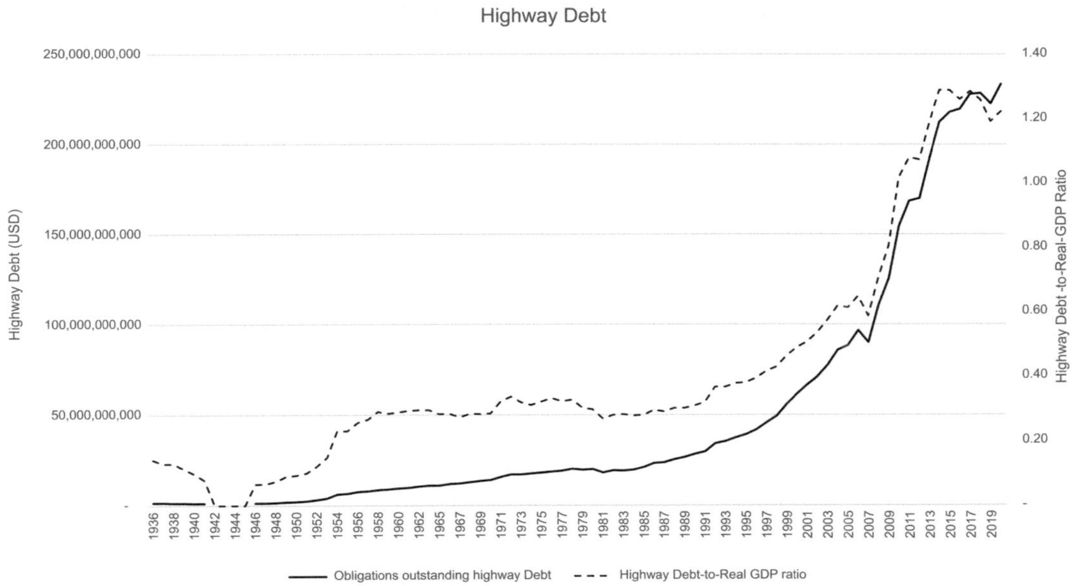

Fig. 4.6 Highway debt and relationship to GDP. Source of data (US Department of Transportation, 2021). The source for real GDP (Amadeo, 2022)

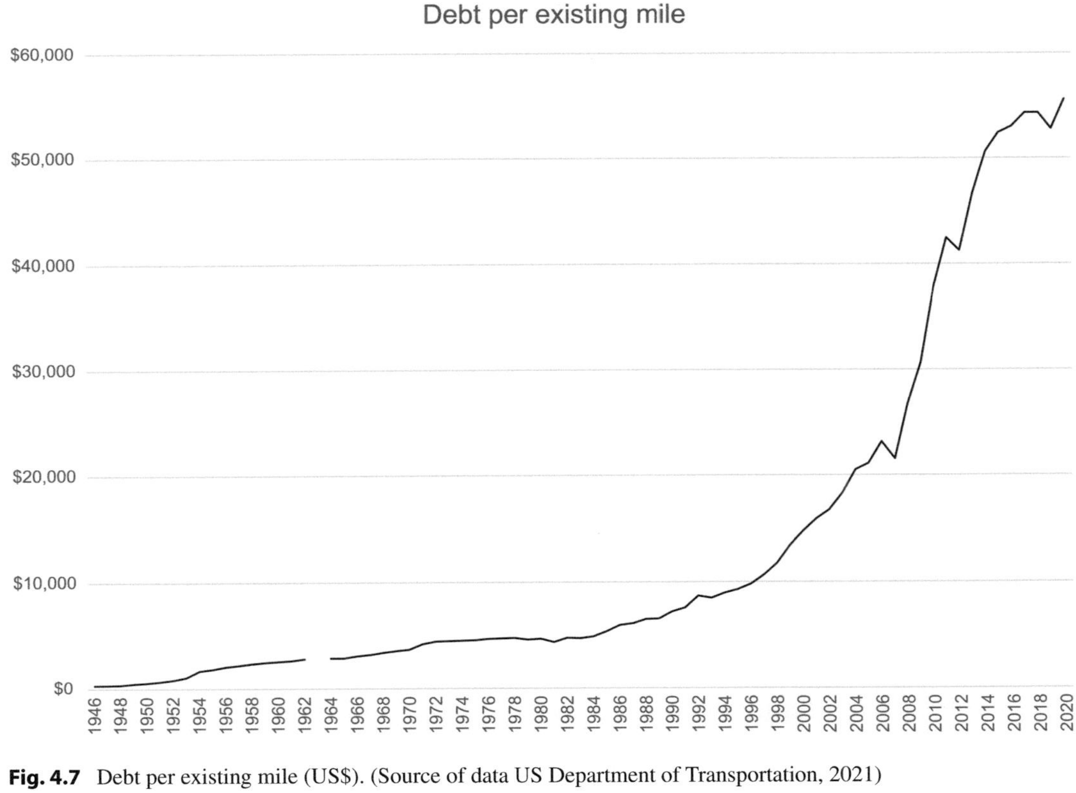

Fig. 4.7 Debt per existing mile (US$). (Source of data US Department of Transportation, 2021)

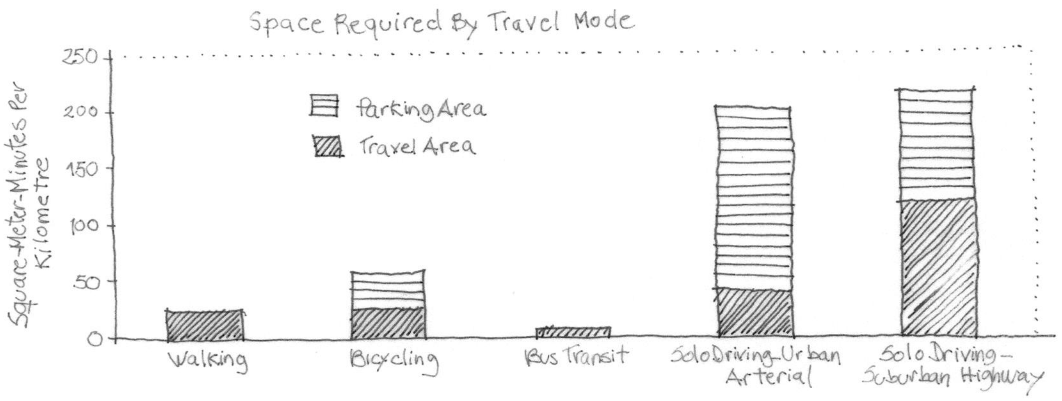

Fig. 4.8 Land taken up by various forms of transport based on a 20 min commute and 8 h parking. (Based on Victoria Transport Policy Institute, 2021)

car is another insidious land use that also affects equality. Shoup (2018:6) notes that: "People who cannot afford to own a car pay more for their groceries to ensure that richer people can park for free when they drive to the store". Parking spaces are normally impervious so increased run-off can lead to flooding. They are also generally of a dark asphalt which heats the run-off water and this, as well as associated pollutants, can interfere with the eco-systems of nearby streams or ponds (Kirkpatrick et al., 2018:136). At home the car may be given a built space—the garage—made of materials similar to those of the main house, even though the car is designed to be weatherproof. It might make more sense to convert garages to housing and park cars on the street (Brown et al., 2020).

Transportation Dependence and Radical Monopoly

The mass production of ICE vehicles created a faster mode of transportation than feet, donkeys, horses or bicycles and was imposed on the built environment with consequences that are currently accepted by most societies. Given climate change mitigation involves reducing the use of fossil fuel, mass changing to electric vehicles would seem to be a solution, at least for those who can afford them. However, Ivan Illich (1974) proposed that above a certain threshold, transportation speed increases energy consumption and this has a profound impact on equity, and hence on SDG 10 Reduced Inequalities. The threshold of speed proposed by Illich is 15 mph (24 kph), a comfortable speed for a beginner bicycle rider, above which Illich proposed there is a detrimental impact on society from transportation.

> High speed is the critical factor which makes transportation socially destructive. A true choice among practical policies and of desirable social relations is possible only where speed is restrained. Participatory democracy demands low-energy technology, and free people must travel the road to productive social relations at the speed of a bicycle. (Illich, 1974:12)

As noted above the early predominance of ICE vehicles on the streets became an issue for people who walked as crossing the road became a problem. Illich's critique was that the addiction to speed, viewed as a sign of progress, displaced or marginalised all the other transportation modes and made the population dependent on having or not having a car. In transportation, too, winners take all. Illich saw this change not only as an attempt to conquer a market and accumulate more money for some but as a problem for ensuring a healthy democracy.

> Industry dominates traffic as soon as daily life comes to depend on motorized trips…This profound control of the transportation industry over natural mobility constitutes a monopoly much more pervasive than either the commercial monop-

oly Ford might win over the automobile market, or the political monopoly car manufacturers might wield against the development of trains and buses. (Illich, 1974:12)

One particular effect of speed in society can be seen in the built environment. With Ford's mass production of vehicles, speed became a sign of progress and a way of shortening the distances within a territory, but only for the people who could afford to buy a car. For the rest, life remained the same but felt more miserable because they were not part of the new mobility. As noted above, cities were developed to benefit the relatively wealthy, not the relatively poor. With more cars and increasing speed came the need for new and more expensive roads and motorways, paid for in part by road users and in part by the taxes of all. This is illustrated by the current situation in the USA. The money for federal highways and investment in transport infrastructure comes from the Highway Trust Fund and in 2021 76% of its money came from road users mostly through tax on fuel and the remaining 24% from the US Treasury General Fund, hence from other taxes (American Road and Transportation Builders Association, 2023). More critically, Ditzik et al. (2015:2) noted that: "Most walking and bicycling takes place on local streets and roads that are primarily paid for through property taxes and other general local taxes". However, the share of these activities on local roads is small compared to the space given up to motorised vehicles. In the past, US road users paid even less of the cost of roads.

> The mythology of transportation finance in the United States is that motorists pay for what they get and get what they pay for when it comes to gas taxes and roads. This myth dates back to the early debates about how to finance highway construction in the first decades of the 20th century. At the time, cities and towns paid for the cost of building and maintaining streets out of general tax revenue or assessments on property owners, while state expenditures on highways were minimal and federal spending on transportation was nearly nonexistent. In 1921, the first year the federal government tracked public spending on highways, local governments accounted for 70 percent of all spending on roads, more than half of which came from property taxes and assessments, and only about one-sixth of which came from charges assessed to drivers. (Ditzik et al., 2015:6)

Henry Ford's wealth came from selling cars but this wealth was subsidised by the tax payer through the building of roads.

Roads and Property Values

One factor in a good location when it comes to buying or renting a dwelling is accessibility as "…a house with easy access to roads and public transportation will be more desirable than one tucked away and [that] can only be accessed by one route" (Struyk, 2023). However, a study of the effect on property of a new highway found that it was increasing traffic flow that tended to lower property values (Carey, 2001:iv). This leads to a potential situation where those without the wealth to buy a car also find themselves in the cheapest housing on a busy traffic route, suggesting accessibility to the network of streets and highways also creates inequality in the built environment.

The Future Must Be the Bicycle

Ford contributed to building Illich's radical monopoly through the reduction in the diversity of mobilisation to the point of dependence on a single object—the car—and with this, the rearrangement of society. In effect ICE vehicles, originally created as a means of transportation, became all pervasive. This is the radical part of the change when the population is pushed to become dependent on an industrial product whose sales will only benefit a few. This mechanism did and still does make wealth accumulation possible for leaders of the automotive industry. When everyday life is disrupted by a product (any industrial output), new habits are created around its use. However, Illich proposed that some empower autonomy, and others enslave you. The bicycle frees you because it neither challenges your autonomy nor replaces your capacity to run or walk but helps to cover longer distances at a low energy cost.

Most importantly, adapting to changes produced by industrial outputs, like the effect of cars on the built environment, leads to their continual consumption because the built environment no longer works unless most people have cars. The criticism of the transportation industry made by Illich was an excuse to attack the excess of industrialisation, or as he termed it over-efficiency, which creates inequalities. One essential point of the criticism is that there are limits to production and consumption above which marginal returns emerge. In the case of transportation, the adverse effects can be seen in the social configuration of the built environment. Ford, and the other leaders of the automotive industry created speed for a few and further distances between amenities for the rest of the population. Not only that, as shown in Fig. 4.3 the advent of ICE vehicles destroyed the public transport system, increasing, even more, the dependence on the private car at the cost of diminishing public transport for everyone. Illich criticised architects and planners for not challenging these industrial outputs. Instead, they adjusted the habitat to internalise the problems produced by cars while never questioning their existence, assuming that cars were an inevitable need or condition for progress.

> These futuristic architects would house and occupy people in self-sufficient units of towers interconnected by tracks for high-speed capsules. Soleri, Doxiadis, or Fuller would solve the problem created by high-speed transport by identifying the entire human habitat with the problem. Rather than asking how the earth's surface can be preserved for people, they ask how reservations necessary for the survival of people can be established on an earth that has been reshaped for the sake of industrial outputs (Illich, 1974:15)

The electric car may be seen as an essential part of climate change mitigation but it is still a car and its effects on the built environment will be no different from those of a car running on fossil fuels. For Illich the only way to reduce inequity in the built environment was to return to the bicycle for all, whether electric or human powered, supplemented by public transport systems for long distance and freight travel. This would free up much inner-city land for housing currently given over to roads and parking spaces for the car, and would also mean that this method of densification would not increase the area of impervious surfaces, which in turn increases the risk of flooding in the face of climate change. However, globally 42% of households already own at least one bicycle (Oke et al., 2015) so a lot more money remains to be made from selling electric cars than mobilising the world by bike.

Wealth, Cars, Sustainable Cities and the SDGs

Through the development and impact of Ford's cars on the built environment, this chapter shows how the link between SDG 8 (Decent Work and Economic Growth) and SDG 9 (Industry Innovation and Infrastructure), assumed positive in the SDGs can, in turn, be found to be negative if the impact of wealth accumulation is taken into account. Industrial innovation in the case of the car does not necessarily imply decent jobs, and although economic growth might be useful for macroeconomics, it does not cascade down to affect everyone positively. In the short term, it can soften unemployment numbers and create millionaires, but this is done with tax-payers' money, re-purposing the city infrastructure to the benefit of a few. The process is done with the complicity of consumers and their faith in progress, which is an obstacle to reducing inequality (SDG 10 Reduced Inequalities) and creating a sustainable city or community (SDG 11 Sustainable Cities and Communities). Modern reliance on the car is also a stumbling block for SDG 7—Affordable and Clean energy—since it takes much less energy to power a bicycle than a car. SDG 12—Responsible Consumption and Production—might also suggest that the age of the car is over since the planet could not cope if everyone had access to a motorised vehicle. The problem is that having less tends not to make certain people rich.

References

Abarro, J. M. E., Gavan, J. N. L., Loresca, D. E. D., Ortega, M. A. A., Esparcia, E. A., & Paraggua, J. A. D. R. (2023). A tale of nickel-iron batteries: Its resurgence in the age of modern batteries. *Batteries, 9*(7), 383.

Amadeo, K. (2022). *U.S. GDP by year, compared to recessions and events.* The Balance. Retrieved from https://www.thebalancemoney.com/us-gdp-by-year-3305543#citation-7, 13 November 2023.

American Road and Transportation Builders Association. (2023). *How is transportation infrastructure paid for in the U.S.?* Retrieved from https://www.artba.org/about/faq/#:~:text=The%20major%20source%20of%20funding,federal%20investment%20in%20highway%20improvements, 4 October 2023.

Anon. (2005). *Henry Ford (1863–1947).* Retrieved from https://www.forbes.com/2005/07/29/ford-autos-assembly-cx_0729bizmanford.html?sh=340efbb126b4, 17 November 2022.

APTA. (2021). *Public transportation fact book Appendix A: Historical tables excel.* American Public Transportation Association. Retrieved from https://www.apta.com/wp-content/uploads/2021-APTA-Fact-Book-Appendix-A.xlsx, 13 November 2023.

Archi-Monarch. (n.d.). *Radburn theory.* Retrieved from https://archi-monarch.com/radburn-theory/, 20 November 2023.

Barrow, H. (2018). *Henry Ford's Plan for the American Suburb: Dearborn and Detroit.* Cornell University Press.

Birch, E. L. (1983). Radburn and the American planning movement. In D. A. Krueckberg (Ed.), *Introduction to planning history in the United States* (pp. 122–151). Routledge.

Bomey, N. (2019). *Why Tesla 'has a problem appealing to women': Electric cars, Elon Musk may be off-putting.* Retrieved from https://www.usatoday.com/story/money/cars/2019/07/08/tesla-electric-cars-women-elon-musk/1616632001/, 18 September 2023.

Brown, A., Mukhija, V., & Shoup, D. (2020). Converting garages into housing. *Journal of Planning Education and Research, 40*(1), 56–68.

Buder, S. (1990). *Visionaries and planners: The Garden City movement and the modern community.* Oxford University Press.

Burclaff, N. (2021). *Standard oil established.* Retrieved from https://guides.loc.gov/this-month-in-business-history/january/standard-oil-established#:~:text=By%20the%20late%201880s%2C%20Standard,laws%20and%20prohibiting%20train%20rebates, 17 September 2023.

Cage, F. (2022). *The long road to electric cars.* Retrieved from https://www.reuters.com/graphics/AUTOS-ELECTRIC/USA/mopanyqxwva/, 20 September 2023.

Carey, J. (2001). *Impact of highways on property values: Case study of the Superstition Freeway Corridor.* Arizona Department of Transportation.

Carlier, M. (2023). *U.S.: Annual car sales 1951–1922.* Retrieved from https://www.statista.com/statistics/199974/us-car-sales-since-1951/, 5 November 2023.

Chicca, F., Vale, B., & Vale, R. (2023). *The environmental impact of cities.* Routledge.

Creative Commons. (n.d.-a). "1916 Baker and Rauch & Lang Electric Cars" by aldenjewell is licensed under CC BY 2.0.

Creative Commons. (n.d.-b). "File:Sonnenschiff in Freiburg-Vauban 4.jpg" by Andreas Schwarzkopf is licensed under CC BY-SA 3.0.

Daugherty, G. (2016). *Who was really the first American billionaire.* Retrieved from America's First Billionaire Isn't the Guy We Thought | Money 15 September 2023.

Ditzik, T., Weissman, G., & Baxandall, P. (2015). *Who pays for roads?* Retrieved from https://publicinterest-network.org/wp-content/uploads/2015/05/Who-Pays-for-Roads-vUS-1.pdf, 4 October 2023.

Doland, K., & Peterson-Withorn, C. (2022). *Forbes World's Billionaires List. The Richest in 2022.* Retrieved from https://www.forbes.com/billionaires/, 26 October 2022.

Doland, K., & Peterson-Withorn, C. (2023). *Forbes World's Billionaires List. The Richest in 2023.* Retrieved from https://www.forbes.com/billionaires/, 14 September 2024.

Eckert, K. (n.d.). *Ford Homes.* Retrieved from https://sah-archipedia.org/buildings/MI-01-WN138, 2 October 2023.

EIA. (2023). *Petroleum and other liquids. U.S. Field production of crude oil. U.S. Energy Information Administration.* Retrieved from https://www.eia.gov/dnav/pet/hist/LeafHandler.ashx?n=pet&s=mcrfpus2&f=a, 13 November 2023.

Eli, S., Hausman, J. K., & Rhode, P. (2023). *The Model T. National Bureau of Economic Research* (Working paper 31454). Retrieved from https://www.nber.org/system/files/working_papers/w31454/w31454.pdf, 20 September 2023.

Esch, E. D. (2018). *The color line and the assembly line.* University of California Press.

EVDB. (2023). *10 cheapest electric cars in New Zealand.* Retrieved from https://evdb.nz/cheapest-electric-cars-nz#, 20 September 2023.

Federal Highway Administration (FHWA). (2023). *The vehicle fleet.* Retrieved from https://www.fhwa.dot.gov/ohim/onh00/onh2p3.htm, 17 September 2023.

Flammang, J. M. (2022). *Poor men's dreams—and the reality.* Retrieved from https://www.tirekick.com/Clunkers-Chapter%202-Model%20T.htm, 18 September 2023.

Ford Motor Company (2020a). *The model T.* Retrieved from https://corporate.ford.com/articles/history/the-model-t.html#:~:text=More%20than%2015%2C000%2C000%20Model%20T's,old%2C%20its%20legacy%20is%20timeless, 20 September 2023.

Ford Motor Company. (2020b). *Highland Park.* Retrieved from https://corporate.ford.com/articles/history/highland-park.html, 27 September 2023.

Ford Piquette Avenue Plant. (2023). *Our History.* Retrieved from https://www.fordpiquetteplant.org/about/history/, 27 September 2023.

Harmon, A. (1999). *Gates Hits $100 Billion Mark, More or Less.* Retrieved from https://www.nytimes.com/1999/07/17/business/gates-hits-100-billion-mark-more-or-less.html, 14 September 2023.

Hedges and Company. (2023). *US vehicle registration statistics.* Retrieved from https://hedgescompany.com/automotive-market-research-statistics/auto-mailing-lists-and-marketing/, 20 September 2023.

Historic England. (2023). *The Spirella Building, associated fountain, lamp standards and steps.* Retrieved from https://historicengland.org.uk/listing/the-list/list-entry/1347670?section=official-list-entry, 27 September 2023.

Holland, E. (2019). *Ford's miracle metal.* Bloomberg Businessweek 4628. 35.

Huu, D. N., & Ngoc, V. N. (2021). Analysis study of current transportation status in Vietnam's urban traffic and the transition to electric two-wheelers mobility. *Sustainability, 13*, 27.

Illich, I. (1974). *Energy and equity*. Calder & Boyars.

International Energy Agency. (2022). *Global EV outlook 2022.* Retrieved from www.iea.org. 15 April 2024.

International Energy Agency (IEA). (2023a). *Electric vehicles.* Retrieved from https://www.iea.org/energy-system/transport/electric-vehicles, 20 September 2023.

International Energy Agency (IEA). (2023b). *Global EV outlook 2023: executive summary.* Retrieved from https://www.iea.org/reports/global-ev-outlook-2023/executive-summary, 20 September 2023.

Jacobs, J. (1961). *The death and life of great American cities*. Random House.

Kilgore, G. (2023). *How many electric cars are there in the United States? We found out.* Retrieved from https://8billiontrees.com/carbon-offsets-credits/cars/how-many-electric-cars-in-the-us/, 17 September 2023.

Kirkpatrick, E., Davis, A., & Pijanowski, B. (2018). The environmental impacts of parking lots. In D. Shoup (Ed.), *Parking and the city* (pp. 133–140). Routledge.

Kirsch, D. A. (1997). The electric car and the burden of History: Studies in automotive systems rivalry in America, 1890–1996. *Business and Economic History, 26*(2), 304–310.

Lazdowski, Y. J. (2020). *Persistence and vigilance: A view of Ford Motor Company's accounting over its first fifty years*. Emerald Publishing Company.

Martin, C. (1990). The British Tabulating Machine Company, 1907–1919. In M. Campbell-Kelly (Ed.), *ICL: A business and technical History* (pp. 24–46). Oxford University Press.

Massey, D. S., & Denton, N. A. (1993). *American apartheid segregation and the making of the underclass.* Harvard University Press.

McCalley, B. W. (2007). *Model T Ford Sales.* Retrieved from https://www.mtfca.com/encyclo/fdsales.htm, 20 September 2023.

McClelland, L. F., Reed, P. S., & Wallace, E. B. (2005). Revisiting Radburn: "Where art and nature combine to make good living conditions"'. *New Jersey History, 12*(1–2), 39–107.

McMahon, B. (2016). *The Ford Century in Minnesota.* University of Minnesota Press.

Miller, M. (1983). Letchworth Garden City eighty years on. *Built Environment, 9*(3/4), 167–184.

Model T Ford Club of America (MTFCA). (2023). *1922.* Retrieved from https://www.mtfca.com/model_t_encyclopedia/1922/, 18 September 2023.

National Museum of American History. (n.d.). *A Streetcar City.* Retrieved from https://americanhistory.si.edu/america-on-the-move/streetcar-city, 20 September 2023.

New York Public Library. (2023). *Automobile Catalogs 1903–1915.* Retrieved from https://www.nypl.org/collections/nypl-recommendations/guides/auto-catalogs-vol-29, 8 November 2023.

O'Malley, J., & Slowic, P. (2024). *Life-cycle greenhouse gas emissions of U.S. sedans and SUVs with different powertrains and fuel sources*. Retrieved from ID-180-—US-GHGs_brief_final.pdf (theicct.org), 16 August 2024.

O'Neill, A. (2022). *Population of the United States 1610–2020.* Retrieved from Population of the United States 1610-2020 | Statista, 15 November 2023.

Oberoi, M. (2023). *How much Tesla stock does Elon Musk own and what other companies does he control?* Retrieved from https://www.nasdaq.com/articles/how-much-tesla-stock-does-elon-musk-own-and-what-other-companies-does-he-control, 5 November 2023.

Oke, O., Bhalla, K., Love, D. C., & Siddiqui, S. (2015). Tracking global bicycle ownership patterns. *Journal of Transport and Health, 2*(4), 490–501.

Orlando, A. W. (2022). *Keeping races in their places: The dividing lines that shaped the American City.* Routledge.

Our World in Data. (n.d.). *Registered vehicles per 1000 people, 2017.* Retrieved from https://ourworldindata.org/grapher/registered-vehicles-per-1000-people, 20 September 2023.

Pappas, T. (2023). *Electric vehicle makers are failing to attract female buyers.* Retrieved from https://www.carscoops.com/2023/02/women-buy-much-fewer-evs-than-men-do/, 18 September 2023.

Partridge, J. (2021). *Electric cars 'will be cheaper to produce than fossil fuel vehicles by 2027'*. Retrieved from https://www.theguardian.com/business/2021/may/09/electric-cars-will-be-cheaper-to-produce-than-fossil-fuel-vehicles-by-2027, 20 September 2023.

Pastor, A.B., Canniffe, E. and Jiménez, C.J.R. (2023). Learning from Letchworth and Welwyn Garden City: Garden cities' policies for the development of existing settlements in the contemporary world. *Land Use Policy 132*. 15pp.

Peterson-Withorn, C. (2017). *From Rockefeller to Ford, see Forbes' 1918 ranking of the richest people in America.* Retrieved from https://www.forbes.com/sites/chasewithorn/2017/09/19/the-first-forbes-list-see-who-the-richest-americans-were-in-1918/?sh=68412c7a4c0d, 20 September 2023.

Piketty, T., Yang, L., & Zucman, G. (2019). Capital accumulation, private property, and rising inequality in

China, 1978-2015. *The American Economic Review, 109*(7), 2469–2496.

Ren, Y., Sun, X., Wolfram, P., Zhao, S., Tang, X., Kang, Y., Zhao, D., & Zheng, X. (2023). Hidden delays of climate change mitigation benefits in the race for electric vehicle deployment. *Nature Communications, 14*, 3164. 12pp.

Richardson, J. (2018). *38% of American cars were electric in 1900.* Retrieved from https://cleantechnica.com/2018/02/25/38-percent-american-cars-electric-1900/, 17 September 2023.

S & P Global Mobility. (2023). *Electric vehicles have a gender issue with women.* Retrieved from https://www.spglobal.com/mobility/en/research-analysis/women-not-buying-electric-vehicles.html#:~:text=All%20these%20brands%20lag%20the,above%2Daverage%20in%20that%20aspect. 18 September 2023.

Shoup, D. (2018). Introduction. In D. Shoup (Ed.), *Parking and the City* (pp. 1–56). Routledge.

Sovacool, B. K. (2009). Early modes of transport in the United States: Lessons for modern energy policymakers. *Policy and Society, 27*(4), 411–427.

Stein, C. S. (1949). Toward new towns for America. *The Town Planning Review, 20*(3), 203–282.

Struyk, T. (2023). *The factors of a "Good" Location.* Retrieved from https://www.investopedia.com/financial-edge/0410/the-5-factors-of-a-good-location.aspx. 4 October 2023.

Swift, M. (2022). *How much the average family needs to earn to live comfortably in New Zealand.* Retrieved from https://www.newshub.co.nz/home/money/2022/08/how-much-the-average-family-needs-to-earn-to-live-comfortably-in-new-zealand.html, 20 September 2023.

The Henry Ford. (2017). *Ford Ambulances for the Red Cross.* Retrieved from https://www.thehenryford.org/explore/blog/ford-ambulances-for-the-red-cross, 8 November 2023.

The Henry Ford (2022). *1914 Detroit electric model 47 Brougham, personal car of Clara Ford.* Retrieved from https://www.thehenryford.org/collections-and-research/digital-collections/artifact/209957/, 16 November 2022.

The Henry Ford. (2023). *Electric cars for well-to-do women.* Retrieved from https://www.thehenryford.org/collections-and-research/digital-collections/expert-sets/11023/, 8 November 2023.

The Insurance Institute for Highway Safety. (2023). *Fatality Facts 2021: Males and females.* Retrieved from https://www.iihs.org/topics/fatality-statistics/detail/males-and-females, 20 September 2023.

The White House. (2022). *President Biden's economic plan drives America's electric vehicle manufacturing boom.* Retrieved from https://www.whitehouse.gov/briefing-room/statements-releases/2022/09/14/fact-sheet-president-bidens-economic-plan-drives-americas-electric-vehicle-manufacturing-boom/, 16 November 2022.

Transportation History. (2021). *1903: A Chicago dentist becomes the first owner of the trailblazing Ford Model A.* Retrieved from 1903: A Chicago Dentist Becomes the First Owner of the Trailblazing Ford Model A – Transportation History, 18 September 2023.

United States Bureau of Labor Statistics. (1913). *Union Scale of Wages and Hours of Labor 1907–1912.* Retrieved from https://fraser.stlouisfed.org/files/docs/publications/bls/bls_0131_1913.pdf?utm_source=direct_download. 4 October 2023.

University of Massachusetts Amherst. (n.d.). *Letter from City Housing Corporation to W.E.B. Du Bois.* Retrieved from https://credo.library.umass.edu/view/full/mums312-b047-i508, 3 October 2023.

US Department of Transportation. (2006). *Celebrating 50 years: The Eisenhower Interstate Highway system.* Retrieved from https://www.transportation.gov/testimony/celebrating-50-years-eisenhower-interstate-highway-system, 4 January 2025.

US Department of Energy. (2014). *The History of the Electric Car.* Retrieved from https://www.energy.gov/articles/history-electric-car, 15 September 2023.

US Department of Energy. (n.d.). *All-electric vehicles.* Retrieved from https://www.fueleconomy.gov/feg/evtech.shtml, 18 September 2023.

US Department of Transportation. (2021). *Federal Highway Administration. Highway Statistics.* Retrieved from https://www.fhwa.dot.gov/policyinformation/statistics.cfm?_gl=1*17zuw5s*_ga*NDA5OTIyMzU3LjE2OTk4NDMzNzM.*_ga_VW1SFWJKBB*MTY5OTg0MzM3Mi4xLjEuMTY5OTg0NTI2OS4wLjAuMA, 13 November 2023.

Victoria Transport Policy Institute. (2021). *Transportation cost and benefit analysis II—Roadway land value.* Retrieved from https://www.vtpi.org/tca/tca0507.pdf, 3 October 2023.

Vitta, S. (2021). Electric cars—assessment of 'green' nature vis-à-vis conventional fuel driven cars. *Sustainable Materials and Technologies, 30*, 27.

Weigand, L., McNeil, N., & Dull, J. (2013). *Cost analysis of bicycle facilities: Cases from cities in the Portland OR region.* Portland State University. Retrieved from https://activelivingresearch.org/sites/activelivingresearch.org/files/Dill_Bicycle_Facility_Cost_June2013.pdf, 26 September 2023.

Wells, C. W. (2007). The road to the model T: Culture, road conditions, and innovation at the Dawn of the American motor age. *Technology and Culture, 48*(3), 497–523.

Wiebenson, D. (1969). *Tony Garnier: The Cité Industrielle*. Studio Vista.

Wilson, K. (2022). *Worth the Watt: a brief history of the electric car, 1830 to present.* Retrieved from https://www.caranddriver.com/features/g15378765/worth-the-watt-a-brief-history-of-the-electric-car-1830-to-present/, 16 November 2022.

Worldometer. (n.d.). *GDP per Capita.* Retrieved from https://www.worldometers.info/gdp/gdp-per-capita/, 20 September 2023.

Worstall, T. (2012). *The story of Henry Ford's $5 a day wages: It's not what you think.* Retrieved from https://www.forbes.com/sites/timworstall/2012/03/04/the-story-of-henry-fords-5-a-day-wages-its-not-what-you-think/?sh=2d5e1870766d, 17 November 2022.

The Cost of Buildings

5

Who Pays for Buildings?

Chapter 4 showed how an idea for a new product—the car—not only made a few people very wealthy but also changed the built environment for everybody. The built environment is not just a product of ideas but a product of money; without money ideas cannot become buildings. It, therefore, seems odd that SDG 11 Sustainable Cities and Communities does not deal with money. Architects, planners, urban designers and developers may be held responsible for the modern environment, but someone has to supply the money to make it. This leaves the question of who pays. Chapter 4 has already shown that, probably unwittingly, taxpayers contribute to the roads whether or not they use them directly. The purpose of this chapter is to explore where the money comes from for some significant public buildings, both in the past and in the present, to see whether who pays has changed. The buildings chosen are partly dictated by the available information but also because most remain significant, even in a modern landscape. The chapter ends by looking at the relationship between houses and wealth.

The funding of a cathedral in both medieval and more modern times is compared first. In the medieval period, when most of the great European cathedrals were constructed, the catholic church was wealthy since it had a monopoly on salvation, and it "over a long span functioned as a 'multinational' bureaucracy of enormous economic power" (Ekelund et al., 1989).

This is followed by looking at the building of a stadium in both the ancient and modern worlds. There are some subtle connections between the church and sport. In medieval times, the only rest day was Sunday, which was a day for going to church and collecting together with a single purpose. Modern times for many in the developed world have 2 days of rest, giving the possibility of football on Saturday and church on Sunday. In some countries such as Argentina, Sunday sees church in the morning and football in the afternoon, irrevocably linking the two events. Perhaps more simply, many stadia have been likened to cathedrals. In Bilbao, St Mames Stadium has been called a cathedral of football (Visit San Mames, n.d.), a distinction also awarded to Wembley Stadium by the famous footballer Pelé (WembleyStadium, n.d.).

The third comparison concerns structures for defence. Spending on defence has always been with us. The average for the world (145 countries) in 2022 was 1.98% of GDP. Obviously, defence spending rises in times of conflict with Ukraine ranked first at 33.55% of GDP while Russia ranked 12th at just over 4% (The Global Economy, 2023). Castles, therefore, form the third pairing by looking at the costs of the Welsh castles of Edward I as these are well detailed in the royal records. A comparison is made with the cost of a relatively modern military installation.

E. J. García, B. Vale, *Wealth and the Built Environment*, Sustainable Development Goals Series,
https://doi.org/10.1007/978-3-031-81954-4_5

Both examples are now redundant. Castles were also places of residence not only for the garrisons defending them but for those sheltering within their walls during times of siege. Later, as discussed below, the domestic function of a castle became as important as its military function. The phrase "My home is my castle", dates to the reporting by Sir Edward Coke of a case from 1604 that "the house of every one is to him as his (a) castle and fortress", meaning that if a house owner harmed an intruder in defence of himself or his property it was not a felony, and leading to the statement "domus sua cuique est tutissimum refugium" (every man's home is his safest refuge) (Coke, 1826: 186). This leads to a brief discussion of the changing funding and cost of houses over time, not because houses are expensive but because they are ubiquitous.

Cathedrals

At first sight, the wealth of the church implies that finding money for buildings should not have been a problem. Piketty (2020: 94) notes in regard to France, where many of the great medieval cathedrals stand: "For more than a millennium, from the fifth or sixth century to the eighteenth or nineteenth century, the church owned a significant share of all property, and especially land…typically a quarter to a third". Piketty goes on to argue that property is the basis of economic law and capitalism, and thus, the great church buildings are a reminder of this. However, those setting out to build a great medieval cathedral did not necessarily benefit from the wealth contained in land ownership, as building money was often hard to come by. Finding money to build a cathedral was problematic because, unlike a parish church, a cathedral has no regular congregation to support it. The task of raising money to build a cathedral was and still is, one of persuading people to part with their money for a project that they may never live to see finished. Yet these huge buildings still dominate the cities in which they stand and continue to attract tourists even in a world of high-rise structures. Their sheer size suggests that equally huge efforts must have been made to raise the money to build them.

New churches were also a way to raise money. Pilgrimages allowed ordinary mortals to see the treasures of the church, but they paid for the privilege. At St. Denis, the new choir of 1144 that heralded the start of gothic building in Europe was designed to allow for the easy circulation of pilgrims and to raise money for the new work. The man in charge, Abbé Suger, fortunately, had a good business head:

> Apart from obtaining numerous royal donations and privileges (the most important of which were the extension of the Abbey's local jurisdiction and the concession of the big annual fair known as the "Foire du lendit") and from securing private benefactions of all kinds, Suger was a great hand at discovering forgotten claims to lands and feudal rights. (Panofsky-Soergel, 1979: 8)

Suger was from lowly origins, and the abbey to which he was sent became his life, including making the abbey church a building to the glory of God. To this end, he raised money in every way possible, even to the extent of asking "royal, princely and episcopal visitors of the Abbey to donate the stones of their very rings for the adornment of a new altar frontal (apparently by divesting himself of his own ring in their presence and thereby forcing them to do likewise" (Panofsky-Soergel, 1979: 33). A visiting Abbot noted the small dimensions of Suger's cell and apparently said "This man puts all of us to shame; he builds, not for himself, as we do, but only for God" (Panofsky-Soergel, 1979: 31).

The great medieval cathedrals were also built in a period when the architect was a respected figure whose services were sought after and who could turn down a job if unappreciated, as happened at the abbey of St. Trond in Belgium. Here, the putative architect stated, "If you do not like my excellent design [*id quod bene concepi*], I suggest you get yourself another architect" (Wilson, 1992: 142). Kemper (2019) suggests that each new cathedral had to be somehow better than the one before: "Every new cathedral needed a superlative quality—biggest, tallest, most elaborate—a bishop would have a difficult time building if he promised anything less". By fulfilling these goals, the new building would attract more people and, hence, more money. As usual, wealth builds wealth at the expense of taking more resources from the natural environment. Given this background, it seems useful to explore how money was raised for these enormous buildings.

Utrecht Cathedral

Fortunately, Wim Vroom (2010: 285–418) has completed a detailed examination of all the papers held by Utrecht Cathedral (Fig. 5.1) that relate to its building and financing over a period of some 275 years up to its virtual completion just before the reformation in 1517 (Vroom, 2010: 465).

What remains of the cathedral today are the choir and transepts and the tower at the west end. The incomplete nave, missing its vaults, which had been replaced with a wooden ceiling, as well as its side chapels and other abutments, collapsed in a storm that produced a tornado in 1674 (Vroom, 2010: 298). The tower remained the tallest man-made structure in the Netherlands until the construction of radio communication towers early in the twentieth century (Emporis, 2022). The sources of money to fund the building of this gothic cathedral after a fire in 1253 had damaged an earlier building are numerous. The building fund for the new work was started through a

Fig. 5.1 Utrecht Cathedral with missing nave. (Source: Creative Commons (n.d.-a))

process whereby when a canon died (a canon was a member of the cathedral's administrative body called the chapter), the new canon had to wait 3 years before enjoying the income generated by his prebend (the prebend was the canon's stipend and could come from land owned by the cathedral). The income from the first year was used to pay any debts of the deceased with any left going to his descendants, while that of the next 2 years went into the fabric or building fund. Although changes were made to this in terms of numbers of years and shares of the income the principle stayed the same. A new canon also had to make admission payments, some of which were diverted to the building fund. However, these contributions from the chapter of canons were not large. "With a maximum share in the fabric's income of 20 percent, and usually much less, chapter contributions did not keep the economy of the fabric afloat during the building campaign" (Vroom, 2010: 317).

Bequests were another source of building money, with most coming from the cathedral chapter, although two architects of Utrecht cathedral left money in their wills for the building fund. Jacob van der Borch, who died in 1475, having spent 30 years working on the cathedral, left the equivalent of the wages of two unskilled workers for a year, while Cornelis de Wael, the architect from 1467–1505, left money that would pay for five unskilled workmen for a year (Vroom, 2010: 324). Vroom also points out that legacies tended to be donated when building work was very active, so it was building activity that generated the bequests. Many people preferred to donate in kind rather than money, so would pay for windows, reliquaries and paraments (ornamental hangings and ecclesiastical vestments) to embellish the building, especially as it was possible to be seen to be making such a donation in perpetuity by being shown depicted in the donated stained glass of a window. Such donations came from both men and women (Hediger, 2019: 239–250).

Most of the money for building Utrecht Cathedral was somehow extracted from the pockets of the faithful through various schemes. These ranged from collection boxes inside the cathedral for donations, through pilgrimages to view the relics owned by the cathedral and the subsequent donations made in thanksgiving, to papal indulgences whereby money could be paid in lieu of some penance, such as making a pilgrimage, just like buying carbon offsets before you fly. Charging outsiders to come into the diocese to raise money for their own cause was another way in which money from the faithful found its way into the fabric fund. The bishop would also send people into the diocese to promise "pardons and other spiritual benefits to all those who supported the cathedral building works" (Vroom, 2010: 333). In lieu of money, pardons could also be issued for days of unskilled work, such as transporting building materials. It was a never-ending quest to raise money for building, which is why building work often went in phases. The same struggle to raise money was repeated in other European cathedrals (Vroom, 2010: 419–456), and still seems to be going on as the story of a modern New Zealand cathedral demonstrates.

Cathedral of St Paul, Wellington

Unlike a medieval cathedral where the church already owned the land, a site had to be purchased for the Anglican cathedral in Wellington (Fig. 5.2). After a couple of earlier attempts to buy land, the current site was eventually purchased in 1943 for the sum of £136,354. At that time: "The sum of 250,000 pounds, which had been raised to build the Cathedral before the war, would now only build about half of it" (Wellington Cathedral of St. Paul, n.d.). The original fund to build the cathedral had been established in 1918 by Bishop Sprott but the site was then elsewhere in the city, "and many of the donors gave their contributions expressly for the purpose of a building on that site" (Press, 1937). This entailed going to the Supreme Court to ensure the donations could be used for the cathedral on the new site. Money had been raised for the pre-WWII building fund through cajoling people that the new cathedral would be a fitting memorial for the victory of WWI (Fig. 5.3). Government stock and war bonds were an acceptable substitute for cash

Fig. 5.2 1919 fund raising advertisement in Wanganui Herald. (Based on Papers Past (n.d.))

and bequests and their donation was also encouraged (Wanganui Herald, 1919). At this stage the architect was Frank Peck, then living in Nelson, and who had arrived from England in 1915 (Blain, 2002: 51). Peck had been articled in Lincoln, where he prepared measured drawings

Fig. 5.3 Cathedral of St. Paul, Wellington. (Source Robert Vale)

of the cathedral. He set up his own practice in 1897 designing breweries and factories but eventually turning to "the less profitable routine of church work" (Anon, 1912).

Peck's design for St Paul's Wellington (Fig. 5.2) although to be built in reinforced concrete rather than stone because of the earthquake risk, drew on French gothic rather than that of the cathedral at Lincoln where Peck had been articled, and was described in the press as "a poor, mean reproduction of a French gothic church of the Chartres type" (Evening Post, 1917).

Peck was eventually replaced as architect by Cecil Wood from Christchurch, although in 1929 Peck did design Nelson Cathedral, which was never completed to his design with work stopping in 1932 (Venture New Zealand, 2022). As in Utrecht the nave at Nelson was given a wooden ceiling, but fortunately this ferro-cement structure is still extant. Cecil Wood (1878–1947) may have become an architect because his uncle by marriage was the well-known UK architect Richard Norman Shaw. In 1893, the 15-year old Wood was articled to another English architect emigré, Frederick Sprouts, for the sum of £100 (Helms, 1996: 13) (approximately NZ$22,500 in 2022) (Reserve Bank of New Zealand, 2022) (see Chap. 10 for further discussion of the cost of architectural education). In 1894, Wood started the 4-year part-time Diploma of Architecture under Samuel Hurst Seager having gained a drawing scholarship to the School of Art in Christchurch (Helms, 1996: 16). Wood also undertook the New Zealand equivalent of the grand tour, which is to spend time working overseas, and from 1901–1906 worked in the UK (Helms, 1996: 21, 31). When appointed in 1937 to design the Cathedral of St. Paul, Wood again travelled overseas to Europe and the USA to look "at the latest methods of concrete construction and modern church design" (Helms, 1996: 188, 193).

Although designed by Wood in 1938, the laying of the foundation stone by Queen Elizabeth II only occurred in 1954, although work did not start until 2 years later, when a contract was signed for the first section for £372,036 (Blain, 2002: 77) (approximately NZ$21.5 million in 2022). The cathedral was finally finished in 2000 (Maclean, 2015) after a massive amount of fund raising, including persuading the government that part of the building qualified as a civic structure so that it could thus gain some NZ$800,000 from the Lottery Grants Board to help complete the tower at the west end (Blain, 2002: 118). This meant the project spanned only 82 years unlike the 275 of Utrecht Cathedral. Cecil Wood died in 1947, so joined the ranks of many medieval architects who never saw their cathedral visions completed.

Just as at Utrecht fund-raising from ordinary people was the way forward, although there was some reluctance post war, at least in Lower Hutt in 1945 where only 25 people turned up to a fund-raising meeting organised by the mayor (Hutt News, 1945a). A meeting a month later in December 1945 where the film *The Battle of Britain* was shown, did much better with a near full attendance at the Red Cross Hall, and where it was hoped the new cathedral would be "a fitting thank offering for the peace that had returned to the world" (Hutt News, 1945b). However, when the Hutt Valley Electric Power Board donated £100 to the cathedral, a letter to the editor stated that this was a misuse of Power Board money, which would be better spent on small churches in the suburbs (Hutt News, 1946). This suggests there is a link between the scale of a building, the resources demanded, the process of fundraising money and the beneficiaries. The decision to cluster money and resources in big buildings can only be made by depriving others of similar benefits and buildings. The opportunity cost of making big buildings is that you can have only a few of them, which following Brecht's ideas (see Chap. 1), means that they create an unequal distribution of money and resources. This goes against the goals of both SDG 10 (Reduced Inequalities) and SDG11 (Sustainable Cities and Communities).

As in Utrecht, bequests were also a means of raising money. The wills of Major and Mrs. Tatum show that Major Tatum left £500 for the installation of a memorial window in the new cathedral, while under Mrs. Tatum's will, after bequests, the residual money would pass to the cathedral building fund, and this was expected to be £20,000 (Otaki Mail, 1943). Towards the end of the building in 1995, a public benefactor, Denis Griffin, agreed to give NZ$500,000 if NZ$300,000 could be raised and upped this by a further NZ$100,000 if the same amount could be raised again, and this was achieved (Blain, 2002: 115–116). The architect who completed the nave and west end was Miles Warren, who had once been articled to Wood.

Perhaps Bishop Holland best expressed who paid for the cathedral at a 1938 fund raising meeting in Masterton. "Enthusiasm, he said, would not build a cathedral unless it also opened the pockets of the people. He said that £300 had been promised from Greytown, and four individuals in Carterton had promised £657" (Evening Post, 1938).

The fact that money diverted for a new building was not helping the poor was noted in medieval times (Vroom, 2010: 70–73). In his 1125 *Apologia* St. Bernard of Clairvaux stated: "The church is resplendent in her walls and wanting in her poor. She dresses her stones in gold and lets her sons go naked. The eyes of the rich are fed at the expense of the indigent" (Hassall, 1996). A similar sentiment was noted regarding the plan to build a cathedral for Wellington. In a 1917 letter to the Dominion newspaper, one Anglican reader wrote:

> The proposed cathedral is, I am glad to think, a good deal in the air, and will be for some time to come, but it disgusts me to hear talk about spending money on a building for which there is no direct use. There are plenty of ways of serving God by helping man just now open without resort to this. The fact of the matter is Wellington wants to be even with Christchurch about a cathedral. There is no worship of God in putting up useless buildings. (Dominion, 1917)

This argument continues. The recent award-winning Bishop Selwyn Chapel attached to the Holy Trinity Cathedral in Auckland looks expensive with its gold ceiling and glass walls. "However, some critics might argue that it required too large an investment – seemingly emphasised by its Italian gold-leaf ceiling – better spent on feeding the poor. But, I disagree because, frankly, individuals are building extravagant houses across the country: no questions asked" (Harvey, 2016). Regardless of the ethical points raised by the expensive construction of the Bishop Selwyn Chapel (see Chap. 9 for the link between ethics and wealth), it illustrates how decisions in the built environment cluster resources while depriving others of them. It seems that unless wealth is questioned as a concept, it is not possible to tackle either SDG 1 (No Poverty) or SDG 10 (Reduced Inequalities).

Why Do People Give?

So what motivates people to pay for a building they might never use or even see finished? In the Middle Ages, the promise of eternal salvation and remission of potential time spent in purgatory from paying for an indulgence would seem to give some benefit to the giver, at least in the next world, if not this. The motivations for giving money to build a new cathedral in Wellington are less clear, although the church, in its fund raising, played heavily on the idea of the building as a memorial of the war, initially of WWI and then WWII, so people might have been motivated by being thankful for their survival and that of family members. However, what both the Utrecht and Wellington cathedral building campaigns show is a dogged persistence in asking people for money in every way possible. In a literature review study of why people give, Bekkers and Wieping (2011) found that actively asking for money enhanced the chance of receiving and the more times people encounter a chance to give, the more likely they are to part with their money, something that those funding the vast cathedral building projects clearly embraced. Even the poor will give money for a cause, hence the tradition of the widow's mite and the idea that even a little helps. This contrasts with those wealthy enough to be able to borrow money, since debt can be used to create wealth (Anderson, 2017: 113–128), something those too poor to borrow can never experience.

The Colosseum and Wembley

The sports industry generates 1% of global GDP or annually some US$700 billion (ISCID, n.d.). Sports also have a long history, stretching back to the Olympic Games of Ancient Greece and have led to the construction of special structures to support sporting activities. Olympia, where the games were held as part of a religious festival from 776 BC to 393 AD, was in part a series of shrines to the God Zeus and in part a series of buildings and spaces for events such as poetry reading, boxing and javelin throwing (Penn Museum, 2022; Ancient-Greece.org, 2022). Roman sports also had their origins in religion (Dunkle, 2014).

> In its earliest years, the primary purpose of the spectacle was to win the good will of the gods, especially in times of crisis, by providing them with good entertainment. This is not to say, however, that the human audience was not taken into consideration. The attraction that spectacles held for the Romans is evident from the steady and abundant proliferation of religious festivals requiring shows made possible by a professionalized entertainment industry.

Many Greek amphitheatres followed the contours of the land, while the Romans built free-standing amphitheatres for drama and gladiatorial contests and circuses for chariot racing. Being a gladiator did not necessarily guarantee a long life as the median age of a gladiator from tombstones was 22.5 years compared to 31 for other males, although the figure for men generally is lowered by frequent infancy death, so 48 becomes the average age of males who survived until 17 (Hopkins & Beard, 2005: 87–88). However, being a charioteer could prove a very lucrative profession. Whereas in the fourth century AD a wage freeze introduced by Emperor Diocletian restricted the wages of working men from teachers to farm labourers to 200 sesterces or less a day, in his 24 years as a charioteer in the second century AD, Appuleius Diocles accumulated over 35,000,000 sesterces, 4000 a day (Matz, 2019: 3). This information comes from a stone monument erected on his retirement from racing aged 42. To be a senator Emperor Augustus brought in a rule such that it was necessary to own property of 1,000,000 sesterces (Dio, 1917: 351). The wealth of Appuleius Diocles has been compared to modern sports men, suggesting he was the wealthiest ever at about US$15 billion, whereas Michael Jordan, the highest paid modern athlete is worth US$1.85 billion and over his career Christian Ronaldo netted US$725 million (Law, 2018). Prowess in sport has thus always been linked to wealth accumulation.

As in Rome, modern sports are supported by large and costly stadia and other buildings. Much is made of the idea that a new and architecturally exciting stadium will improve the local economy by attracting visitor money (Watson, 2022). However, in a study of stadia for professional teams in the USA, Noll and Zimbalist (1997) note that before 2006 "approximately $7 billion coming from public sources such as tax-exempt bonds, will be spent on new structures for professional sports teams. These facilities actually have minimal or even a negative effect on overall economic activity and employment". Nevertheless, modern stadia have become important icons in the city, which often push the boundaries of design and engineering, leading to failures such as the 2010 collapse of the Minneapolis Metrodome roof under snow loads (Mann, 2021) or the 1999 collapse of the retractable roof of the Miller Park stadium in Milwaukee during construction, which tragically killed three workers (Stavropoulos, 1999). Because of their design and high cost, sports stadia seem worthy of further investigation.

The Colosseum

After Emperor Nero committed suicide in AD 68, civil war followed with Vespasian, a Roman general charged with supressing the Jewish rebellion, eventually being declared emperor a year later. The sacking of Jerusalem had yielded a treasury of "silver, gold, ivory and precious gems" (Hopkins & Beard, 2005: 27). Nero had previously taken a large area of Rome for his golden palace and pleasure grounds that included a lake that was to be the site of Vespasian's new Flavian amphitheatre (named after the emperor's dynasty), later in medieval times called the Colosseum. "In building the Colosseum Vespasian was dramatically making the point that the profits of Roman military success belonged, in part at least, to the common people of Rome" (Hopkins & Beard, 2005: 32). The new building was also seen as a reward for those who had borne Vespasian's austerity campaign which had been established to refill the treasury that Nero had drained (Matz, 2019: 75). The Colosseum opened in AD 80, a year after Vespasian's death and when his son Titus, also a former Roman general who had been part of supressing the Jewish rebellion, was emperor.

The designers of the Colosseum are not known, but their skill in draining not only the site but also in disposing of the rainwater that fell on the seats and ran down into the arena area is to be admired (Hopkins & Beard, 2005: 142). The elliptical design of the Colosseum heralded the shape of many modern stadia, but whereas the latter are often more of a spilt circle with side walls inserted, the Colosseum (Fig. 5.4) was a true ellipse covering about five acres. It had external diameters of 615 ft. and 510 ft. (187 m and 155 m), while those for the internal arena were 281 ft. and 176 ft. (86 m and 54 m). The external walls were 160 ft. (49 m) high incorporating 80 numbered arches that formed the entrances (Matz, 2019: 76).

The elliptical shape probably derived from the amphitheatre. What in Greece had been a terraced hillside, in Rome had become a free-standing building with the arc of tiered seating facing a stage that closed off the long side, thus creating an interior where the focus was all on the central area (Brown, 1968: 28). The inspiration for the elliptical form may have come from two extraordinary wooden amphitheatres built in 52 BC for funeral games. Possibly set side by side, they were the scene of two separate dramatic performances in the morning, and through a pivoting system were moved together to form

Fig. 5.4 The Colosseum, Rome. (Source: Creative Commons (n.d.-b))

an enclosed space for gladiatorial combats, with some spectators apparently remaining in their seats during the move (Dodge, 2014).

The seating in the Colosseum was basically a hierarchy, with the most important people in the seats nearest the arena, rising up to the top where the slaves, non-citizens, and women sat, although the Vestal Virgins were down the front. The advantage of being up at the top once all the stairs were surmounted was being under the cloth shade sails, whereas those in the senatorial seats could well have "baked" in the mid-day sun as it has been estimated that the linen sails could only shade about half of the arena. It seems there was no charge for attendance although tickets were distributed beforehand (Hopkins & Beard, 2005: 106–111). The seating capacity, given the Romans were smaller than modern humans, has been estimated at around 45,000 (Matz, 2019: 77). The cost of the building at the time is not known but an estimate for the foundations alone (Hopkins & Beard, 2005: 148) came out at £28.5 million. Another estimate for building the whole structure today including the Travertine marble facings was US$435.3 million (approximately £374 million) (Sork, 2018).

Wembley Stadium

Completed in 2007 after the old heritage listed stadium was demolished in 2002, Wembley Stadium has been described as "the most expensive stadium that has ever been built as a cost of £798 million (roughly US$1.57 billion)" (CPG Europe, 2022). The original stadium with its renovations to host the 1948 Olympic Games, was listed in 1976 but demolished with the consent of the Heritage Authorities (Kiuri and Teller, 2015). The site had previous involvement with extremes. During excavations for the original Wembley Stadium, the foundations of the 1890 Tower for

London were uncovered. Known as the Watkin Tower after one of its promoters this would have been the highest man-made structure in the world had it been finished (Lewis, 2025). The idea behind it was to have a tower in London that would eclipse the Eiffel Tower in Paris by being both 45.8 m taller and situated on high ground rather than next to the river. A competition was held, and the winning entry (Fig. 5.5) was by Stewart, McLaren and Dunn of London (Stewart was a civil engineer and McLaren and Dunn architects (Lynde, 1890: 83)). Its estimated cost was £352,222, equivalent to £28.9million in 2017. Only the first 49 m were built after the company building it went into liquidation, not least because of the marshy ground that led to movement in the foundations, and even this "stump" was eventually blown up in 1904 (Green, 2012). More curious is the cheaper (£155,080, equivalent to £12.7million in 2017) competition entry of the cable-stayed tower of Findlay, Rendel (engineers) and Halsey Ricardo (architect) (Fig. 6.5), given the cable-stayed roof of the new Wembley arena, although there appears to be no link between the new Wembley Stadium roof and the unplaced design in the 1890 competition.

Apart from the ill-fated Watkin Tower, the new Wembley Stadium also sits on the land occupied by previous Wembley Stadium built as part of the 1924–1925 British Empire Exhibition. When built, the stadium with its famous twin towers was claimed to be "bigger than the Coliseum in Rome" (Crow, 2018: 12). The original intention was to demolish the stadium after the exhibition but "businessman and civil servant Sir James Stevenson suggested the stadium stay open, as football had been played on the grounds where it was built since the 1880s. Soon after, King George V officially opened the stadium and the pitch was dubbed the 'hallowed turf'" (WembleyStadium, 2022). Sir James Stevenson had chaired the committee organising the British Empire Exhibition. The architects were John Simpson and Maxwell Ayrton, working with engineer Owen Williams and at the time the reinforced concrete stadium cost £750,000 (the equivalent of £30.8 million in 2017) (Crow, 2018: 12) (Fig. 5.6). The big money comes with the

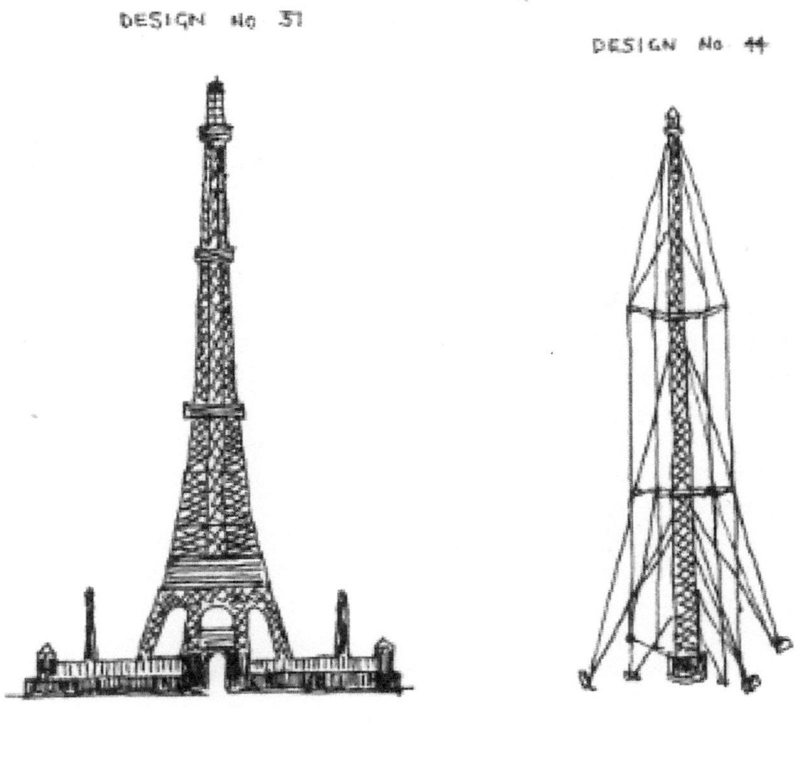

Fig. 5.5 Left: Winning Design. (Source: Lynde (1890: 82)); Right: Cable Stayed Design. (Source: Lynde (1890: 96)). Both designs were 1200 ft. high

Fig. 5.6 The old Wembley Stadium. (Source: Creative Commons (n.d.-c))

new Wembley Stadium of 2007, costing £798 million (equivalent to £1093 million in 2017 as a comparison with the old Wembley Stadium). With its iconic arch as a substitute for the twin towers, it was designed by Foster and Partners in conjunction with Populous (Foster + Partners, 2022) (Fig. 5.7). Norman Foster is currently the richest architect in the world with a net worth of some US$200 million (see Chap. 3), although this does not put him in the ranks of the richest in the world. Apart from the arch, which partly supports the roof, and the arena, the building encompasses conference facilities, restaurants and other food outlets, along with 2618 toilets (Crow, 2018: 131).

Seating 90,000 the rough estimate in terms of building for the price per seat at £9000 compares with nearer £10,000 for a seat at the Colosseum if built today, though the seats at Wembley are red plastic and those of the Colosseum were of stone or wood for the upper seats. The cheapest seat for a 2021 World Cup qualifying match (England versus Andorra, which England won 5–0) at Wembley was £25 for an adult in the family enclosure, which would be a seat at the higher level, while the most expensive seat was £65 (Koobit, 2022). For the 2022 FA Cup Final (Chelsea versus Liverpool, which Liverpool won on penalties) adult tickets ranged from £45–250 (Khairi, 2022). Tickets at the Colosseum were free and either distributed by someone in authority, and consequently wealthy, or via subscription to a society that would then give out free tickets.

The new Wembley Stadium belongs to the Football Association, the body that oversees English football, through its subsidiary company, Wembley National Stadium Limited (WNSL). The FA, as a not-for-profit organisation, makes its money from owning the rights to international and FA Cup matches, with broadcasting rights as its main source of income (Bolvin, 2022), and profits are fed back into football. Unlike the Colosseum, ticket sales along with letting out the facilities are vital for the new stadium since finance had to be borrowed to complete it, with public money raised through taxes only going so

Fig. 5.7 The new Wembley Stadium. (Source: Creative Commons (n.d.-d))

Table 5.1 Money for Wembley Stadium

Stakeholder	Money	Purpose
Football association (operating through fully owned subsidiary Wembley National Stadium ltd.)	"Facilities agreement with a group of commercial banks as part of the original financing package" (amount unknown)[a]	Balance of money required
Department of culture, media, and sport	£20 million	Improvements to non-stadium infrastructure
Sport England (responsible for sport section of lottery funds)	£120 million	Purchase of 1924 Wembley Stadium
London development agency (promotes economic renewal)	£21 million	Towards project

[a]GLA (Greater London Authority, 2022)

far. When the rebuilding of the stadium was initiated in 2002, it "received £20 m from the Department of Culture, Media and Sport, and another £21 m from the London Development Agency, at which point safeguards were introduced to protect the public money invested in the project" (The Guardian, 2003). The same source also mentions a lottery grant of £120 million that "could have been lost when the project ran into financial trouble for failing to raise enough commercial funding". Lottery money could be regarded as voluntary taxation. Despite the wealth of the FA, taxpayers' money was still needed to ensure the project would go ahead. Table 5.1 sets out the initial stakeholders in the project and what they funded, where known (WembleyStadium, 2022).

The stadium thus needs a revenue stream from sales of tickets and hiring out its other facilities to cover the balance of the finance needed for the project. In 2018, there was a proposal to sell it for

£600 million to Shahid Khan, the owner of Fulham Football Club, so that the money could be used to improve grassroots football in the UK (Conn, 2018). The deal fell through because of division amongst those making the decision. However, the cancelling of events because of COVID-19 saw the stadium in trouble. For the 12 months up to July 2020, WNSL made an operating loss of £27.7 million, whereas the loss in the previous 12 months had been £2.5 million. Turnover also went from £100.7 million to £35.5 million. In addition, "with an estimated £15m in annual costs that come with keeping a landmark building top of the range—from maintaining the roof to upgrading the wifi—the stadium is never going to be an earner for the governing body" (MacInnes, 2021). In its turn, the Colosseum has seen much spent on it over the years, including significant rebuilding in the eighteenth century and reconstruction of some of the missing south side in the 1830s (Hopkins & Beard, 2005: 166, 171), but as the same authors note (2005, 122), "it is a well-known axiom among archaeologists that the more famous a monument is, the less likely any of its original structures are to survive—the more likely it is to have been restored, rebuilt and, more or less imaginatively, reconstructed".

The owners of Wembley Stadium, through its subsidiary company, the Football Association, would seem to be able to bear its continuing costs as in 2021, its turnover was £443 million with a £132 million profit, up from turnover of £335 million and profit of £9.8 million the previous year (The Football Association, 2022), although the money put into the new stadium is money that will fail to reach grass roots football. Overall, the money in football is huge despite some of the best players coming from very poor backgrounds (GhanaSoccernet, 2021), but then most gladiators who fought in the Colosseum were probably not wealthy, being either condemned prisoners, slaves or prisoners of war (Dunkle, 2013: 30–31). The architecture of sports spectacles then and now is both grand and expensive, but fortunately, as in Rome, those who succeed can emerge wealthy, and at least now, those who fail do not die.

Castles and a Modern Military Installation

Money spent on defence is money for which the hope is it will never be used. Globally, in 2021 5.9% of government spending was on the military (Da Silva et al., 2022). Buildings that were once important militarily have now become heritage items. For these reasons, it seems useful to compare expenditure on castles with a more modern military physical installation.

The Welsh Castles of Edward I

A medieval monarch had two options when it came to preparing for an attack, he could pay for an army or build a castle, but the money it was possible to raise through taxation could seldom do both (Brauer & van Tuyll, 2008: 46). The castle once built and paid for, at least paid for in theory, only required normal maintenance whereas an army needed constant feeding, clothing and housing. The fact Europe is littered with substantial castles or castle remnants suggests that money spent on a castle was seen as better value for the castle owner than the alternative of spending money on an army. Castles also gave an expressive opportunity for those asked to design them. It, therefore, seems unsurprising that William the Conqueror arrived in England with prefabricated wooden castles, setting one up at Pevensey and then another at Hastings as strongholds before embarking on the battle that gave him the throne (Great Castles, n.d.).

The stone castle was far superior to its wooden predecessor, but such castles required labour for building and labour had to be paid. Table 5.2 sets out the cost of the ring of castles build by Edward I in his campaign to subdue the Welsh, including the equivalent number of cows that could be purchased for the same money at that time. "From the very outset; castles and campaigning went hand in hand for Edward in Wales; it was not a matter of building only once victory had been achieved" (Prestwich, 2010: 4). Often called Edward's ring of iron (Lewis, 2022; Williams, 2011), Siddall (2021: 133) states the castles were

Table 5.2 Approximate costs of selected Welsh castles of Edward I

Castle	Date	Cost £[j]	Approximate 2017 equivalent million £[k]	Cow purchase equivalent[k]
Aberystwyth	1277–1289[b]	3900	2.7	8666
Rhuddlan	1277–1285[c]	9500	6.6	21,111
Bluith	1277–1282[d]	1700	1.2	3777
Flint	1277–1284[e]	7000	4.9	15,555
Harlech	1283–1290[f]	9–10,000	7.0	21,111
Caernarfon[a]	1283–1292[g]	12,000[g]	8.8	26,666
Conwy	1283–1292[h]	15–19,000	12.5	37,777
Beaumaris	1295–1298[i]	9000	6.6	20,000
Total			49.3	154,663

[a]Dates for phase one. The second building phase for slightly more money lasted from 1295–1323, with additions until 1330, although never finished (Cartwright, 2019b). Brauer and van Tuyll (2008: 54) state over 12 years the total cost was £16,000–27,000
[b]Welsh Government (2022a)
[c]Welsh Government (2022b)
[d]CastlesFortsBattles.co.uk (2019)
[e]Welsh Government (2022c)
[f]Cartwright (2019a)
[g]Cartwright (2019b)
[h]Cartwright (2019c)
[i]Thomas (2009)
[j]Brauer and van Tuyll (2008: 54)
[k]National Archives (n.d.)

more “symbols of power, superiority and dominance rather than means of protecting the Welsh coasts and their hinterland”. Prestwich (2010: 1) also notes Edward’s conquest led to the loss of the Welsh ruling dynasty and the introduction of English administration and laws, including English taxation, taxes being an important source of income for building on this scale, although Brauer and van Tuyll (2008: 45) note that: “Rulers might tax, but neither economic reality, administrative capacity, nor tradition would allow collections anywhere near as extensive or methodical as today”.

As a comparison, the total UK military expenditure budget for 2017 was £36,209.5 million (Statista, 2022). In 2017, the UK population at the end of June was 66.04 million (Office for National Statistics, 2018), making a military expenditure of £548 for each woman, man and child. Obviously, a direct comparison is not possible as the castles in Table 5.2 were not the only expenditure on defence at the time as other castles were being built, repaired and renovated. Taylor (2003: 119) shows that for the period 1277–1304 the money spent on 10 Welsh castles, including all those in Table 5.2, amounted to £78,267 11 s 6d. Over 27 years, this amounts to £2898 1 s per annum, or the approximate 2017 equivalent of £2.1 million. Broadberry et al. (2011: 22) estimate the English medieval population to be 4.43 million in 1279, 4.75 million in 1290 and 4.69 million in 1315. Taking an average population of 4.6 million gives an expenditure per annum of £0.46 or just over 9s per head (just under 50p in modern money) for this period, a vast difference with the cost of modern defence. Moreover, the castles have become a lasting presence in the landscape, with Conwy (Fig. 5.8), Caenarfon (Fig. 5.9), Harlech and Beaumaris accorded world heritage status (Geography and Technology, Welsh Government, 2022), something that could not be said of the physical presence of UK’s current defence force.

To build these Welsh castles took stone and wood and workers. The former were shipped in from quarries by boat although recycled stone from earlier local structures was also used (Sidall, 2021: 151). Workers were impressed from around England. In 1282–83, workers came from as far north as Northumberland and as far west as Somerset and Dorset and most counties in between, with Norfolk and Suffolk sending 100

Fig. 5.8 Conwy Castle. (Source: Creative Commons (n.d.-e))

Fig. 5.9 Caernarfon Castle. (Source: Creative Commons (n.d.-f))

diggers and 20 carpenters and Lincolnshire 150 diggers, 40 carpenters and 20 masons (Taylor, 2003: vi).

The architect of many of the Welsh castles was Master James of Saint George (c. 1235–1308). Edward I had probably met him in Savoy, where the king stayed on his return from the crusades in the Holy Land (Siddall, 2021: 138). Master James of Saint George was known as "ingeniator and mazun", the latter being translated as "builder in stone" (Taylor, 1950). Taylor goes on to explain the problem in translating "ingeniator" as the word "is not capable of concise translation into modern terminology, for it embodies something of all those functions which to-day are shared between the architect, the contractor and the resident superintendent of works". Master James of Saint George was attached to the court and in 1278 at the start of the castle building was paid 1 s per day whilst at court and an additional 1 s per day when away dealing with building (Taylor, 1950). His basic wage was the 2017 equivalent of £12,667 per annum, which would purchase 40 cows at the time. In 2021, the median salary for a partner/director in an architectural firm was £64,313 (RIBA, 2022), and this would purchase 71 dairy cows priced at £900/cow (based on 2-year old Hereford (Gov.UK, 2022)). However, the wages received by Master James of Saint George were increased. "On 20 October 1284, the king made a grant for life to Master James de Sancto Georgio, king's sergeant, of 3s. a day from the wardrobe, and at his death Is. 6d. a day to his wife Ambrosia if she survives him. He also received 8 marks yearly for his summer and winter clothing", making him the highest paid master craftsman in royal employment in the medieval period (Taylor, 1950). A mark was an allowance of 160d (13 s 4d) (MacLeod, 1863: 459). Taking the salary as 3 s/day and his clothing as a fringe benefit would mean the latter formed about 10% of salary. As a comparison in USA, the average fringe benefit rate is 30%, but for most people this would include health insurance (Tanski-Phillips, 2020). What it does show is that not much has changed in terms of remuneration and conditions of employment for skilled workers.

The work of Master James of St George can be recognised from the rounded towers of the castles (Fig. 5.8). Caernarfon Castle (Fig. 5.9) is different with its faceted towers and walls. Taylor (1950) notes that although Master James of St George was paid for work on this castle, from 1295–1308 the master mason was Walter of Hereford, and it may be his hand that is revealed in the design. However, he was only paid 1 s a day with an annual clothing allowance of 3.5 marks, compared to the 3 s and 8 marks of Master James of St George. One curiosity is that women also worked on the castles in various roles and were paid the same as men in the same role (Siddall, 2021: 141), something that seems far in advance on the modern situation. Strange that medieval England for all its faults was in accord with SDG 5 Gender Equality.

The money for the architects as well as for the materials and labour for the castles came from taxes, which were collected in both money and kind. Brauer and van Tuyll (2008: 70) make this clear.

> As landholdings of monarchs became less important, rulers became quite creative. The English kings exploited archaic land taxes, managed vacant bishoprics, pocketed profits from minting, collected every conceivable feudal due, taxed the laity for the crusades, and laid the foundation for the shift to universal taxation. This pattern—modernization of taxation due to war—can be seen in France and Spain as well. In France, the ordinary revenues grew from 86,000 to 400,000 livres tournois between 1180 and 1290, supplemented by a variety of special levies to fund wars and crusades. The Spanish monarchy achieved similar results, further stretching its money through devaluation. Monarchies grew in power as well as wealth.

Siddal (2021: 133) also notes that the coastal locations of castles meant that not only was it easy to ship in building materials, but the location made it easy to collect taxes on all goods coming into the port. A study of medieval taxation in Sweden from 1320–1550 (Retsö & Söderberg, 2015) showed how the level of taxes could change dramatically according to royal demand. From 1320–1349 in Sweden, the median tax paid per taxpayer was equivalent to 32 gm of silver or 16 kilos of butter, representing 2% of the value of a farm. For the period 1365–1424, this rose to 177 gm silver or 105 kilos of butter, representing 15% of the value of the farm, which means a tax on the capital. People coped with taxes by forming into groups, whereby the better off paid a higher proportion of the total tax than those less well off. Although tax evasion, as in the present, was practised, after the Black Death (1347–1351) and the subsequent loss of population, rents fell. These lower rents for farms meant it was possible for those in charge to raise taxes, suggesting a keen interest was paid as to how much it might be feasible to raise in taxation.

The church also raised a tax in the form of the tithe, or a tenth of the produce of the parish. "The tax was intended to sustain parish priests, maintain church buildings and provide alms for the poor, but religious corporations could appropriate parish tithes to their own purposes" (Dodds, 2005). Some splendid architecture still celebrates this further tax on the luckless medieval peasant in the form of the tithe barn (Fig. 5.8). Mortimer (2009: 47) makes the point that food was much more valuable to those living in medieval times compared to now, whereas labour and land were cheap, so taxes based on produce hurt (Fig. 5.10).

Željava Air Base

Sitting between Croatia and Bosnia Herzegovina, the underground Željava Air Base beneath Mount Pljesivica outside Bihac was constructed from 1948–1968 at a cost of US$6 billion but by 2007 no longer had any military use (Atlas Obscura, 2023; BBC Monitoring European, 2007). In 1948, the population of the Socialist Federated Republic of Yugoslavia was approximately 15.8 million, rising to 21.5 million by 1971 (Demographic Research Centre, 1974: 11). The cost per head of population thus varies between US$380 per head, assuming all was constructed in 1948, and US$280 per head if all built in 1971.

> The designer of Zeljava, Dragoslav Sobotka, baring his soul to Professor Radmila Tonkovic, said that the total length of the underground tunnels, galleries, and ancillary chambers was 3,500 metres. The facility had four entrances, which gave out onto two main runways. The mountain concealed also three galleries of 400 metres, 500 metres, and 350 metres in length; each was 8 metres high and 20 metres wide. The facility could sustain life in total isolation from the rest of the world for 30 days. (BBC Monitoring European, 2012).

Dragoslav Sobotka (1914–2008) was an engineer and colonel, and Željava airfield was his greatest work (People/Dragoslav Sobotka, n.d.). His father had been a railway engineer and was the designer of the famous Šargan Eight, which was

Fig. 5.10 Great Coxwell Tithe Barn, Oxfordshire. (Source: Creative Commons (n.d.-g)

a way of allowing the track to rise in a short distance, now part of a heritage railway (Pejović, 2020). Also, like Edward I's castles, and despite land mines and other hazards within the underground complex, it is possible to have a tourist experience and visit the external parts of the Željava site for US$55.72 per person (Trip Advisor, 2022). Currently it costs £11.10 (US$13) for an adult to visit Conwy Castle (Welsh Government, 2022d).

Who Pays for Architecture?

The above is a silly question since the answer is the client, but as the paired examples above show the client also has to find the money from somewhere, unless they are already very rich. The wealth of the Catholic Church in the Middle Ages was brought about through tithes, not paying taxes to the crown and limiting the power of the nobility to concentrate family wealth through marriage by forbidding marriage between cousins (Piketty, 2020: 94), however this did not necessarily mean that money was immediately available for the great cathedrals and churches of the period. Tithes were levied to support parish churches, not cathedrals. Money for building cathedrals came from the people through donations made either in thanksgiving, fear of eternal damnation, or somewhere in between. At the same time, these buildings paid for by the people became the major icons of the city, helping to attract people to these centres both then, as pilgrims, and now as tourists (Chicca et al., 2023: 229). The fact the cathedrals were the largest buildings by far in the city reinforced the power of the church, as its architecture made this hard to forget. The architects of these great buildings were respected men of moderate wealth, sometimes devoting a working life to a building they would never see finished. The curious fact is that a relatively modern cathedral, such as that of St. Paul in Wellington, was also financed by the people. Nor is this situation unique. When it came to the rebuilding of Coventry Cathedral after WWII, the gothic structure having been severely bombed and damaged in 1940, part of the £825,000 came from the War Damage Commission, a government body, but part still came from public subscriptions (Campbell, 1992).

The military buildings examined were financed through general taxation, together with some impressment of labour in the case of the castles of Edward 1. The expenditure of money for defence raised through taxation is, to an extent, money used as a deterrent rather than for a specific purpose. The hope is never to go to war but maintaining an army "just in case" is thought to be prudent. Modern warfare tends to demand trained people and stores of weapons rather than built environment, whereas the castles of the thirteenth century were a cheaper option for keeping the peace than maintaining a fighting force as a deterrent. Some of this architecture has survived as testament to those who designed and built the fortresses. The modern example of the air base in the mountain has lasted as well but had a very much shorter useful life, as well as being more expensive. In the end, both examples have become touristic ruins, a fate that tends to happen to large buildings when societies collapse (Garcia & Vale, 2021: 289–290).

Only in the example of stadia does a different means of building finance emerge. The Roman Colosseum was paid for through the spoils of war as a reward to a people who had been suffering under a regime of austerity. Could the money have been spent in a different way? Juvenal, writing in the second half of the first century AD coined the phrase *panem et circenses* (bread and circuses) to reflect the fact that Romans had lost interest in being active citizens, and that their rulers kept them "in a complacent and inert state by continually plying them with handouts of free food and lavish, violent shows" (Aldrete, 2021). This raises the issue of whether there is a parallel with the interest in modern sport. Raising money to build the new stadium at Wembley was only partly funded through taxation, with the most money raised in a loan by the Football Association, and that loan is financed by the money raised through operating the building, possibly at the expense of the FA's profits not going to grassroots football. As the FA raises money through broadcasting rights, it is the people who pay to watch who ultimately have a hand in supporting the new stadium. However it is viewed, it is the people who ultimately pay for great public architecture.

The examples above could be construed as grand buildings, and few architects manage to design such buildings during their working lives. A 2020 European study found that the design of houses remained important for many architects: "Private housing remains the most important sector for architects; almost 9 out of 10 architectural practices have undertaken some housing design work in the past year and the sector generates 54% of average practice turnover" (Architects' Council of Europe, 2020). Who pays for housing then becomes a further question.

Housing

Shelter is a basic right under the four requirements of traditional Buddhism, along with food, clothing and medicine. These are seen as fundamental needs that all people should have met and without which life becomes precarious (Reinert, 2018). SDG 11 (Sustainable Cities and Communities) target 11.1.1 is a measure of those with inadequate housing as the slum conditions in the world illustrate that life can be precarious even with some form of shelter however rudimentary. Nevertheless, in some developed countries not only is homelessness an issue, for example within USA some 500,000 people are homeless every night (Ding, et al. 2021), but what has post WWII been thought of as a human right—the right to own property, including a house (UN, 1948)— for some in the developed world is increasingly becoming a right they may never experience. Despite this, perhaps the real need is to be decently housed rather than own a house. Should decent housing be provided by government or should housing be left to the market?

House Ownership in the UK

According to the UK National Archives (2016) in 1918, renting was much more common than home ownership as 77% of households were renting. Only in 1971 were renters and owners both 50% of total households, with a peak in ownership of 69% reached in 2001, which has since fallen to 64%. Since most people buy a house with a mortgage, the ability to buy a house depends both on house price and earnings. Figure 5.11 compares the 1845–2019 cost of ownership of a house with income for the UK (Lamont, 2023).

What Fig. 5.11 demonstrates is that house ownership in the mid-nineteenth century was for many an even more remote dream than at present. Lamont (2023) gives three reasons for the drop in the ratio of house price to earnings: between 1815 and 1911 UK housing stock doubled; houses were smaller and built on smaller plots; and earnings rose more quickly than house prices fell because of the first two reasons given. The median house price to median income ratio, or median multiple, is a measure of housing affordability and a ratio of 3 or less is considered indicative of housing affordability (interest.co.nz, 2022). However, like all simple measures, reality is generally more complicated. Most houses are bought via a mortgage and in a pre-COVID-19 period of falling interest rates globally (Berkely Economic Review, 2020) proportionally less income was needed to service the loan so there was more money available for buying houses, house prices rose and so did the median multiple. What also emerges is that as populations grow and cities constrain urban sprawl, housing affordability drops, not least because land is less readily available (Ball et al., 2014).

Housing and what it costs is inextricably tied up with land. Land, in turn, is linked to sustainability as within a finite boundary, whether that of a city, nation or planet, land from somewhere has to supply the resources needed to sustain the way of living. As we know, the developed world is mostly living beyond the resources available within national limits, and humanity as a whole is annually consuming more than the renewable resources of a single planet (Earth). Land availability and cost is, therefore, a critical element in supplying housing people can afford, whether renting or buying.

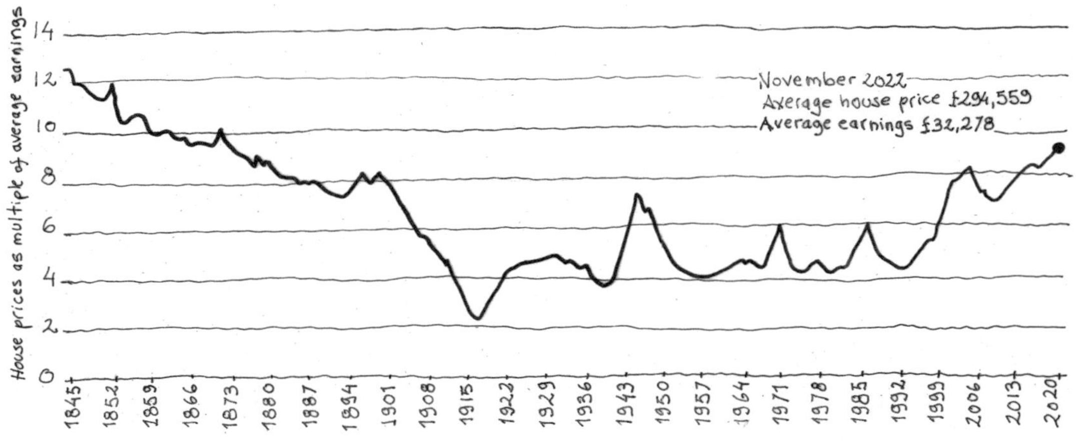

Fig. 5.11 UK house prices and average earnings based on Bank of England. (Based on Lamont 2023)

The last century in developed countries including the UK has seen land use controlled through planning processes, which effectively means political control of how much land is available. Hartmann and Gerber (2018: 5) note: "Many European cities have reached their maximum spatial expansion and are focusing on reshaping their urban core or industrial areas. Even where economic growth is still a given, the actors of land use planning struggle with this new focus on the redevelopment of pre-used plots". The problem here is that such redevelopment generally involves sorting out complications of land ownership. It also implies densification and often multi-storey housing. As populations continue to grow, the idea of having a single-family house on its own fenced plot—the idea of home is my castle—becomes much less viable.

Housing and Wealth

The reason for including some discussion of housing within the context of wealth and the built environment is that owning a house is seen as part of wealth accumulation (Andrews & Sánchez, 2011). Being tied to a mortgage forces available income into paying for a house, the value of which can rise with time as populations grow and land becomes scarcer. Most other normal household expenditures do not result in an asset that increases in value. Perhaps more importantly, ownership confers a stability that is more certain than renting, since tenures can be terminated, although in most societies such terminations are regulated. In fact, one of the definitions of slum housing is that it is housing with no guarantee of tenure (Chicca et al., 2023: 209). Nevertheless, in New Zealand where around one third of households rent (Stats NZ, 2019), "In 2018, the most common reason given by renters who had moved to another rental in the last five years was because their landlord ended their tenancy" (Stats NZ, 2020). The same source shows that most rentals in NZ are also privately owned (83.5%) rather than being some type of social housing for rent. Broadly this situation has come about because of the sale of state-owned houses, although the history of state housing supply and sale in New Zealand is complicated, with changes occurring according to which shade of government—right or left—is in power (Ministry for Culture and Heritage, 2014). The real issue is that private rentals are part of wealth accumulation unavailable to those doing the renting. Those who own a house for renting out, which assumes they have another house for their own use, will benefit from a rise in house prices. This is not the case with rental housing provided by the public sector. However, public sector housing for rent is a relatively new phenomenon.

The original European housing crisis came about as a result of the industrial revolution, the rapid increase in population that came with this, and the depopulation of the countryside as people migrated to the cities for work. Denby (1938: 24) makes it clear that this situation had not been foreseen.

> It is not surprising that this vast problem found the municipal authorities inexperienced and unprepared. With new towns springing up and old ones doubling and trebling in a generation, the public authorities had more than enough to do to increase the public services, such as water-supply, sanitation, transport and police, without also undertaking the provision of cheap housing. Besides, it was assumed that the demand for dwellings would be satisfied by private enterprise…

Private enterprise in housing provision at this time related to land ownership. Being a wealthy land owner did not automatically mean that you would provide good housing for those living on your urban land. Through the enclosure acts, the Duke of Norfolk owned a considerable area of Sheffield, an industrial town in the north of England. On this land "the Duke of Norfolk's slums are built—streets of "back-to-back" houses, small, dirty, vile-smelling, and productive of a death-rate of 40 per 1000" (Anon, 1912). The real problem resided in the level of wages. Unless you could earn enough to live in a decent house, the slum was the only other option. The wages of UK agricultural labourers were so low by the end of the nineteenth century that they could never pay the economic rent on a new cottage (Swenarton, 1981: 31). At this time only those earning more money, such as the skilled artisan or the middle classes, could attempt to become house owners through joining a building club. Through subscription and the interest earned on this pool of money and with the benefit of bulk savings it was possible to become a home owner (Chapman & Bartlett, 1971: 238). Business owners keen to attract and keep skilled workers might also build decent houses for them to rent (Chapman & Bartlett, 1971: 236).

One problem is that house ownership makes for a less mobile population and hence can lead to higher unemployment (Andrews & Sánchez, 2011). This is why in the past housing has at times been tied to employment, as it gives those doing the employing guaranteed labour. Here the wealth lies with the employers who can afford to supply houses for rent for their workers, a famous example in the UK being housing for Cadburys.

> The Cadburys, who moved from crowded Birmingham to rural Bournville, five miles out, in 1879, employed then 300 hands, and never dreamed of the present expansion of their business, which now requires 3,600 hands, including 2,400 girls. Surrounded by a model village, which is national property, here is a factory under truly ideal conditions, from which numerous illustrations of industrial betterment will be drawn…(Meakin, 1905: 68).

As much as the garden city movement or the development of the car may have led to the split between home and workplace and the inevitable commute (see Chap. 4), the desire for improved working conditions and housing may also have had a hand in this split, as other manufacturers moved factories to the edge of settlements, noting that this was also where land was cheaper.

Only in public sector housing provided by the state through taxes is individual wealth accumulation avoided. Unfortunately, because of neoliberal economics and the selling off of local authority owned houses in the UK the waiting list for social housing is long (Shelter, 2023) and there is a similar situation in New Zealand where public housing has dropped from 6.5% in 1951 to a current 3.5% of stock (Te Ora, 2023), forcing people into private renting, and without a good income, rentals can be less than satisfactory, even in a developed country. Before the end of 2023, the New Zealand Labour government was making efforts to provide affordable housing to people. Under Kainga Ora, the housing department of the government in New Zealand, KiwiBuild was "part of a national strategy to address New Zealand's housing challenges through partnering with private developers, government agencies, local councils, infrastructure providers, Māori and mana whenua" (KiwiBuild, 2023). An individual home buyer was eligible for a house at the time of writing (October 2023) if they had a salary of no more than NZ$120,000 (before tax), while for two or more buyers, like a couple, they needed to make less than NZ$200,000 (before tax) (Kiāinga Ora, 2023). The deposit was 5% for a first-time buyer loan and to be eligible under the scheme a three-bedroom house in Auckland would have needed to cost no more than NZ$860,000, which equates to finding a deposit of NZ$43,000. This was all wound down when the Labour government were voted out of power in November 2023.

However, what happened with households that had earnings slightly above the minimum required by KiwiBuild? They would need to borrow money from the bank and acquire a mortgage to buy a house. The bank will lend a certain amount of money depending on wealth and income. This money is directly related to income, expenses and saving capacity. Banks or intermediate agencies can demand a deposit of funds ranging from 10% to 20% of the price of the house. In Auckland, the average price of a house is just over NZ$1.2 million (Quotable Value, 2023). The ANZ bank demands first home buyers pay 20% of the cost of the house as a deposit, which, for an average price of 1.2 million, would be NZ$240,000. The average household salary (2022, the nearest date available to 2023) was NZ$89,036 after taxes (Stats NZ, 2023). For a household with two working members this would be NZ$178,072. However, people are also expected to save for retirement from this same income. Very roughly various sources suggest saving 10% or 15% of annual salaries to have a safe retirement. Assuming a couple is saving 10%, NZ$8904 each, it will take them 13.5 years to save the money for the bank deposit on a house, without any saving for retirement. For households with salaries above the median, which are already paying rent and cannot apply to KiwiBuild, the real challenge is to have the capital for the deposit. The mortgage is less problematic since it replaces the rent already being paid. The bank deposit is the problem, and it is directly linked with the wealth of a person, not merit in having a good job and salary or saving capacity. People working in jobs advertised in the market will have a hard time trying to save NZ$240,000 while paying rent. Why are banks asking for a deposit from people who have paid rent for many years? Isn't the record as a good tenant and a solid job proof enough of the capacity to pay for a house loan? In this situation, the middle class is trapped because they are too wealthy to receive the benefits of the poor and too poor to comply with rules designed for the rich and privileged.

The implication of not owning a house can be felt in everyday life. Two people with the same salary can live completely different lifestyles and have extra savings depending on their wealth, what they own, and what they have. A good income can provide a safe retirement and lifestyle only if it is backed up by wealth. Wealth is the key to ownership in the built environment, and housing ownership is the most critical asset in the wealth of most ordinary people. Without owning a house, retirement can be miserable. This wealth gap weakens society and creates a divide.

Housing Now and Then—It Was Ever So

Centuries ago renting grew out of tenant farmers renting land for subsistence from the landowner. In a study of cruck-framed peasant houses, mainly in four counties in the English midlands (Buckinghamshire, Leicestershire, Oxfordshire and Warwickshire), Dyer (2013: 106) first noted that a peasant was often a small land holder with sufficient land to feed a family and to trade any surplus or someone working as an artisan or in mining or fishing: "As small-scale landholders, peasants occupied a subordinate position in society, paying rents to their lords and taxes to the state". However, for some at least this did not mean living in abject poverty, as their houses reveal. Clark (2013) states that the land-holding peasant family would be able to generate the income to build a house of at least three bays, so in many respects was better off than those middle-income New Zealand households discussed above. The fact that so many houses have survived attests to how well they were built and often the adjoining buildings for housing animals and storing crops were equally well constructed (Dyer, 2013: 109). "The actual costs of finished buildings, where these are known from documentary sources, range from £2 to £11, with £4 as the median figure" (Anon, 2013). Dyer (2000: 164) notes peasants could borrow to raise the sum required for building "and many of them must have spent years paying off the cost of a dwelling or barn. Pleas of debt in manorial court rolls show that facilities for borrowing cash existed in many peasant communities, and there are occasional references to mortgages". The 1380 median value of £4 for a house equates to a little under £2500 in 2017 money, and that sum in 1380 could have purchased 3 horses, or 6 cows, or 200 days of skilled trades labour. In 2022, £2500 would buy you 4 dairy cows in New Zealand (Trademe, 2022a) and 1 horse (Trademe, 2022b) but only 0.5% of a dwelling in the same country based on the median house price of NZ$900,000 (Interest.co.nz, 2022). Obviously, the huge increase in human population has had an equally huge effect on the price of obtaining a house.

The cost of land has an effect on housing costs. The cost breakdown for a developer's 180m^2 two storey house in Auckland, New Zealand on a 300m^2 plot with a total cost of NZ$1.175 million before COVID-19 (probably 2017) was land 21%, civil/infrastructure costs 11%, materials 19%, labour 14%, contingency/prelims 8%, government taxes/consents 4%, holding costs 6%, professional fees 5%, GST (goods and services tax) 12%, carpark-marginal. In Wellington and Christchurch, where land is much cheaper, materials account for 23.5% and 22.7% and labour for 17.6% and 18.6% respectively (Deloitte, 2018: 33, 39). Such a house would be framed in timber although the methods used and the finishes would be different from a fourteenth century timber peasant's house. The only relevant costs for the latter would be labour and materials, and the small sum for professional fees for the modern house, which would include those of the architect, would be subsumed in the labour costs of the peasant's builder. Then as now not all labour for building a house is necessarily skilled. The peasant commissioning the house might save money by digging the foundations (Dyer, 2013: 107), but the same author continues that "even the least skilled jobs might be done by specialists". Assuming that the cost of the peasant house was split equally between materials and labour, the 100 days of skilled labour represent 12.5 weeks of work assuming an 8-day week. The current build time for a modern house in NZ is 10–12 months including all planning and consents with a build time of 6–8 months (Stonewood Homes, 2022). In fairness, the modern NZ house at 180m^2 is more than twice the size of the average peasant house and will have a more complex fit-out. Dyer (2000: 138) states that 84% of 113 peasant houses investigated in Worcestershire "were either of three bays, 15 ft by 45 ft (c. 4.6 m × 13.8 m), or of two bays, 15 ft by 30 ft (c. 4.6 m × 9.2 m), with three bays in a majority. The remaining sizes were of one bay (2 per cent), four bays (11%) or five or six bays (4%)". Looking at the neighbouring counties of Gloucestershire, Staffordshire and Warwickshire, 80% of peasant houses were of two or three bays. Woodforde (1970: 74) gives the standard bay as between 12 ft. and 15 ft., "a standard allowance based on the room needed for a team of four oxen." Using the Worcestershire data, a three-bay house would have a ground floor area of 63m^2 with a similar area for the first floor assuming it was completely floored over (Dyer, 2000: 140), although the first-floor area would be compromised by the roof slope (Fig. 5.12). This gives a peasant house area of an approximate maximum of 120m^2 overall, and proportionally a two-bay house would have a maximum area of 80m^2. However, this is floor area and although people were smaller then, Dyer (2000: 148) notes that "Many cruck buildings failed to survive...not because they were flimsy in construction, but because their lack of height made them difficult to convert to two-storey dwellings in the sixteenth and seventeenth centuries".

Whether the modern New Zealand timber house will last as long as some of the cruck framed peasant cottages is a moot point. What is evident is that to own a house then as now,

Fig. 5.12 Cruck frame house. (Source: Creative Commons (n.d.-h))

required a level of income that reflected the ability to borrow the money required. The difference is that houses were handed down through the generations so that families rather than individuals owned property and hence wealth. This works in the present where the money from house ownership is handed to the next generation. However, in a country like New Zealand which relies on skilled immigrants there may be no such wealth to be handed down, and such families may never have the chance to own a house and hence accrue the wealth that comes with this.

As ever, the buildings of the very poor have not survived. A history of Dartford suggests the town's poor "tended to live in single-cell buildings about 3 metres (10 feet) square, with internal partitioning to divide them into a living area and a bedchamber" (Dartford Town Archive, n.d.). As with modern slums, these structures would have been self-built of materials to hand—small wood, mud and straw. The tragedy is that housing has never been seen as a right in the western world and unless the state provides for those who have very little, slum-like conditions of overcrowding and lack of tenure will continue (Pryor, 2014). Meanwhile, those who own or are buying a house have wealth.

Perhaps what this chapter has raised is the issue of how public money should be spent and how the built environment reflects spending priorities. A large public building should always raise the question of whether this is the best use of the available money, especially when that money is raised through taxes. The problem is that it is usually those who have wealth who make the decisions.

References

Aldrete, G. S. (2021). "Bread and circuses": Ancient Rome, modern science fiction, and the art of political distraction. *Film and History, 51*(2), 4–20.

Ancient-Greece.org. (2022). *The Olympic Games.* Retrieved from https://ancient-greece.org/culture/olympic-games.html, 6 September 2022.

Anderson, T. J. (2017). *The value of debt in building wealth.* Wiley Online books.

Andrews, D., & Sánchez, A. C. (2011). The evolution of homeownership rates in selected OECD countries: Demographic and public policy influences. *OECD Journal: Economic Studies (2011/1)*, 207–243.

Anon. (1912). *A Duke's Slums.* Colonist. Issue 13511, Sept 12th. 2.

Anon (2013). *Peasant Houses in Midland England.* Retrieved from https://archaeology.co.uk/articles/peasant-houses-in-midland-england.htm, 15 June 2022.

Architects' Council of Europe. (2020). *The Architectural Profession in Europe 2020—ACE Sector Study.* Retrieved from HYPERLINK. https://www.ace-cae.eu/activities/publications/ace-2020-sector-study/, https://www.ace-cae.eu/activities/publications/ace-2020-sector-study/, 13 October 2023.

Atlas Obscura. (2023). *Zeljava Airbase.* Retrieved from https://www.atlasobscura.com/places/zeljava-airbase, 13 October 2023.

Ball, M., Cigdem, M., Taylor, E., & Wood, G. (2014). Urban growth boundaries and their impact on land prices. *Environment and Planning A, 46*(12), 3010–3026.

BBC Monitoring European. (2007). *Bosnian, Croatian CGS agree Bihac airfield no longer has "military importance".* BBC Worldwide Limited.

BBC Monitoring European. (2012). *Serbian paper details military "underground fortresses" across former Yugoslavia.* BBC Worldwide Limited.

Bekkers, R., & Wiepking, P. (2011). A literature review of *empirical* studies of philanthropy: Eight mechanisms that drive charitable giving. *Nonprofit and Voluntary Sector Quarterly, 40*(5), 924–973.

Berkely Economic Review. (2020). *The Long Decline of Global Interest Rates.* Retrieved from https://econreview.berkeley.edu/the-long-decline-of-global-interest-rates/, 19 September 2022.

Blain, M. (2002). *Wellington Cathedral of St Paul.* Victoria University Press.

Bolvin, S. (2022). *How does the FA get its money?* Retrieved from https://sport-net.org/how-does-the-fa-get-its-money/, 5 October 2022.

Brauer, J., & van Tuyll, H. (2008). *Castles, battles and bombs: How economics explains military history.* University of Chicago Press.

Broadberry, S., Campbell, B.M.S., & van Leeuwen B. (2011). *English medieval population: Reconciling time series and cross sectional evidence.* Retrieved from https://warwick.ac.uk/fac/soc/economics/seminars/seminars/conferences/venice3/programme/english_medieval_population.pdf, 24 August 2022.

Brown, F. E. (1968). *Roman architecture.* Studio Vista.

Campbell, L. (1992). Towards a new cathedral: The competition for Coventry Cathedral 1950–51. *Architectural History, 35*, 208–234.

Cartwright, M. (2019a). *Harlech Castle.* Retrieved from https://www.worldhistory.org/Harlech_Castle/, 23 August 2022.

Cartwright, M. (2019b). *Caernarfon Castle.* Retrieved from https://www.worldhistory.org/Caernarfon_Castle/, 24 August 2022.

Cartwright, M. (2019c). *Conwy Castle.* Retrieved from https://www.worldhistory.org/Conwy_Castle/, 24 August 2022.

CastlesFortsBattles.co.uk. (2019). *Bluith Castle.* Retrieved from http://www.castlesfortsbattles.co.uk/central_wales/builth_castle.html, 24 August 2022.

Chapman, S. D., & Bartlett, J. N. (1971). The contribution of building clubs and freehold land society. In S. D. Chapman (Ed.), *The history of working-class housing* (pp. 219–246). David and Charles.

Chicca, F., Vale, B., & Vale, R. (2023). *The environmental impact of cities.* Routledge.

Clark, D. (2013). The medieval peasant house—Towards a new paradigm? *Vernacular Architecture, 44*(1), 1–5.

Coke, E. (1826). *Reports of Sir Edward Coke, Knt. In Thirteen Parts.* (Reprint) Lawbook Exchange, 2002. (Original) J. Butterworth and Son.

Conn, D. (2018). *Wembley will not be sold by the FA after Shahid Khan withdraws offer.* Retrieved from https://www.theguardian.com/football/2018/oct/17/wembley-stadium-deal-off-fa-shahid-khan, 13 September 2022.

CPG Europe (Construction Products Group). (2022). *Wembley Stadium.* Retrieved from https://www.flowcrete.eu/en-gb/case-studies/wembley-the-worlds-greatest-stadium/#:~:text=The%20new%20Wembley%20Stadium%2C%20located,largest%20seating%20capacity%20in%20Europe. 5 September 2022.

Creative Commons. (n.d.-a). *Sergey Ashmarin, CC BY-SA 3.0.* https://creativecommons.org/licenses/by-sa/3.0, via Wikimedia Commons

Creative Commons. (n.d.-b). "The Colosseum, Rome" by kevinpoh is licensed under CC BY 2.0.

Creative Commons. (n.d.-c). "Wembley Stadium Twin Towers—Shut down!" by Merv Payne is licensed under CC BY-SA 2.0.

Creative Commons. (n.d.-d). "Wembley Stadium (34)" by Martin Pettitt is licensed under CC BY 2.0.

Creative Commons. (n.d.-e). "Conwy Castle and car park from Town Walls—geograph.org.uk -1,723,358" by David Dixon is licensed under CC BY-SA 2.0.

Creative Commons. (n.d.-f). "Caernarfon Castle from outside" by dgoomany is licensed under CC BY-SA 2.0.

Creative Commons. (n.d.-g). "Great Coxwell Tithe Barn" by Fearless Fred is licensed under CC BY 2.0.

Creative Commons. (n.d.-h). “Cruck cottage in Wick” by Philip Halling is licensed under CC BY-SA 2.0.

Crow, M. (2018). *Images of the Past Wembley Stadium.* Pen and Sword Books Ltd.

Da Silva, D.L., Tian, N., Béraud-Sudreau, L., Marksteiner, A. & Linag, X. (2022). *Trends in World Military Expenditure 2021.* Retrieved from https://www.sipri.org/sites/default/files/2022-04/fs_2204_milex_2021_0.pdf, 28 August 2022.

Dartford Town Archive. (n.d.). *Homes of the poor: Dartford's dismal hovels.* Retrieved from https://www.dartfordarchive.org.uk/medieval/buildings_p.shtml, 23 June 2022.

Deloitte. (2018). *Cost of residential housing development: A focus on building materials.* Retrieved from https://www2.deloitte.com/content/dam/Deloitte/nz/Documents/Economics/nz-en-DAE-Fletcher-cost-of-residential-housing-development.pdf, 21 June 2022.

Demographic Research Centre. (1974). *Population of Yugoslavia.* Retrieved from http://www.cicred.org/Eng/Publications/pdf/c-c55.pdf, 28 August 2022.

Denby, E. (1938). Europe re-housed. Allen & Unwin.

Ding, H., Loukaitou-Sideris, A., & Wasserman, J. L. (2021). Homelessness on public transit: A review of problems and responses. *Transport Reviews, 42*(2), 134–156.

Dio, C. (211–233 AD; 1917). *Roman History.* Retrieved from http://penelope.uchicago.edu/Thayer/E/Roman/Texts/Cassius_Dio/54*.html#26, 11 September 2022.

Dodds, G. (2005). Managing Tithes in the Late Middle Ages. *The Agricultural History Review, 53*(2), 125–140.

Dodge, H. (2014). Amphitheatres in the Roman World. In P. Christesen & D. G. Kyle (Eds.), *A companion to sport and spectacle in Greek and Roman antiquity.* John Wiley and Sons. Chapter 37 (unpaginated).

Dominion. (1917). *Anglican Cathedral.* Vol. 10, Issue 3140. 19 July 1917. p. 6.

Dunkle, R. (2013). *Gladiators: Violence and spectacle in ancient Rome.* Routledge.

Dunkle, R. (2014). Overview of Roman spectacle. In P. Christesen & D. G. Kyle (Eds.), *A Companion to Sport and Spectacle in Greek and Roman Antiquity.* John Wiley and Sons. Chapter 25 (unpaginated).

Dyer, C. (2000). *Everyday life in medieval England.* Bloomsbury Publishing.

Dyer, C. (2013). Documentary evidence. In N. Alcock & D. Miles (Eds.), *The Medieval Peasant House in Midland England* (pp. 105–152). Oxbow Books.

Ekelund, R. B., Hébert, R. F., & Tollison, R. D. (1989). An economic model of the medieval church: Usury as a form of rent seeking. *Journal of Law, Economics and Organization, 5*(2), 307–331.

Emporis. (2022). *Domtoren Utrecht.* Retrieved from https://www.emporis.com/buildings/111232/domtoren-utrecht-utrecht-netherlands, 30 August 2022.

Evening Post. (1917). *The Cathedral.* Vol. 44, Issue 10, 23 July 1917, p. 2.

Evening Post. (1938). *Visit to Masterton: Address by the Bishop.* Vol. 126, Issue 28, August 1938, p. 12.

Foster + Partners. (2022). *Wembley Stadium.* Retrieved from https://www.fosterandpartners.com/projects/wembley-stadium/, 7 September 2022.

Garcia, E., Vale, B., & Vale, R. (2021). *Collapsing gracefully: Making a built environment that is fit for the future.* Springer.

Geography and Technology, Welsh Government. (2022). *World Heritage Sites in Wales.* Retrieved from https://lle.gov.wales/catalogue/item/WorldHeritageSites/?lang=en#:~:text=Wales%20currently%20has%20four%20world,Canal%20in%20north%2Deast%20Wales, 25 August 2022.

GhanaSoccernet. (2021). *From poverty to power, footballers who grew up poor—South American edition.* Retrieved from https://ghanasoccernet.com/from-poverty-to-power-footballers-who-grew-up-poor-south-american-edition, 14 September, 2022.

Gov.UK. (2022). *Livestock prices, finished and store.* Retrieved from https://www.gov.uk/government/statistical-data-sets/livestock-prices-finished-and-store, 26 August 2022.

Great Castles. (n.d.). *Hastings Castle.* Retrieved from https://great-castles.com/hastings.html, 19 August 2022.

Greater London Authority. (2022). *MD2644 Wembley National Stadium Limited—Financing Facility.* Retrieved from https://www.london.gov.uk/decisions/md2644-wembley-national-stadium-limited-financing-facility, 8 September 2022.

Green, A. (2012). *Catalogue of the 68 Competitive Designs for the Great Tower for London (1890).* Retrieved from https://publicdomainreview.org/collection/catalogue-of-the-68-competitive-designs-for-the-great-tower-for-london-1890, 5 September 2022.

Hartmann, T., & Gerber, J.-D. (2018). Land, scarcity, and property rights. In J.-D. Gerber, T. Hartmann, & A. Hengstermann (Eds.), *Instruments of land policy: Dealing with scarcity of land* (pp. 3–7). Routledge.

Harvey, J. (2016). *Gold sanctuary: Bishop Selwyn Chapel.* Retrieved from https://architecturenow.co.nz/articles/gold-sanctuary-bishop-selwyn-chapel/, 11 September 2022.

Hassall, P. (1996). *Bernard of Clairvaux: Apology.* Retrieved from http://web.mit.edu/aorlando/www/SaintJohnCHI/Church%20History%20Readings/Bernard%20Apology.pd 18 August 2022.

Hediger, C. (2019). Female donors of medieval stained-glass windows. In B. Kurmann-Schwarz & E. Pastan (Eds.), *Investigations in medieval stained glass.* Brill.

Helms, R. M. (1996). *The architecture of Cecil Wood.* PhD Thesis. University of Canterbury.

Hopkins, K., & Beard, M. (2005). *The Colosseum.* Harvard University Press.

Hutt News. (1945a). *Cathedral Appeal.* Vol. 19, Issue 12. 7 November. p. 5.

Hutt News. (1945b). *Cathedral Appeal: Lower Hutt Campaign.* Vol. 20, Issue 2. 12 December. p. 9.

Hutt News. (1946). *Power Board's Donation*. Vol. 21, Issue 2. June 1946. p. 9.

Interest.co.nz. (2022). *Median multiples*. Retrieved from https://www.interest.co.nz/property/house-price-income-multiples, 19 September 2022.

ISCID (n.d.). *How much money does the sports industry spend each year?* Retrieved from https://www.icsid.org/business/how-much-money-does-the-sports-industry-spend-each-year/, 30 August 2022.

Kemper, N. (2019). *Building a Cathedral*. Retrieved from https://theprepared.org/features-feed/2019/4/28/building-a-cathedral, 30 August 2022.

Khairi, N. (2022). *FA Cup 2022 final tickets: Prices for Liverpool* vs *Chelsea decider & how to buy*. Retrieved from https://www.goal.com/en-us/news/fa-cup-2022-final-tickets-prices-for-liverpool-vs-chelsea/blt-d52793aa9da75553, 13 September 2022.

Kiānga Ora. (2023). *KiwiBuild*. Retrieved from https://kaingaora.govt.nz/home-ownership/kiwibuild/, 18 October 2023.

Kiuri, M., & teller, J. (2015). Olympic stadiums and cultural heritage: On the nature and status of heritage values in large sport facilities. *International Journal of the History of Sport, 32*(5), 684–707.

KiwiBuild. (2023). *About us*. Retrieved from https://www.kiwibuild.govt.nz/about-kiwibuild/about-us/. 18 October 2023.

Koobit. (2022). *Category 5 (Family Enclosure)*. Retrieved from https://www.koobit.com/england-v-andorra-e2007/tickets/category-5-family-enclosure-6108, 13 September 2022.

Lamont, D. (2023). *What 175 years of data tell us about house price affordability in the UK*. Retrieved from https://www.schroders.com/en/uk/private-investor/insights/markets/what-174-years-of-data-tell-usabout-house-price-affordability-in-the-uk/, 11 February 2023.

Law, E. (2018). *The highest paid athlete in history was a Roman Charioteer*. Retrieved from https://theculturetrip.com/europe/italy/articles/the-highest-paid-athlete-in-history-was-a-roman-charioteer/, 11 September 2022.

Lewis, M. (2022). *The 10 'Ring of Iron' Castles Built by Edward I in Wales*. Retrieved from https://www.historyhit.com/edward-i-ring-of-iron-castles-in-wales/, 28 August 2022.

Lewis, R. (2025). *Wembley Stadium*. Encyclopedia Britannica. Retrieved from https://www.britannica.com/place/Wembley-Stadium, 6 February 2025.

MacInnes, P. (2021). *Wembley the headline act at Euro 2020 but ailing finances cast a shadow*. Retrieved from https://www.theguardian.com/football/2021/jun/12/wembley-the-headline-act-at-euro-2020-but-ailing-finances-cast-a-shadow, 13 September 2022.

Maclean, C. (2015). *St. Paul's Cathedral*. Retrieved from https://teara.govt.nz/en/photograph/13388/st-pauls-cathedral, 17 August 2022.

MacLeod, H. D. (1863). *A dictionary of political economy*. Longman, Brown, Longman and Roberts.

Mann, R. (2021). *Remember when the Metrodome's roof collapsed 5X before it was taken down?* Retrieved from https://www.theweathernetwork.com/ca/news/article/this-day-in-weather-history-december-12-2010-metrodome-roof-collapse, 6 September 2022.

Matz, D. (2019). *Ancient Roman Sports, A-Z: Athletes, venues, events and terms*. McFarland and Company Ltd.

Meakin, B. (1905). *Model factories and villages*. T. Fisher and Unwin.

Ministry for Culture and Heritage. (2014). *The State steps in (and out)*. Retrieved from https://nzhistory.govt.nz/culture/we-call-it-home/the-state-steps-in-and-out, 16 October 2023.

Mortimer, I. (2009). *The time traveller's guide to medieval England*. Vintage Books.

Noll, R. G., & Zimbalist, A. (1997). Sports, jobs and taxes. *Brookings Review, 15*(3), 35.

Office for National Statistics. (2018). *Population estimates for the UK, England and Wales, Scotland and Northern Ireland: mid-2017*. Retrieved from https://www.ons.gov.uk/peoplepopulationandcommunity/populationandmigration/populationestimates/bulletins/annualmidyearpopulationestimates/mid2017, 24 August 2022.

Otaki Mail. (1943). *Bequests to Charities*. 29 March. p.2.

Panofksy-Soergel, G. (1979). *Abbot Suger* (2nd ed.). Princeton University Press.

Papers Past. (n.d.). *Wanganui Herald, Volume LIII, Issue 15745, 18 February 1919, Page 3*. Retrieved from https://paperspast.natlib.govt.nz/newspapers/WH19190218.2.17.2?end_date=31-12-1938&items_per_page=10&query=cathedral+for+wellington+proposed+use+of&snippet=true&start_date=01-01-1917, 20 August 2022.

Pejović, B. (2020). *A century since the construction feat in Šargan*. Retrieved from https://www-presslider-rs.translate.goog/vek-od-graditeljskog-podviga-na-sarganu/?_x_tr_sch=http&_x_tr_sl=sr&_x_tr_tl=en&_x_tr_hl=en&_x_tr_pto=sc, 30 August 2022.

Penn Museum. (2022). *The Games*. Retrieved from https://www.penn.museum/sites/olympics/olympic-origins.shtml, 6 September 2022.

People/Dragoslav Sobotka. (n.d.). *Dragoslav Sobotka*. Retrieved from https://peoplepill-com.translate.goog/people/dragoslav-sobotka?_x_tr_sl=sr&_x_tr_tl=en&_x_tr_hl=en&_x_tr_pto=sc, 29 August 2022.

Piketty, T. (2020). *Capital and ideology*. Harvard University Press.

Press. (1937). *Cathedral for Wellington: proposed use of old fund*. Vol. 73, Issue 22233. 26 October. p.7.

Prestwich, M. (2010). Edward I and Wales. In D. M. Williams & J. R. Keynon (Eds.), *The impact of the Edwardian Castles in Wales* (pp. 1–8). Oxbow Books.

Pryor, N. (2014). *Slum warning over Auckland CBD*. Retrieved from https://www.rnz.co.nz/news/national/253620/slum-warning-over-auckland-cbd, 19 October 2023.

Quotable Value. (2023). *QV House Price Index*. Retrieved from https://www.qv.co.nz/price-index/, 18 October 2023.

Reinert, K. A. (2018). The Wisdom of Need: Basic Goods Provision in Buddhist Economic Ethics. *Journal of Buddhist Ethics, 25*, 607–636.

Reserve Bank of New Zealand. (2022). *Inflation calculator.* Retrieved from https://www.rbnz.govt.nz/monetary-policy/about-monetary-policy/inflation-calculator, 30 August 2022.

Retsö, D., & Söderberg, J. (2015). The late-medieval crisis quantified. *Scandinavian Journal of History, 40*(1), 1–24.

RIBA. (2022). *Salary guide 2022—architects and architecture.* Retrieved from https://jobs.architecture.com/staticpages/10290/salary-guide-2022-architects-and-architecture/, 26 August 2022.

Shelter. (2023). *14,000 social homes lost last year, as over a million households sit on waiting lists.* Retrieved from https://england.shelter.org.uk/media/press_release/14000_social_homes_lost_last_year_as_over_a_million_households_sit_on_waiting_lists#:~:text=In%202022%20there%20were1%2C206%2C376,lettings%20and%20tenancies%2C%20Table%20600, 18 October 2023.

Siddall, R. (2021). *Natural Stone and World Heritage: The Castles and Town Walls of King Edward in Gwynedd.* CRC Press.

Sork, A. (2018). *What famous ancient structures would cost to build today.* Retrieved from https://www.bigrentz.com/blog/cost-build-ancient-structures#:~:text=A%20261%2C36%2D%20square%20feet,costs%20at%20around%20%2422%20million, 11 September 2022.

Statista. (2022). *Military expenditure of the United Kingdom (UK) from 2009–2019.* Retrieved from https://www.statista.com/statistics/547441/military-expenditure-total-uk/, 24 August 2022.

Stats, N.Z. (2019). *Renting* vs *owning in NZ.* Retrieved from https://www.stats.govt.nz/infographics/renting-vs-owning-in-nz/, 16 October 2023.

Stats, N.Z. (2020). *Housing in Aotearoa: 2020.* Retrieved from www.stats.govt.nz., 16 October 2023.

Stats, N.Z. (2023). *Household income and housing-cost statistics: Year ended June 2022.* Retrieved from https://www.stats.govt.nz/information-releases/household-income-and-housing-cost-statistics-year-ended-june-2022/#:~:text=average%20annual%20household%20disposable%20income,increased%20from%20%2450%2C164%20to%20%2450%2C561, 18 October 2023.

Stavropoulos, P. (1999). *Rush to complete new stadium blamed for deaths of three US construction workers.* Retrieved from https://www.wsws.org/en/articles/1999/07/milw-j23.html, 6 September 2022.

Stonewood Homes. (2022). *How Long Does it Take to Build a House?* Retrieved from https://stonewood.co.nz/how-long-does-it-take-to-build-house/#:~:text=So%2C%20in%20total%2C%20once%20Building,before%20handover%20can%20take%20place. 19 October 2023.

Swenarton, M. (1981). *Homes fir for Heroes*. Heinemann.

Tanski-Phillips, M. (2020). *What in the world is a fringe benefit rate?* Retrieved from https://www.patriotsoftware.com/blog/payroll/fringe-benefit-rate/, 27 August 2022.

Taylor, A. J. (1950). Master James of St George. *The English Historical Review, 65*(257), 433–457.

Taylor, A. (2003). *The Welsh Castles of Edward 1.* Bloomsbury Publishing.

Te Ora, E. (2023). *The 'true wait list' for state housing could be hundreds of thousands of people.* Retrieved from https://www.thepost.co.nz/a/nz-news/350041747/true-wait-list-state-housing-could-be-hundreds-thousands-people, 18 October 2023.

The Football Association. (2022). *FA Reports and Financial Statements.* Retrieved from https://www.thefa.com/about-football-association/what-we-do/financial-statements#:~:text=During%20the%202020%2D21%20season,the%20National%20Football%20Centre%20Limited., 14 September 2022.

The Global Economy. (2023). *Military spending, percent of GDP—Country rankings.* Retrieved from https://www.theglobaleconomy.com/rankings/mil_spend_gdp/#:~:text=Military%20spending%2C%20percent%20of%20GDP%20%2D%20Country%20rankings&text=The%20average%20for%202022%20based,countries%20where%20data%20are%20available. 11 October 2023.

The Guardian. (2003). *£120 m Wembley grant 'could have been lost'.* Retrieved from https://www.theguardian.com/uk/2003/jun/06/football.wembleystadium, 7 September 2022.

The National Archives. (2016). *A century of home ownership and renting in England and Wales (full story).* Retrieved from https://webarchive.nationalarchives.gov.uk/ukgwa/20160107120359/http://www.ons.gov.uk/ons/rel/census/2011-census-analysis/a-century-of-home-ownership-and-renting-in-england-and-wales/short-story-on-housing.html, 14 October 2023.

The National Archives. (n.d.). *Currency converter: 1270–2017.* Retrieved from https://www.nationalarchives.gov.uk/currency-converter/#currency-result, 23 August 2022.

Thomas, J.L. (2009). *Beaumaris Castle.* Retrieved from http://www.castlewales.com/beaumar.html, 23 August 2022.

Trademe. (2022a). *Dairy cattle.* Retrieved from https://www.trademe.co.nz/a/marketplace/business-farming-industry/farming-forestry/livestock/dairy-cattle, 21 June 2022.

Trademe. (2022b). *Horses.* Retrieved from https://www.trademe.co.nz/a/marketplace/sports/equestrian/horses-ponies/horses/search?search_string=horses, 21 June 2022.

Trip Advisor. (2022). *Underground military airport - #zeljavatour.* Retrieved from https://www.tripadvisor.co.nz/AttractionProductReview-g670545-d19991555-Underground_military_airport_zeljavatour-Rakovica_Plitvice_Lakes_National_Park_Cen.html, 28 August 2022.

United Nations (UN). (1948). *Universal Declaration of Human Rights*. Retrieved from https://www.un.org/en/about-us/universal-declaration-of-human-rights, 14 September 2022.

Venture New Zealand. (2022). *Nelson Cathedral*. Retrieved from https://venturenewzealand.co.nz/place/nelson-cathedral, 30 August 2022.

Visit San Mames. (n.d.). *Why is San Mames Stadium called the cathedral of football?* Retrieved from https://visit-sanmames.com/en/blog/san-mames-stadium-bilbao-called-cathedral-of-football/, 9 September 2022.

Vroom, W. (2010). *Financing Cathedral Building in the Middle Ages: the generosity of the faithful*. Amsterdam University Press.

Waganui Herald. (1919). *Advertisements Column 2*. Vol. 53, Issue 15745, 18 February. p. 3.

Watson, L. (2022). *Why does Christchurch need a stadium, anyway?* Retrieved from https://www.nzherald.co.nz/nz/why-does-christchurch-need-a-stadium-anyway/U4QIGURXHU2KKQCR4KJOKVM3SM/, 6 September 2022.

Wellington Cathedral of St. Paul. (n.d.). *Our History*. Retrieved from https://www.wellingtoncathedral.org.nz/our-history, 16 August 2022.

Welsh Government. (2022a). *Aberystwyth Castle*. Retrieved from Aberystwyth Castle | VisitWales, 23 August 2022.

Welsh Government. (2022b). *Rhuddlan Castle*. Retrieved from https://cadw.gov.wales/visit/places-to-visit/rhuddlan-castle, 23 August 2022.

Welsh Government. (2022c). *Flint Castle*. Retrieved from https://cadw.gov.wales/visit/places-to-visit/flint-castle, 23 August 2022.

Welsh Government. (2022d). *Conwy Castle*. Retrieved from https://cadw.gov.wales/visit/places-to-visit/conwy-castle, 28 August 2022.

WembleyStadium. (2022). *Our Story*. Retrieved from Our Story (wembleystadium.com), 7 September 2022.

Williams, G. (2011). The iron ring of Edward Longshanks. *Medieval Warfare, 1*(2), 38–42.

Wilson, C. (1992). *The Gothic Cathedral*. Thames and Hudson.

Woodforde, J. (1970). *The truth about cottages*. The Country Book Club.

6 Luxury

Understanding Luxury

Luxury seems a long way from SDGs, like No Poverty (SDG1) and Zero Hunger (SDG2). Still, any discussion of wealth needs to tackle luxury because it challenges responsible consumption and production (SDG12) and even the idea of Sustainable Cities and Communities (SDG11). Luxury might be good for economic growth, but it must come at a resource use cost. A pattern here is that actions whose starting point is highly dependent on accumulating wealth, justified as economic growth or industrial innovation, tend to benefit two goals—SDG 8 Decent Work and Economic Growth, and possibly SDG 9 Industry, Innovation and Infrastructure—at the expense of negatively impacting all the rest.

In its essence, a luxury is something that does not happen in the everyday world. You might indulge in the luxury of extra time in bed on a non-work day, or perhaps the luxury of a hot bath instead of the fast daily shower. Since luxury is, therefore, contextual, it is also relative to the referent used to make the comparison between having something and wanting something else. The context for the emergence of luxury is an unfair comparison between what is at hand or available (the have) against what is desired and found in more expensive lifestyles or in richer environments (the want). The origins of that desire and feelings are multiple, from personal drivers, such as the wish to belong to an elite to gain respect, to external factors, such as the simple luxury of ready access to water in the desert. Associating luxury with a desire is not a caprice. In Elizabethan English, luxury meant lust and even adultery, as Shakespeare's lines from Act 4 of *Much Ado About Nothing* show: "She knows the heat of a luxurious bed. Her blush is guiltiness, not modesty". The desire to have more than you really need, which is fundamental to the modern idea of luxury, naturally works against any idea of sustainability, which can simply be defined as living within the means at hand (Garcia & Vale, 2017: 15), whilst luxury is the desire for something outside the means at hand. This idea is hardly new. In Plato's *The Republic,* Socrates discussed the implication of luxury with Glaucon by arguing that if everybody wants more than just "the bare necessities of houses, clothing and shoes", society would become much more complex with the need for many more occupations, including an army. To support this accumulation of luxuries "the territory which was formerly enough to support us will now be too small". The consequence is that "we shall have to cut a slice off our neighbour's territory. And if they too are no longer confining themselves to necessities and have embarked on the pursuit of unlimited material possessions, they will want a slice of ours too" (Plato, 2007; 123), an observation that reflects modern global society and the challenge luxury poses to the necessity of achieving SDG 11 Sustainable Cities and Communities.

E. J. García, B. Vale, *Wealth and the Built Environment*, Sustainable Development Goals Series,
https://doi.org/10.1007/978-3-031-81954-4_6

Debates about luxury were very prevalent during the Age of Enlightenment (1685–1815) because it was tied to the idea of economic growth and its consequences of creating more stuff (Hont, 2006: 380). On one side were those who looked back to the Greeks and the idea of luxury leading to the need for more territory, while on the other were those prepared to accept economic growth and the idea of having more stuff but to work out "how to make it politically and morally benign" (Hont, 2006: 380). At the time these debates made it difficult to define luxury, and when in the 1750s Diderot included it in his Encyclopédie, he struggled to define the term.

> We say, without any of us being mistaken, of an infinity of objects of all sorts, that they are luxurious …but what is this luxury that we attribute so infallibly to so many objects? Here is a question which will not be answered with any precision except as a result of a discussion which those who demonstrate the most accuracy in the application of the word luxury have not had, are not perhaps even in a position to have. (Goodman, 2003: 75)

Others saw the production of goods as part of economic growth, goods that were bought not just by the rich but also by the middle class, as leading to inequality in society, since there was inevitably a group of people who could not afford these new goods—the poor. Rousseau writing in 1750–55 thought luxury "brought about a continuous process of moral corruption. It ruined taste, substituted effeminacy for courage, and drew the population from the countryside to capital cities" (Berg & Eger, 2003: 21). Luxury then became associated with the built environment through the necessity of having cities both as a means of housing labour to produce goods and as a means of housing those who would buy the goods, with everyone in close proximity. However, while economic growth remained the goal, as it still does for modern politics, the production of goods over and above those that might be thought necessities for life is inevitable. The architect Le Corbusier noted the consequences of this when it came to the design of the Radiant City.

> It is recognised that industry is submerging us with products that are irrelevant to our happiness… Modern industry, organised according to the laws of supply has inundated us with useless consumer goods…Free competition has been forced to invent advertising, the travelling salesman, the prospectus, product prizes, etc: all those competing have to summon up all their energy to cancel out the efforts of their competitors. (Le Corbusier, 1933: 68)

More recently Hill (2005: 124) goes further by suggesting that although it might be human nature to want luxuries, unfortunately this "brings disorder and instability into the world, and to counter those explosive forces there must be a powerful state apparatus".

What Is a Modern Luxury Item?

When it comes to choosing an intended purchase, luxury has become associated with something that costs more than its regular equivalent, even though the buyer may not be aware of why the luxury product is different from the non-luxury one, apart from the price and possibly the name associated with the product. In some circumstances buying the name may be the most important aspect of obtaining a luxury. This is illustrated by the expensive Karl Lagerfeld villas discussed in more detail later in this chapter. The site plan (Fig. 6.1) reveals that one very expensive villa is seriously compromised in terms of privacy, being surrounded by roads and paths but this has not stopped it from being a very expensive house. The first villa was sold off-plan for approximately €15 million (US$16.5 million) (EFE, 2022). This must be an example of buying a name as much as a house.

As a further example, luxury cars are defined as cars with enhanced performance, interior comfort and safety features. "In order to be considered a luxury car, the vehicle must have high-end features that go above and beyond the average necessities" (Certified Autoplex, 2022). Although an engine, transmission chain, wheels, something to sit on and a body to keep out the rain might be thought of as the necessities for a car, essentially the luxury car has more money spent on aspects of these even though the same objective is to be

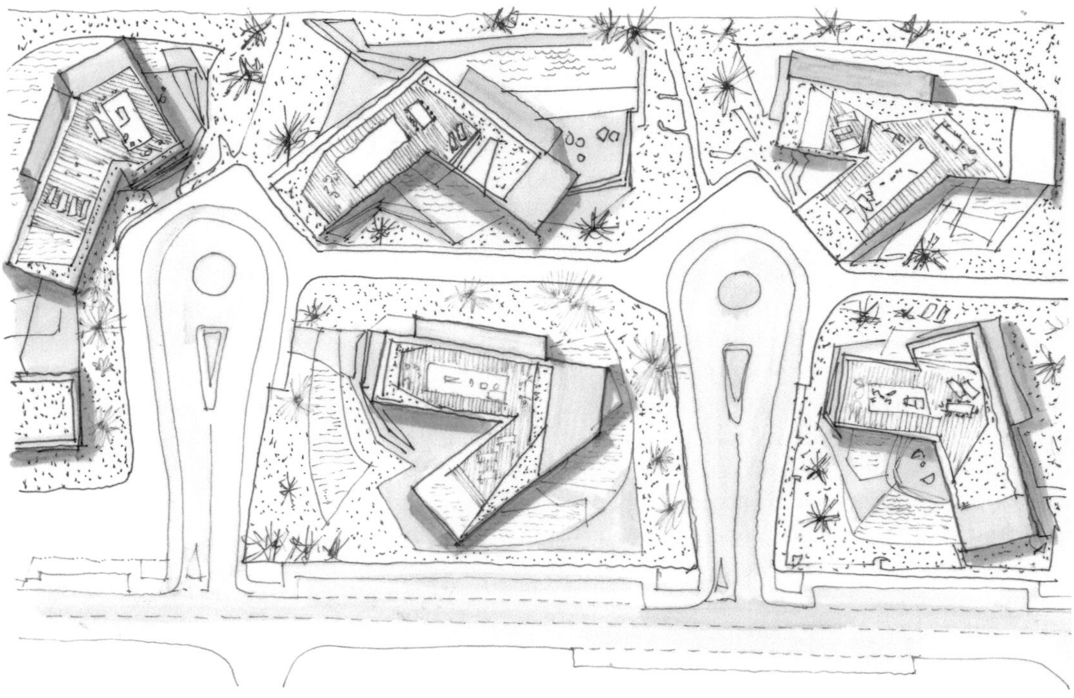

Fig. 6.1 Karl Lagerfeld Villas, Marbella: Site Plan. (Based on Karl Lagerfeld (n.d.))

able to drive from A to B. Kelley Blue Book editors are more specific about the luxury car (KBB Editors, 2021): "The driving experience should be one of the most prominent differences between mainstream car brands and luxury marques. People buy luxury cars to feel rich, pampered, and prestigious". It also seems that men are very much more likely to buy a luxury car than women, at least in a study of the city of Douala in Cameroon (Robert, 2019) (see Chap. 4 for discussion of gender difference in buying electric cars).

Luxury could also be construed as something as simple as having a soak in a hot bath, rather than the daily shower, although it turns out the daily shower could also be seen as an unnecessary luxury for some people, since all that is required for a sedentary person in a temperate climate is to shower perhaps once or twice a week (Batliwala, 2016). Shmerling (2021), a doctor, goes further:

> ...daily showers do not improve your health, could cause skin problems or other health issues—and, importantly, they waste a lot of water. Also, the oils, perfumes, and other additives in shampoos, conditioners, and soaps may cause problems of their own, such as allergic reactions (not to mention their cost).

To top it off, shower gels also have luxury brands, even though all you are buying is soap in water, rather than using the soap in a bar, but selling water is a great way to make a profit. On the supermarket shelves shower gels are more expensive than a bar of soap, so hence appear more luxurious. This is the key to luxury; it costs more and so demonstrates that you have money.

This means that luxury is contextual. It depends on choosing between one version of an object and a seemingly superior and more expensive version of the same type of object. Sometimes, it is even a case of where the product was bought. In New Zealand, the same company owns two major supermarket chains—New World and Pak'nSave. New World is the "aspirational" supermarket compared to the more affordable Pak'nSave. The imported cheese Castello is more expensive in New World than in Pak'nSave even though it is the same cheese, brand and product. There is perhaps a building parallel here since a house built with the same materials and to the same size would be valued differently depending on its location and surrounding amenities. In the supermarket example, Pak'nSave advertises the fact that its lower prices are related to its "no frills" or non-luxury approach to the supermarket interior. Still, few people shop in a supermarket to admire its interior.

Inherent in the idea of luxury or a luxury product is the desire to have an object that costs more. Where there is a choice, shopping at New World in the example above shows that you have the money to do this and hence indicates your status in the locality. The origins of the desire for luxury are multiple, from personal drivers (belonging to an elite, gaining respect through ownership) to external factors (such as the luxury of having water in the desert or clean air in a polluted area). If as suggested above, sustainability is living within the means at hand, then luxury could be viewed as going against sustainability since the assumption is that more resources will have gone into the luxury product.

We have already written about electric cars and wealth in this book. Still, it is worth considering whether electric cars are a luxury because they generally cost more than the equivalent petrol car with the same characteristics. Because their effect on the environment overall is less than that of the equivalent petrol car, although the electric car demonstrates a level of wealth it is also showing the spending of that wealth in a way that lessens environmental impact. However, to someone living in Nairobi and commuting to work by bicycle, an electric car, or indeed any car, would seem to be a luxury since it demonstrates a level of acquired wealth, whereas, for someone living in a slum in Dhaka, even a bicycle might seem a luxury. This means that luxury is relative to context and that, within that context, luxuries are only available to those with sufficient wealth. For those in the developed world who can afford luxuries, this raises an ethical issue since unnecessary luxuries compromise the environment through their unnecessary use of resources. Hence, the luxuries for the few affect the environment for all (see Chap. 9 for further discussion of ethics).

Luxury and Wealth

In 1899, the Norwegian-American economist Thorstein Veblen dedicated an entire book to the analysis of what he called the "leisure class" a social elite that owned so much that they needed to do little or nothing for living. This was part of Tevye's dream when he sang in the musical *Fiddler on the Roof* that, if he were rich, he wouldn't need to work hard. Veblen explains the process:

> As wealth accumulates, the leisure class develops further in function and structure, and there arises a differentiation within the class. There is a more or less elaborate system of rank and grades. This differentiation is furthered by the inheritance of wealth and the consequent inheritance of gentility. With the inheritance of gentility goes the inheritance of obligatory leisure; and gentility of sufficient potency to entail a life of leisure may be inherited without the complement required to maintain a dignified leisure. (Veblen, 1994: 48)

The problem with such a leisure class is that time must be passed somehow, given the lack of need to work, and one way of passing the time is to consume. Since people normally only drive one car at a time, buying a very expensive and luxurious car helps to use up some of the money. However, having several very expensive, luxurious cars and constructing a building to house them will use up even more. There is also considerable increased environmental impact in the materials and energy that went into making the multiple cars owned and the building to house them. The link between wealth, consumption (ownership of goods) and luxury is thus fixed.

Wealth and ownership are deeply linked, as it is impossible to accumulate what is not owned. Veblen (1994: 17) proposed that the "motive that lies at the root of ownership is emulation". There is a utility in consumption, without which being wealthy becomes less attractive. Being a wealthy person implies having an excessive amount of time to dedicate to leisure activities and indulgent activities that people from the lower classes cannot have, such as extended overseas trips. "Throughout the entire evolution of conspicuous expenditure whether of goods or of services or human life, runs the obvious implication that in order to effectually mend the consumer's good fame it must be an expenditure of superfluities. In order to be reputable, it must be wasteful" (Veblen, 1994: 60). Waste here is not used in the general sense of the term as something no longer wanted, but in the sense of whether expenditure on it is really necessary, that is "whether it serves directly to enhance human life on the whole—whether it furthers the life process taken impersonally" (Veblen, 1994: 61).

As Kapferer (2017: 25) much more recently notes: "Luxury growth comes mainly from the newly rich, especially those in emerging economies that are enjoying great growth...the newly rich crave love, power, and status through conspicuous consumption". These attitudes have been extremely useful for those promoting luxury products, whether directly or through social media, by pushing people to buy more stuff, without which they will not be part of an acceptable group or society. A study of young adults in the USA found that Facebook use tended to undermine subjective well-being in terms of how those using it felt and how satisfied they were with their lives (Kross et al., 2013). More recently, analysis of the social media use of respondents to an extensive and nationally representative time use survey in China found "that social media browsing does sizable damage to the users' subjective well-being, and relative income matters in the negative effect of social media browsing... [being] more pronounced for low-income people than for high-income people" (Bao et al., 2021). This suggests that not being able to afford the necessary items to be socially acceptable (according to Facebook) is detrimental to how people feel about themselves. In this way, luxury becomes an aesthetic requisite to be socially acceptable (see Chap. 8). Certain luxury products are things people are manipulated to want but cannot have. A parallel can be seen here with architecture in that websites like ArchDaily and Dezeen seldom feature small, affordable buildings for poor people, thus influencing the idea that what passes for news-worthy architecture tends to be large and expensive and, hence, inevitably more environmentally detrimental.

Luxury and Necessities

These ideas need to be compared to Adam Smith's eighteenth-century discussion of necessities and luxuries, where he allows that custom may make what had formerly been a luxury into a necessity, examples given by him being leather shoes and linen shirts. Linen shirts were formerly worn by the rich but were now required by a day labourer, as without such a shirt, he would be deemed to be in poverty. In contrast, beer and wine were luxuries because life could go on without their consumption (Smith, 1776: Book V, ch.II, part II). In a global economy, custom may no longer dictate what is a luxury, this being a task left to the global media and the advertisements that support them. Wilkinson and Piketty (2018: 103–104) note that the consumption of luxury goods is more important in less equal societies and that spending on advertising increases with greater income inequality. Ubel (2017) also notes that in the USA "people in states with relatively high income inequality were more likely to search for luxury brands than those in other states". His reason for this is that the less wealthy need to spend money on luxury brands to give the illusion of having wealth to achieve a status in society, whereas those with money are immune to such behaviour.

One problem with luxuries is that once you have them, they become part of everyday life and no longer fit the definition of luxury as something that happens occasionally. This vicious circle can lead to greater expenditure to satisfy the feeling of treating oneself to a luxury product. Since cost and the spending of money form an approximate proxy for environmental impact, given that wealthier nations with more money to spend tend to have higher ecological footprints (Chicca et al., 2023: 200–210), luxury is not good for the environment. This problem of acquiring a luxury item that no longer becomes a luxury can be seen in houses. As architects of houses, we are selling the same product over and over again; houses were and will be houses. Designers of the architecture of wealth work on what people do not need but might want. The prevalence of en-suite bathrooms is an instance. A bathroom is part of a house that is only used infrequently, perhaps 30–60 min a day in a family house (Khajehzadeh et al., 2016), and multiple bathrooms will inevitably be used less. In New Zealand, a small en-suite bathroom next to the main bedroom might be found in some 1970s houses (BRANZ, n.d.) and having a house with two bathrooms would then have been considered a luxury. Now, the en-suite has to be as large as the main bathroom, if not larger, and luxury can only be had by having multiple en-suite bathrooms, all of which come at an environmental cost. Design for the wealthy has thus created the desire for multiple bathrooms, some of which can be the most expensive room in the house in terms of finishes (Vora, 2021).

You can buy items that are labelled "luxury" in everyday stores, such as luxury shower gels, so you can still treat yourself occasionally. Still, these shower gels cost more than regular ones, and as noted above, soap is even cheaper. This has led to such branding of more exclusive goods as "superluxury" or "überluxury" as a means of satisfying the need to be seen to be able to afford luxury. Batat (2019: 6) states that being seen as consuming luxurious goods sends "tacit social signs as well as explicit messages, visible or invisible, directly or indirectly, to other social actors who can decode them by referring to luxury objects and lifestyles displayed by individuals". The problem for the environment is that as wealth increases, so does the demand for more luxury. Pajur (2017: 23) notes that in eighteenth-century Tallinn, the consumption of luxury items that moved from the courts to urban societies was thought to be good for economic development. Since economic development is always a goal, as highlighted in the SDGs (see Chap. 1), people now have to be made to want luxury items. How luxury makes money is evident in the modern Hermès silk scarf. Whereas the Hermès scarf may not use any more raw resources than a silk scarf of similar size without the Hermès brand name, it will cost approximately ten times as much, which may account for the fact that Hermès is said to be the most profitable luxury brand (Kapferer, 2017: 27). What does go into a Hermès scarf is time because, unlike the high street silk scarf which is computer printed, the Hermès scarf is screen printed and engravers have to translate the complex design to the screens (dotLuxury, n.d.). Thus, the luxury item which costs more potentially has an increased environmental impact through the involvement of the resources bound up in these extra steps in manufacture.

In the past, when goods were hand-made, they were expensive, but people had few possessions because of this. The development of industrialised production meant that energy to run machines substituted for handwork, but this also made goods cheaper, at least while energy was cheap and plentiful. Because machine production is so much faster, the desire for having more goods had to be created. Luxury products have in part turned this around, in that the inclusion of more handwork makes the product more expensive, and hence only available to the wealthy. In the nineteenth century, William Morris was desperate to return industrial production to craftwork but his own firm could not make such handprinted fabrics and handblocked wallpapers available to all (Cody, 1988).

> When, in 1876, a wealthy industrialist whose house Morris's company was redecorating asked him why he was pacing up and down and room and muttering to himself, Morris replied "It's only that I spend my life in ministering to the swinish luxury of the rich." All through the 1880s and '90s, after his conversion to a militant Socialism, Morris was denouncing capitalism and the swinish upper classes while his company decorated their houses at considerable expense.

However, luxury goods are not quite the same as expensive hand-crafted ones. A room decorated with expensive hand blocked Morris papers will probably not be changed for many years. Luxury brands, however, do change, so the real luxury is to own the latest version, whether a Hermès scarf or a Louis Vuitton suitcase, the latter developed so "The process of travelling tomorrow could be a little more elegant and a little more seamless and a little more enjoyable" (Louis Vuitton, n.d.).

This has now also happened with cars as not only are new models introduced every 5–10 years but in between cars are given a "facelift", meaning "often they incorporate a subtle re-sculpting of the headlamps, grille and front bumper" (WhichCar, 2019). In contrast, Ford sold the Model T for 19 years (see Chap. 4).

This does not limit resource consumption in the manner of the handcrafted product that lasts a long time, because there has to be a design input as well as marketing and fashion shows to send the message that luxury is now this year's expensive product, not last year's. Even when someone buys the latest version of a smartphone, the feeling and status of owning it tend to only last the time it takes for a new model to be released. This means that luxury products cannot be sustainable. Many luxury brands can trace their ownership back to Bernard Arnault. At the time of writing this (October 2023), he and his family are at the top of the Forbes list, worth US$211 billion (LaFranco & Peterson-Withorn, 2023), suggesting luxury goods make a lot of money for someone. Arnault's company LVMH (Louis Vuitton Moët Hennessy) owns many fashion brands and deals with wines and spirits, fashion and leather goods, perfumes and cosmetics, and watches and jewellery (Forbes, 2022a, b), all arguably products without which life would still be possible as many people in the world make evident, while those who can show possession of such products underline their status in society.

Kapferer (2017: 25–26) also notes that the newly rich "want their success to be visible, such that they are frequent buyers of mansions and gorgeous flats all over the world, as well as high-end furnishings to fill these homes", and this is where luxury starts to impact on the built environment. The following sections explore further the impact of luxury and buildings, first through the involvement of architects in designing for the luxury brands and what might be defined as luxury buildings either because of their temporary nature or because there is no apparent need for them, and then through the effect the search for luxury has on both permanent and temporary buildings.

Architects and Luxury Brands

How many museums are a reasonable number for any city? Museums (see Chap. 9 for an ethical take on museums) are there to create venues for tourists. Chicca et al. (2023: 229–238) created a link between such urban icons and increases in tourism, pointing out the significant environmental impact of the latter. The LVMH group recently commissioned a new museum of contemporary art in Paris that features the art collection of Bernard Arnault, who is behind the Fondation Louis Vuitton. The architect was Frank Gehry and the museum was completed in 2014 (Fig. 6.2).

Earlier Gehry was the architect of the Bilbao Museum and hence of the Bilbao effect (Chicca et al., 2023: 232–233), whereby new architectural icons encourage visitors and hence city regeneration. Paris, however, is hardly short of visitors. Before the effect of COVID-19, Paris was the third most visited city after London and Bangkok and also ranked third for visitor spending of US$17.02 billion in 2014 after London (US$19.27)) and New York (US$18.57) (Hedrick-Wong & Choong, 2014: 3, 5). According to Paris-Walking-Tours.com (n.d.) there are 140 museums in Paris, not apparently including the Fondation Louis Vuitton, suggesting the city is hardly short of museums to visit. The Fondation Louis Vuitton has decorative curved glass sails that reflect in the surrounding ponds: "With two stories [sic] above ground, and one below, set within an excavated hollow and bordered by a pool, the 126,000-square-foot [11,706 m^2] building cost more than $135 million. (The Vuitton folks will not say how much more.)" (Giovannini, 2014). Brown and Pes (2018) state the final cost of the building was approximately US$900 million, which is quite a lot more, and gives a cost of around US$77,000/m^2. This should be compared with the highest average 2021 global cost for primary and secondary schools of US$4850/m^2 in New York (Statista, 2023). There must be something wrong with humanity if so much more money is spent on art than on the next generation.

Fig. 6.2 Fondation Louis Vuitton. (Source Creative Commons (n.d.-a))

Perhaps more controversial as noted by Brown and Pes (2018) are "France's generous corporate philanthropy law, which allows a company to deduct 60 percent of the sum it gives to cultural projects and institutions [such as building a museum] from its corporate tax bill". Sutton (2018) also described the claim made by the Front Républicain d'Intervention Contre la Corruption that building the museum had led to tax evasion in that:

> ...the nonprofit branch of the luxury conglomerate LVMH was able to deduct some 60% of the cost of the museum from its taxes, and request tax refunds on some other costs. In all, FRICC claims that LVMH and the Louis Vuitton Foundation received nearly €603 million ($682 million) from the state toward the nearly €790 million ($894 million) construction costs of the museum". The case was dismissed.

It seems the citizens of France are paying heavily for such a luxury, and arguably unnecessary, building. Why was the building so expensive? The sails are of laminated glass and the building structure is wood, steel and stainless steel, claimed to be "a choice of materials with low environmental impact" (Paris Digest, n.d.), but without a life-cycle analysis to support such a claim. The building does collect rainwater used for cleaning the glass and topping up the pond. Geothermal energy is used in the heating and cooling system (Paris Digest, n.d.), but this is, at best, a nod towards making a sustainable building, one specifically designed to attract visitors, especially those from out of town, with the consequent environmental impact of tourism. There were just over a million visitors in 2016 and 1.4 million in 2017 (Statista, 2021). A study of tourism in Barcelona found the "average carbon footprint of a tourist is 111.6 kg CO_2 eq/day and 43 kg CO_2 eq/day for a day-tripper "(Rico et al., 2019). This should be compared to the around 14 kg CO_2 eq/day for the average Spaniard (Ritchie & Roser, 2020). Even if all visitors to Gehry's Fondation Louis Vuitton in 2017 were day trippers with a similar carbon footprint to those in Barcelona, and assuming this was their only visit of the day, this would be annual emissions of 60,200 tonnes CO_2 eq. To put this into perspective, these arguably unnecessary emissions equate to the 2021 annual emissions of 11,577 UK citizens or 602,000 citizens of Madagascar (Our World in data, n.d.).

At the 2014 contemporary arts fair in Paris, the comment was made that museums like the Fondation Louis Vuitton were given such an amazing exterior appearance because they were "a marketing tool" (Bilefsky & Carvajal, 2014). Pritchard (2014) takes a similar view in summing up this building created from selling luxury brands.

> It's a bling edifice for a wealthy brand, with bags of swagger; a nonsensical sculptural building that will be loved and loathed in equal measure—it's either too vulgar, or not enough. It should be expected that an architect famous for sculptural flamboyancy would produce something so theatrical for the richest man in France and the philanthropic arm of his trashy luggage and fashion company.

Harrouk (2021) noted that luxury and fashion brands have been venturing into architecture. This is obvious in the design of stores to sell such brands. The Italian firm Bulgari, now part of the LVMH group, sells luxury jewellery, watches, perfume and leather goods and has a flagship store in Rome at the Via dei Condotti. The façade motif of the flagship store, which first opened in 1905 (LVMH, n.d.), has been stamped on facades around the world (Fig. 6.3). In the latest version in Kuala Lumpur by Dutch firm MVRDV, the motif has been extended to embrace the façade, being created from glass reinforced concrete cut apart with the gaps filled with resin illuminated by amber LED lights, so that the façade glows (MVRDV, n.d.) (Fig. 6.4). What was once a motif of the brand has now become the branded façade.

The technical innovation required to produce the veined façade of the Kuala Lumpur Bulgari store was also required for MVRDV's earlier Crystal Houses in Amsterdam, where a new façade of glass bricks and glass window mouldings was created both to give transparency so as to see into the building interior but also retain the details of the classic Dutch brick façade. Crystal Houses sits in "The Netherlands most expensive shopping street" and was designed to be a "modern flagship store" now revamped and occupied by Hermès (MVRDV, 2017) (Fig. 6.5). Baldwin (2019) saw the project as "a way for Amsterdam to be home to distinctive, upmarket flagship stores without compromising the city's historical character". However, the interweaving of glass and clay bricks at the junction between the old and new facades makes an uncomfortable junction that certainly does not lie within "Bene aedificandi scientia" (see Chap. 7) but feels more like smoke and mirrors.

These two stores for luxury brands designed by famous architects are not isolated examples. Macguire (2011) noted that those shopping for luxury goods wanted the full experience of the brand through a high-end shopping experience, and this generally involves big-name architects. He continued: "In recent years, Prada, Christian Dior, Gucci and Hermès have all briefed some of the world's finest designers—including Doriana & Massimiliano Fuksas, Herzog & de Meuron and Rem Koolhaas—to create spectacular outlets that they hope will embody the essence of their products". This idea of encouraging people to buy goods through more avant garde architecture is not confined to the luxury brands (Fig. 6.6).

The architectural firm LOT-EK created a portable merchandising experience from 24 containers for the sporting goods firm Puma, that could be moved to different sporting locations. As well as retail space there were two decks, offices, storage and a bar area (Basulto, 2008). Because of its use of retrofitted containers and demountable nature, this project appears to have green credentials. Made in China, it arrived by cargo ship in Alicante and was assembled for the start of the 2008 Volvo Ocean race. Once the yachts had left it was disassembled and crossed the Atlantic on a cargo ship before being reassembled in Boston, the home of Puma North America (Anon, 2008). Meanwhile the yachts took the hard route to South Africa, India, China, and then round Cape Horn to Brazil, the USA, Ireland and Sweden. Puma was a sponsor and had a yacht in the race—il Mostro (World Architecture Community, 2008). After Boston, Puma City went to South

Fig. 6.3 Bulgari Manhattan. (Source: Creative Commons (n.d.-b))

Africa and since then to New York and back to Boston (Faverman, 2010). After this the trail is lost. Puma, via its chief marketing officer, were "mum on the cost to create Puma City, simply saying it cost 'a lot'" (Muther, 2009).

What seems to be happening is that architecture is being used as part of branding, the whole aim being to sell more goods. The architecture and architects that contribute to this end, are effectively materializing wealth and the desire of being wealthy. Such expensive buildings are useful to those who pay for them because they move people to "aspire" by consuming. This affects design by defining its role, meaning and ultimate goal. The apparent goal is to build the scene for never ending consumption, and to materialize the apparent benefits of wealth.

This issue has also affected dwellings. Karl Lagerfeld, who died in 2019, was a German fashion designer who headed Chanel for many years. As noted earlier in this chapter, his fashion label has initiated a development of five supposedly sustainable villas in Marbella to celebrate Lagerfeld's passion for architecture (Forbes, 2022a, b) (Fig. 6.7). Apart from the earlier comments on the layout of the five villas (Fig. 6.1), there is a considerable area of road within the site since the people who buy these villas will have cars, although the design assumption is these will be electric and there are also cycle ways that lead straight to the coast from the development.

Work started in October 2022 and the five villas are due for completion in 2024. The architect, Andrea Boschetti of The One Atelier, said: "The units are not super big, between 660 and 900 sqm, intended for private families" (Harrouk, 2021). In 2017, among a selected group of developed countries, the largest average house size was 189 m^2 (Australia), the highest in the world, while Spain averaged 122 m^2, ranked eighth in the cluster (Statista, 2022). Consequently, most Spaniards might view the Karl Lagerfeld villas as

Fig. 6.4 Bulgari Kuala Lumpur. (Based on MVRDV (n.d.))

super big. The architect also stated no "luxury" materials were used, these so-called non-luxury materials being terrazzo, recycled marble and wood and ceramic material that forms the white façades. Looking at recycled marble in the UK, a worktop-sized piece would sell for £588/m^2 (Authentic Reclamation, 2023) while the conventional granite-based composite worktop would be £137/m^2 (Householdquotes, 2023), and these are prices for relatively small pieces of marble. The marble may be recycled, but it is not a cheap material. These statements misunderstand sustainability by making it seem to apply to trendy luxury for the wealthy.

Elsewhere (Klein, 2022) the façade material is stated to be "wood coated with a synthetic ceramic material and white gold dust that has been polished to a gleam". Hopefully it was Fairtrade or fair-mined gold as all other sorts are not good for the environment. Other sustainability credentials are the lush planting, the solar panels on the roofs with geothermal power for the mechanical systems, insulation, "and each villa will be oriented to make good use of the sun and weather" (Klein, 2022), although the latter could be viewed as nothing special since this is what any architect will do for any building that is not portable. Responding to the site conditions is as basic to architecture as kicking a football is to a professional footballer—part of the job. However, it is the sheer size and cost of the villas that belie any notion of their being sustainable, the essence of which is to have less both in the name of equity and not destroying the resources of the planet on which humanity depends, both now and in the future. This raises the issue of whether an expensive item, however it is made, can ever be sustainable. The answer is simple—it depends on how long the item lasts. Research has shown that investing more at the point of construction to make a sustainable building, such as adding additional insulation will normally lead to lower resource use over the life of the building through savings in operation (Mithraratne & Vale, 2004;

Fig. 6.5 Crystal Houses, Amsterdam. (Source: Creative Commons (n.d.-c))

Fig. 6.6 Puma City. (Source: Creative Commons (n.d.-d))

Fig. 6.7 Karl Lagerfeld Villas, Marebella. (Based on Harrouk (2021))

van Hal, 2006). Providing the building does not receive significant changes during its life, such as the frequent new fit outs found in many commercial situations, the more expensive building designed to be sustainable will be just that. This is where fashion plays a role. An expensive luxury garment should last a very long time if well-made and of good fabric, but it will not change as fashion dictates. If used as is, it could be a good investment for the environment. However, luxury brands that bring out new products to encourage consumption inevitably lead to the expensive item being unsustainable. If this product is also large and expendable, the environmental damage could also be large. As noted earlier, Veblen saw how the rise of a leisured class with no need to work led to more time devoted to consumption, so there is an inevitable link between wealth and having to spend the money on goods, and preferably expensive goods that reflect societal status. Alternatively, those with money could choose to spend it on combatting climate change for the greater good, part of which might be asking architects to produce zero energy buildings. As noted by Garcia et al. (2021: 303), dealing with climate change would cost the world's high net worth individuals US$22,600 per annum. Annually, such a sum would buy one haute couture gown (Hush, 2010), 16 nights' accommodation for two in a room at the Ritz Hotel, London, including afternoon tea, breakfast and a bottle of champagne (The Ritz London, 2023), or almost three Barcelona chairs for the living room (Knoll, 2023), arguably all things one could live without, therefore all luxury items.

Luxury brands also affect urban design. Anyone who has spent time air-side in an airport soon realises that the same luxury brands like Prada and Burberry are ubiquitous among the more useful outlets selling pain killers and batteries. Since for most people an airport is a transitory experience, this is perhaps boring but not a big problem. However, it becomes a problem when luxury brands dominate the streetscape, thereby effectively excluding those without the means to acquire them from this part of the urban landscape. Morandi (2017: 114) notes how luxury brands have transformed the centre of Milan, which has long been the site of small workshops producing and selling luxury goods related to fashion and jewellery. Now the centre is the retail hub for modern luxury brands while their manufacture has been pushed out of the centre. This has led to a change in who lives in the centre as formally: "Low-income workers and the low middle-class could find accessible living conditions despite the greater presence of high-class residents, so that in this part of the city some sort of social mixité could persist". This is no longer the case as the centre of Milan is now a place for those with money to shop, or those without to watch them shopping.

Although as shown above, some architects have been keen to work with luxury brand names to produce buildings that reflect the nature of the products, in terms of their unusual but imageable appearances that rely on many hours of painstaking work behind the scenes, the notion of luxury has affected buildings both before the advent of luxury brand names and in other ways, and these ideas are explored in the next section.

Luxury and Its Effect on Buildings

As noted above, Adam Smith had realised that luxury was a moveable concept and what was once seen as a luxury could become if not an absolute necessity, something that most people in the same society saw as essential. The humble window was once such a luxury item because of the cost of glass production. Before the 1700s, windows were literally "wind-eyes" coming from the old Norse word "vindauga" as most were unglazed, framed holes in a wall that could be closed off with the "use of sliding or folding wooden shutters, oiled cloth, paper or even thin sheets of horn" (Wychavon District Council, 2007). Only the wealthiest could afford the small pieces of glass set into lead cames reinforced with metal bars, the largest versions of which were the stained glass windows of churches and cathedrals. These religious windows formed a valuable way of telling the stories of Christianity at a time when many people could not read. The labour-intensive process necessary to produce glass meant that it was an expensive product. Eskilson (2018: 2) describes how in the 1600s in France, crown glass was first produced by blowing a large spun disc from which rectangular panes could be cut, as well as the bullseye in the centre where the blowing rod had been attached, use of the latter being a cheaper way of glazing a window even if it was hard to see out of it. The cost of glass meant that large glazed windows were the prerogative of the rich, as much as the large houses of which they were a part. This did not go unnoticed as in the seventeenth and eighteenth centuries a tax was levied on windows in England and Scotland. In England (and Wales), the tax was introduced in 1695, and the more windows in a house the higher the tax, thus the burden would fall more heavily on the rich. France also introduced a window tax and went one better in 1798 by introducing a tax on doors (Keen & Slemrod, 2021: 19). At the start of the window tax in England and Wales, each existing inhabited dwelling had to pay 2 shillings annually, but for those with more than 10 but less than 20 windows an additional 4 shillings was paid, and for those with 20 windows or more an additional 8 shillings (Institute of Historical Research, 2019). The window tax appeared in Scotland in 1748: "The Window, Commutation, Inhabited House and Consolidated Assessed Taxes were all taxes on householders though in practice only the better off were taxed. The house had to have at least 7 windows or a rent of at least £5 to be taxed" (National Records of Scotland, 2022). There were changes to the taxing of windows but the principle remained the same. Windows were expensive items when building, and are still more expensive than the wall in which they sit.

The window tax had the huge advantage of being easy to administer since it did not require probing into income or entering premises as windows could be counted from the outside. However, it led to bricking up windows or, for new dwellings, building filled-in window openings that could later have windows inserted when the tax was repealed, which only happened in 1851 in both England and Scotland (Fig. 6.8). Thus, those with money could still have big houses and show that they were avoid-

Fig. 6.8 Dr. Johnson's house, City of London. (Source: Creative Commons (n.d.-e))

ing tax, or just pay for having windows. "Of course, there are some instances in which residents by design had numerous windows as a means of displaying their wealth" (Oates & Schwab, 2015).

To complete the story, the tax was disapproved of as it was seen as a tax on health since windows allowed both light and air into the building. The window tax also affected the poor who of necessity rented. Although the tax was levied on the occupant, not the owner, exceptions to this were the large tenements that housed many families, as here the landlord paid the tax as these were considered to be single dwellings. An unscrupulous landlord would block up windows without thought to what it might be doing to the interior conditions (Oates & Schwab, 2015). However, when it comes to dwellings, size, and hence the number of windows it was possible to have along with the use of exotic materials, normally in the interior, was all part of how wealth and status in society could be displayed.

Dwelling Size

The second verse of the hymn *All things bright and beautiful*, with its suggestions that status in society is something ordained by a higher power (The rich man in his castle/The poor man at his gate), may no longer appear in hymn books (Ward, 2004), but dwelling size and location still reflect wealth in the environment. The increasing size of houses in New Zealand, for instance, has been ascribed to the way taxes are organised, as New Zealand has no capital gains tax on dwellings. The interest on savings and money earned through dividends is taxed but not the money that goes into houses. Coleman (2017) notes: "If housing is taxed less than other assets, people have incentives to buy bigger houses, and they will pay more to live close to desirable locations". In this instance, large dwellings are not necessarily a sign of luxury but more one of making the best out of a particular fiscal situation, providing the individual has the means to do this. Ultimately, if you have money, it is easier to find ways to make money and owning a large single-family house sends the message that you are wealthy enough to do this.

In the past, a very large house would accommodate not just the owner but many other people involved in running the house and its surroundings. As an example, Holkham Hall in Norfolk, UK (Fig. 6.9) was constructed from 1734–1765 in the Palladian manner.

The design has been variously attributed to William Kent, Lord Burlington, Thomas Coke otherwise the Earl of Leicester who was the client, and Matthew Brettingham the elder, who acted as clerk of works and was employed to look after the Earl of Leicester's buildings (Salmon, 2013). It is Brettingham's drawings that are of interest here as they note the whole south façade to measure 364 ft. (110 m). From these drawings the section through the central block is three and in places four storeys high. They also show the central area to be 162 ft. from east to west (49 m). This means each of the wings, which are two storeys with a small third storey attic area in the roof space, and that are approximately square in plan, measure 30.5 m × 30.5 m. Very roughly, therefore the floor area of Holkham hall including the four wings is around 12,000m^2. The Holkham archives give the number of people living in the house apart from the owners and excluding the time when there were visitors who would come with their own servants (Purvis, 2023).

> The number of servants working varies greatly over the last 250 or so years depending on the season and who was visiting the hall. In 1759, Lady Margaret records having only 29 indoor and outdoor servants, but by 1881 the 2nd Earl has 28 indoor servants living in the Hall with at least that many again living in Holkham village.

Thomas Coke never lived to see Holkham finished though his wife Lady Margaret did, and their only son had died earlier. The number of people permanently inhabiting this vast mansion were therefore, small—perhaps some 20 people in an area of 12,000m^2, though for the servants this was a place of work and for those at the bottom of the servant hierarchy their only private space might be a room in the attic floor with "a bed, a chair, perhaps a table, a small mirror or a piece of carpet" (Stobart, n.d.). If space is luxury, then Holkham Hall is a very luxurious building and is reputed to have cost £90,000 (Heritage Emergency, n.d.), the equivalent of more than £20 million in 2023. This seems like a bargain compared to the house now worth £40 million near Hamstead Heath, London (Fig. 6.10) with its 10 bedrooms and 7 bathrooms that also took some 10 years to construct at a cost of over £30 million (Ivey, 2021).

When it comes to big and expensive homes in London, Buckingham Palace outshines everything, being reckoned the most expensive dwelling in the world with its 775 rooms that include "19 State rooms, 52 Royal and guest bedrooms, 188 staff bedrooms, 92 offices and 78 bathrooms" noting that Buckingham Palace also serves as the administrative hub as well as the official London home of the UK monarchy (The Royal Household, n.d.). In 1826, George IV and his architect John Nash aimed to turn a more modest but large house into a palace with money obtained from parliament (see further discussion about who pays for buildings in Chap. 5) and the famous state rooms date from this time. However, by

Fig. 6.9 Holkham Hall, Norfolk. (Source: Creative Commons (n.d.-f))

1829, the year before the death of George IV, the costs had risen "to nearly half a million pounds. Nash's extravagance cost him his job" (The Royal Household, n.d.). This is approximately equivalent to £70 million in 2023.

Even older and larger, the Palace of Versailles is not only huge, with 2300 rooms, but was deliberately designed to show the power and wealth of the royal family of France. Growing from a hunting lodge, under Louis XIV it became the centre of the court and the administrative seat of the monarchy. Perhaps most magnificent is the 1678–1684 Hall of Mirrors (Fig. 6.11). This replaced a terrace that separated the apartments of the King and Queen and so was effectively a vast corridor (Château de Versailles, 2023).

"In an era when the Renaissance had jogged France's memory of the cultural splendour of the ancient world, particularly imperial Rome, Louis tasked his team with devising 'how to glorify the king in architecture and décor'" (Pyrcz, 2019). Such glorification often helps to glorify the architect. The Hall of Mirrors was designed by Jules Hardouin-Mansart (he was the nephew of the Mansart who possibly designed the mansard roof). Hardouin-Mansart was described as a "skilled courtier" and rose to become the overseer of the king's buildings: "The king was so pleased with his skills that he bestowed a title on him in 1682 and made him a knight of Saint-Michel in 1693—a distinction conferred notably upon artists". Eventually Hardouin-Mansart became Count of Sagonne (Château de Versailles, 2023). Even in the seventeenth century being involved with expensive, luxury architecture had its rewards for the architect. Ultimately this glorification of monarchy of which the architecture at Versailles was a part led to the French Revolution and the demise of Louis XVI at the hand of the guillotine in 1793. After this the glories of the building of Versailles were given back to the people of France in the form of a museum, which

Fig. 6.10 Mansion near Hampstead Heath. (Based on Ivey (2021))

Fig. 6.11 The Hall of Mirrors, Versailles. (Source: Creative Commons (n.d.-g))

opened in 1796, although many of the contents had been moved to other venues such as the Louvre (Château de Versailles, 2023).

A palace is a big house, therefore showing the owner has money, but not all palaces are luxurious buildings. The 1923 Mrigadayavan Palace at Cha'Am in Thailand, a summer palace of King Rama VI who was advised for health to spend time by the sea, is a large building but it uses the details of vernacular Thai houses, including ceramic pots to collect rainwater, louvered shutters, and raised living areas (Fig. 6.12). It was also built of teak from a demolished palace. The building work was overseen by Ercole Manfredi architect of many buildings in Thailand in that period. Although fitted with a generator for electricity King Rama VI kept candles by his bed in his very modest bedroom so as not to wake everyone by turning on the light in the middle of the night and thereby starting up the noisy engine. Nevertheless, a palace is still a palace.

The Media

Not everyone can live in a palace, but the idea of living in a luxurious environment has penetrated the design of normal dwellings through television programmes such as *The Block*. In this programme, people have to renovate some form of house in the hope of making money when it is sold at the end. The emphasis is on what the programme calls high-end design. Winning rooms created by the contestants in the NZ series have been described by the judges as having "a beautiful luxe feel" (Resene, 2021). One of the NZ Block judges has a 2022 book, Live Luxe, which purports to give a step-by-step guide to designing the perfect home, which, by inference, epitomises luxury (Ferguson, 2022). Why a home should be perfect is a strange goal, not least because the lives of those who live there are constantly changing so what they need from a home must also change. Perfection is also not some-

Fig. 6.12 Mrigadayavan Palace. (Source: Robert Vale)

thing the individual parts of the natural world achieve as no flowers on a plant are ever the same, so there can be no perfect flower, though all might be perceived as beautiful. Nevertheless, the idea of it being possible to be able to create the perfect home has made *The Block* a popular series in both Australia, where it originated in 2003, and in New Zealand. In 2017, an election year in New Zealand, more people watched the final of *The Block* than election debates: "306,476 people in the 25–54 demographic watched the finale of *The Block NZ*, making it that age group's most-watched television event of the year" (Stuff, 2017). In Australia, *The Block* is equally popular with the 2022 Australian finale watched by over 1.7 million viewers (Thomas, 2022). People seem to want to look at luxury even if they cannot necessarily have it, remembering that many viewers may also not have the "luxury" of owning or buying their own home.

Following the New Zealand version of the programme in 2021, a small apartment complex due for completion in 2023 in New Plymouth was being advertised as "high-end" by taking inspiration from *The Block,* with the designers opting for "luxury over budget" (Bayleys Realty Group, 2021). The real estate agents, McGrath, responsible for selling the houses from the 2021 series of *The Block* in Australia, defined a luxury home for the buyers as one "that affords them more space, a home office for their high school children to learn in, and a garden or pool area they can escape to" (Sissian, 2021). Obviously, this comment looks back to the problem of living under lockdown conditions and having to work and school children at home, but statistics show that new houses in both New Zealand and Australia have become large in terms of floor area while households have decreased in size (Khajehzadeh, 2017: 5). More importantly, Khajehzadeh and Vale (2017) note that the extra space in large, new houses in New Zealand is made up of space that may be seldom used, such as additional circulation and extra bathrooms. This is where the idea of luxury also comes in, since it is the possession rather than the use of the particular item that is important, as possession reflects position in society. Ultimately, luxury, whether the Prada handbag or the large house, is conspicuous consumption. Given that humanity is consuming resources beyond the means of the planet to regenerate these annually, having a larger house than is really needed does not seem the path to sustainability or the achievement of SDG 11 Sustainable Cities and Communities. Rather it seems for some architects and designers, sustainability has become the newest form of luxury building.

Disposable Buildings

Buildings take resources and making a building last a long time is a good move towards a sustainable built environment. Perhaps the greatest luxury when it comes to buildings, therefore, are those which are disposable or only have a very short life. Mohammad al-Asad (2005) argues that in places buildings are treated as disposable items like many consumer goods. He gives the rise of the nuclear family as one reason for this as families no longer spend generations in the same building, but move from one dwelling to another as circumstances change, and this change inevitably involves the use of extra resources. Additionally, a short building life is associated with the price of land as in places like Japan the land is very much more valuable than the building on it, leading to short building life. A reduced building life also happens where tax legislation permits depreciation of a building as an asset along with furniture and fittings: "Once a building is completely depreciated, it offers no additional incentives in terms of tax deductions and replacing it by tearing it down or selling it becomes the preferred course of action" (al-Asad, 2005). However, when it comes to considerations of luxury, it is perhaps the very short life architectural pavilions designed for various exhibitions that would seem to be the ultimate in disposable buildings.

The Dutch Pavilion by MDRVD for the 2000 Expo at Hanover was designed to show how artificial eco-systems could be engineered in a six-storey building, with each storey dedicated to a type of Dutch landscape. The aim was to show how more people could live on the available land. Its useful life was 5 months (Shen, 2012: 150), the period of the Expo, although it was designed to last longer. Currently, it is an abandoned shell, so it is hardly a sustainable building, although there are plans by the same designers, MDRVD, for its conversion to offices and student accommodation with the addition of new adjacent buildings (Baldwin, 2020). The Dutch Pavilion demonstrates the problem of temporary buildings in that they waste a lot of resources unless they can find a reuse.

The London Serpentine Pavilions perhaps epitomise how temporary buildings have become associated with wealth. Begun in 2000 as a means of giving architects a chance to create their first structure in the UK, many have since become architectural collectors' items, as from the start they were designed to be moved once their sojourn in Hyde Park outside the Serpentine Gallery was over (Design Diffusion, 2023). The first a clear span tent structure by Zaha Hadid was eventually bought by a farmer for use in the Flambards Theme Park in Cornwall. "The pavilion (which is now known as the Kingsford Centre, hireable for £500) has played host to weddings and discos, comedy nights and concerts" (Wainwright, 2015), so a not very glorious if useful end. The Serpentine pavilions are viewed as concrete expressions of the individuality and creativity of each of the chosen designers, or auteur architecture, without (much) thought to use and context, two ideas which underpin the very nature of architecture. After their initial showing many have been bought by the equivalent of wealthy architectural connoisseurs and transported to new locations. In this way, the 2008 pavilion by Gehry has become part of a permanent architectural exhibition (Otero, 2015).

> It was bought by the Irish developer Partick McKillen and rebuilt at Château la Coste, a winery that has repositioned itself and maximized its market value by juggling its wines with a collection of art and architecture pieces. In approximately an hour and a half, and after paying the admission fee, visitors can enjoy art and architecture in a landscape scattered with works by Ai Weiwei, Jean Nouvel, Frank Gehry, Norman Foster, Renzo Piano, Oscar Niemeyer, Richard Serra, Louise Bourgeois, or Paul Matisse.

Here we have a situation where wealthy, well-known architects are partly responsible for selecting the year's pavilion architect (David Adjaye and Richard Rogers are advisors to the Serpentine Gallery directors) (Design Diffusion, 2023), who then, through sponsorship, create a structure that is sold to the wealthy as an art object. This seems a very long way from the ideas of "firmness, commodity and delight" and even further from the idea that architects might have a responsibility for creating a sustainable built environment. Moreover, it seems that the 2019 pavilion designed by Junya Ishigami used unpaid interns in the midst of all this wealth (Jonze, 2019). This pavilion was supposed to celebrate the roof, being made of many overlapping slates that appear to rise from the ground. According to Ishigami this made the pavilion "part of the landscape rather than making it an independent building" (Jordahn, 2019). Despite the fact that slates belong to the Welsh landscape rather than that of central London, this idea rather works against the nature of a temporary building, unless the whole work is intended as an ironic metaphor. After the showing, the pavilion was bought by Therme Group, who also bought the 2018 pavilion, with "plans to exhibit Ishigami's structure to the public at its resorts around the world" (Waite, 2019). Despite the experience that might be gained working as an unpaid intern on the Ishigami pavilion it is very, very unlikely ever to lead to the chance to design a Serpentine Pavilion. The elite architects who are given this opportunity are just that—a small coterie of elite-named architects. Yet, in education, so many architectural design studios face students with the challenge of designing a museum, a pavilion or an exhibition building. To achieve the SDG goals, there will have to be a big change in the education and practice of those responsible for the design of the built environment.

The crossover between art and luxury has a long history, according to Federica Carlotto, who is the director for a short course on the current global trends when it comes to luxury products. She goes on to say (Sotheby's Institute of Art, n.d.).

> Art and luxury have a long history of influencing each other to create timeless, aspirational experiences…From creative and editorial collaborations with artists like Salvador Dali, Cindy Sherman, Ed Ruscha and Takashi Murakami, to brand and creative directors increasingly drawing on trends and philosophy of art in their work.

What the Serpentine Pavilions demonstrate is the application of elite architects to the production of a luxury building, just as graffiti artist Banksy has teamed up with the Japanese firm MediCom Toy to make a range of Bearbricks. Bearbricks are a limited range of collectible normally plastic bear-shaped toys that are not toys because of the cost (Limn Gallery, 2023). In a way, the Serpentine pavilions are as much architecture as an expensive collectable plastic "teddy" Bearbrick is a toy. The meaning of architecture has been lost just as the essence of the teddy bear as a child's comforter and friend has been left behind in the ideal of making and spending money for the sake of it. Given that SDG 12 is to secure sustainable consumption and production patterns, luxurious goods have no place in the debate.

References

Al-Asad, M. (2005). Disposable buildings: Urban crossroads #36. *Centre for the Study of the Built Environment.* Retrieved from https://www.csbe.org/disposable-buildings, 25 April 2023.

Anon. (2008). PUMA city makes its global debut as the start of the Volvo Ocean Race 2008–2009 in Alicante, Spain. *Leisure and Travel Week.* October 11. 31.

Authentic Reclamation (2023). *Reclaimed marble, granite and slate.* Retrieved from https://authentic-reclamation.co.uk/stock-categories/reclaimed-marble-granite-slate/, 12 April 2023.

Baldwin, E. (2019). *MVRDV reopens crystal houses with new tenant and façade.* Retrieved from https://www.archdaily.com/919767/mvrdv-reopens-crystal-houses-with-new-tenant-and-facade?ad_source=search&ad_medium=projects_tab&ad_source=search&ad_medium=search_result_all, 22 November, 2022.

Baldwin, E. (2020). *MVRDV to transform iconic expo 2000 pavilion into co-working complex.* Retrieved from https://www.archdaily.com/943508/mvrdv-to-transform-iconic-expo-2000-pavilion-into-co-working-complex, 25 April 2023.

Bao, T., Liang, B., & Riyanto, Y. E. (2021). Unpacking the negative welfare effect of social media: Evidence from a large scale nationally representative time-use survey in China. *China Economic Review, 69*, 101650.

Basulto, D. (2008). *PUMA city, shipping container store/LOT-EK.* Retrieved from https://www.archdaily.com/10620/puma-city-shipping-container-store-lot, 23 November 2022.

Batat, W. (2019). *The new luxury experience: Creating the ultimate customer experience.* Springer International Publishing AG.

Batliwala, A.K. (2016). *Is it possible to shower too often?* Retrieved from Is it possible to shower too often? | Stuff.co.nz, 31 October 2022.

Bayleys Realty Group. (2021). *A touch of The Block NZ comes to New Plymouth at new apartment complex.* Retrieved from https://www.bayleys.co.nz/news/residential/a-touch-of-the-block-nz-comes-to-new-plymouth-at-new-apartment-complex, 24 November 2022.

Berg, M., & Eger, E. (2003). The rise and fall of the luxury debates. In M. Berg & E. Eger (Eds.), *Luxury in the eighteenth century: Debates, desires and delectable goods* (pp. 7–27). Palgrave Macmillan.

Bilefsky, D. & Carvajal, D. (2014). *A capital of the arts is forced to evolve.* Retrieved from https://www.nytimes.com/2014/10/28/world/europe/a-capital-of-the-arts-is-forced-to-evolve.html, 21 November 2022.

BRANZ (Building Research Association of New Zealand) (n.d.). *Bathrooms and laundries.* Retrieved from https://www.renovate.org.nz/1970s/kitchens-and-bathrooms/bathrooms-and-laundries/, 9 October 2023.

Brown, K., & Pes, J. (2018). *How much should taxpayers pay for a private museum? A French watchdog group files a claim to find out.* Retrieved from https://news.artnet.com/art-world/louis-vuitton-foundation-fricc-1406812, 21 November 2022.

Certified Autoplex. (2022). *What is considered a luxury car.* Retrieved from https://blog.certifiedautoplex.com/what-is-considered-a-luxury-car/, 31 October 2022

Château de Versailles. (2023). *History.* Retrieved from https://en.chateauversailles.fr/discover/history#louis-xiii-and-versailles1607-1638, 22 April 2023.

Chicca, F., Vale, B., & Vale, R. (2023). *The environmental impact of cities: Death by democracy and capitalism.* Routledge.

Cody, D. (1988). *Morris & Co.* Retrieved from https://victorianweb.org/authors/morris/morisco2.html, 16 November 2022.

Coleman, A. (2017). *Why does New Zealand keep building such massive houses?* Retrieved from https://thespinoff.co.nz/society/15-05-2017/why-does-new-zealand-keep-building-such-massive-houses, 30 November 2022.

Creative Commons. (n.d.-a). "Fondation Louis Vuitton" by Henry_Marion is licensed under CC BY-SA 2.0.

Creative Commons. (n.d.-b) "Bulgari NYC Flagship" by ajay_suresh is licensed under CC BY 2.0.

Creative Commons. (n.d.-c) "Crystal Houses 1 MVRDV" by Jleaguer is licensed under CC BY-SA 4.0.

Creative Commons. (n.d.-d) "Puma City Precedent Study Presentation P.5" by FHKE is licensed under CC BY-SA 2.0.

Creative Commons. (n.d.-e) "Dr Johnson's House—City Of London" by Jim Linwood is licensed under CC BY 2.0.

Creative Commons. (n.d.-f) "Holkham Hall, Norfolk" by Rev Stan is licensed under CC BY 2.0.

Creative Commons. (n.d.-g) "Hall of Mirrors, Versailles" by Dimitry B is licensed under CC BY 2.0.

dotLuxury. (n.d.). *How Luxury is Made: The Hermès Scarf.* Retrieved from https://fashion.luxury/style/made-hermes-scarves/, 14 November 2022.

EFE. (2022). *First Karl Lagerfeld villa in Marbella sells off-plan for €15 million.* Retrieved from https://ww.fashionnetwork.com/news/First-karl-lagerfeld-villa-in-marbella-sells-off-plan-for-15-million,1451673.html, 29 March 2023.

Eskilson, S. (2018). *The age of glass.* Bloomsbury Publishing.

Favermann, M. (2010). *Shipping container architecture: Puma city store near Fenway Park.* Retrieved from https://www.berkshirefinearts.com/10-08-2010_shipping-container-architecture.htm, 23 November, 2022.

Ferguson, S. (2022). *Live Luxe.* Allen and Unwin.

Forbes. (2022a) *LVMH Moët Hennessy Louis Vuitton.* Retrieved from https://www.forbes.com/companies/lvmh-moet-hennessy-louis-vuitton/?sh=1109471e1371, 16 November 2022.

Forbes, A. (2022b). *"Karl Lagerfeld Villas" luxury residential project breaks ground in Marbella.* Retrieved from https://theluxuryeditor.com/news/karl-lagerfeld-villas-luxury-residential-project-breaks-ground-in-marbella/#:~:text=In%20an%20event%20celebrated%20on,the%20third%20quarter%20of%202024.&text=come%20to%20life.%E2%80%9D, 30 November 2022.

Garcia, E., & Vale, B. (2017). *Unravelling sustainability and resilience in the built environment.* Routledge.

Garcia, E., Vale, B. and Vale, R. (2021). *Collapsing gracefully: making a built environment that is fit for the future.* Cham: Springer.

Giovannini, J. (2014). *Paris Envy: Frank Gehry's Vuitton Foundation.* Retrieved from https://lareviewofbooks.org/article/paris-envy-frank-gehrys-vuitton-foundation/, 18 November 2022.

Goodman, D. (2003). Furnishing discourses: Readings of a writing desk in eighteenth-century France. In M. Berg & E. Eger (Eds.), *Luxury in the eighteenth century: Debates, desires and delectable goods* (pp. 71–88). Palgrave Macmillan.

Harrouk, C. (2021). *Is sustainability the new luxury? In Conversation with Andrea Boschetti on Karl Lagerfeld Villas in Marbella.* Retrieved from https://www.archdaily.com/972822/is-sustainability-the-new-luxury-in-conversation-with-andrea-boschetti-on-kl-villas-in-marbella?ad_source=search&ad_medium=projects_tab&ad_source=search&ad_medium=search_result_all, 21 November 2022.

Hedrick-Wong, Y. & Choong, D. (2014). *Mastercard 2014 Global Destination Cities Index.* Retrieved from https://newsroom.mastercard.com/wp-content/uploads/2014/07/Mastercard_GDCI_2014_Letter_Final_70814.pdf. 17 November 2022.

Heritage Emergency. (n.d.). *Holkham Hall.* Retrieved from Holkham Hall—Heritage Emergency, 20 April 2023.

Hill, R. (2005). Architecture, luxury and ethics. In N. Ray (Ed.), *Architecture and its ethical dilemmas* (pp. 123–131). Taylor and Francis.

Hont, I. (2006). The early enlightenment debate on commerce and luxury. In M. Goldie & R. Wokler (Eds.), *The Cambridge history of eighteenth-century political thought* (pp. 377–418). Cambridge University Press.

Householdquotes. (2023). *Composite Worktop Installation Costs: Your 2023 Guide.* Retrieved from https://householdquotes.co.uk/composite-worktop-costs/, 12 April 2023.

Hush, K. (2010). *Who wears Paris haute couture? Very few, apparently.* Retrieved from https://www.smh.com.au/lifestyle/fashion/who-wears-paris-haute-couture-very-few-apparently-20100709-1041u.html, 13 April 2023.

Institute of Historical Research. (2019). *William III, 1695–6: An Act for granting to His Majesty severall Rates or Duties upon Houses for making good the Deficiency of the clipped Money. [Chapter XVIII. Rot. Parl. 7&8 Gul. III. p.5.n.4].* Retrieved from https://www.british-history.ac.uk/statutes-realm/vol7/pp86-94#s1, 1 November 2022.

Ivey, P. (2021). *Britain's most expensive home is one of the most popular properties on the internet.* Retrieved from https://www.standard.co.uk/homesandproperty/luxury/london-s-ps40-million-supermansion-highgate-most-viewed-a137646.html, 20 April 2023

Jonze, T. (2019). *Row over use of unpaid interns by Serpentine pavilion architect.* Retrieved from Row over use of unpaid interns by Serpentine pavilion architect | Architecture | The Guardian, 25 April 2023.

Jordahn, S. (2019). *Serpentine Pavilion designed to be "part of surrounding landscape" says Junya Ishigami.* Retrieved from https://www.dezeen.com/2019/06/19/serpentine-pavilion-2019-junya-ishigami-interview-video/#:~:text=In%20this%20exclusive%20Dezeen%20video,the%20gallery's%20CEO%20Yana%20Peel, 27 April 2023

Kapferer, J.-N. (2017). The end of luxury as we knew it? In J.-N. Kapferer, J. Kernstock, T. O. Brexendorf, & S. M. Powell (Eds.), *Advances in luxury brand management* (pp. 25–41). Palgrave Macmillan.

Karl Lagerfeld. (n.d.). *Immerse in legendary style.* Retrieved from https://karllagerfeldvillasmarbella.com/thevillas/, 30 October 2022.

KBB Editors (2021). *What is a luxury car?.* Retrieved from https://www.kbb.com/what-is/luxury-car/, 31 October 2022.

Keen, M., & Slemrod, J. (2021). *Rebellions, rascals, and revenue: Tax follies and wisdom through the ages.* Princeton University Press.

Khajehzadeh, I. (2017). *An investigation of the effects of large houses on occupant behaviour and resource-use in New Zealand*. PhD Thesis. Victoria University of Wellington.

Khajehzadeh, I., & Vale, B. (2017). How house size impacts type, combination and size of rooms: A floor plan study of New Zealand houses. *Architectural Engineering and Design Management, 13*(4), 291–307.

Khajehzadeh, I., Vale, B., & Isaacs, N. (2016). Time-use in different rooms of selected New Zealand houses and the influence of plan layout. *Indoor and Built Environment, 27*(1), 19–33.

Klein, L. (2022). *First-ever Karl Lagerfeld real estate development model of upscale sustainability.* Retrieved from luxuryportfolio.com/blog/first-ever-karl-lagerfeld-real-estate-development-model-of-upscale-sustainability/, 30 November 2022.

Knoll. (2023). *Barcelona chair.* Retrieved from https://www.knoll.com/product/barcelona-chair, 13 April 2023.

Kross, E., Verduyn, P., Demiralp, E., Park, J., Lee, D. S., Lin, N., Shablack, H., Jonides, J., & Ybarra, O. (2013). Facebook use predicts declines in subjective well-being in young adults. *PLoS One, 8*(8), e69841.

LaFranco, R., & Peterson-Withorn, C. (2023). *Forbes world's billionaires list.* Retrieved from https://www.forbes.com/billionaires/, 11 October 2023.

Le Corbusier. (1933). *The Radiant City*. Faber and Faber.

Limn Gallery. (2023). *Bearbrick.* Retrieved from https://www.limngallery.co.nz/collections/bearbrick-luxury-collectible-toy-nz/banksy, 26 April 2023.

Louis Vuitton. (n.d.). *The new rolling luggage.* Retrieved from https://us.louisvuitton.com/eng-us/magazine/articles/spirit-of-travel-women-the-new-rolling-luggage#, 16 November 2022.

LVMH. (n.d.). *Bulgari—Via dei Condotti.* Retrieved from https://www.lesjourneesparticulieres.com/2022/the-locations/bulgari-via-dei-condotti/, 21 November 2022.

Macguire, E. (2011). *Building the brand: 10 great fashion-architecture hook ups.* Retrieved from https://edition.cnn.com/2011/10/12/living/fashion-store-architecture/index.html, 23 November 2022.

Mithraratne, N., & Vale, B. (2004). Life cycle analysis model for New Zealand houses. *Building and Environment, 39*(4), 483–492.

Morandi, C. (2017). The city as a playground. In M. Paris (Ed.), *Making prestigious places: How luxury influences the transformation of cities* (pp. 113–128). Routledge.

Muther, C. (2009). A portable 'city' visits Fan Pier. *The Boston Globe.* Retrieved from http://archive.boston.com/lifestyle/fashion/articles/2009/04/23/a_portable_city_visits_fan_pier/, 23 November 2022.

MVRDV. (2017). *Crystal house/MVRDV.* Retrieved from https://brickarchitecture.com/projects/crystal-houses-mvrdv, 22 November 2022.

MVRDV. (n.d.). *Bulgari Kuala Lumpur.* Retrieved from https://www.mvrdv.nl/projects/343/bulgari-kuala-lumpur, 21 November 2022.

National Records of Scotland. (2022). *Taxation Record.* Retrieved from https://www.nrscotland.gov.uk/research/research-guides/research-guides-a-z/taxation-records, 1 November 2022.

Oates, W. E., & Schwab, R. M. (2015). The window tax: A case study in excess burden. *The Journal of Economic Perspectives, 29*(1), 163–179.

Otero, M. (2015). Arquitecturas de circulación y acumulación: El remontaje de los pabellones de la Serpentine Gallery. *ARQ (Santiago).*, 100–109.

Our World in Data. (n.d.). *Per capita CO_2 emissions, 2021.* Retrieved from https://ourworldindata.org/grapher/co-emissions-per-capita, 11 October 2023.

Pajur, A. (2017). The fabric of a corporate society: Sumptuary laws, social order and propriety in early modern Tallinn. In J. Ilmakunnas & J. Stobart (Eds.), *A taste for luxury in early modern Europe: Display, acquisition and boundaries* (pp. 21–38). Bloomsbury Publishing Plc.

Paris Digest. (n.d.). *Fondation Louis Vuitton in Paris.* Retrieved from https://www.parisdigest.com/monument/fondation_louis_vuitton_paris.htm, 21 November 2022.

Paris-Walking-Tours.com. (n.d.). *City of architecture and heritage.* Retrieved from https://www.paris-walking-tours.com/cityofarchitectureandheritage.html, 17 November 2022.

Plato (trans. Lee. D.). (2007). *The Republic*. Penguin Books.

Pritchard, O. (2014). Gehry Monster: Fondation Louis Vuitton by Frank Gehry. *The Architects' Journal.* October 31st.

Purvis. L. (2023). *A working world beneath your feet.* Retrieved from A working world beneath your feet—Holkham, 20 April 2023.

Pyrcz, A. A. (2019). Versailles: The ultimate 'French Revs' excursion. *Ágora, 54*(3), 14–21.

Resene. (2021). *Check out these luxurious living rooms from Three's The Block NZ.* Retrieved from https://www.habitatbyresene.co.nz/check-out-these-luxurious-living-rooms-from-threes-the-block-nz-this-week/, 29 November 2022

Rico, A., Martínez-Blanco, J., Montlleó, M., Rodríguez, G., Tavares, N., Arias, A., & Oliver-Solà, J. (2019). Carbon footprint of tourism in Barcelona. *Tourism Management, 70*, 491–504.

Ritchie, H. & Roser, M. (2020). *CO_2 emissions.* Retrieved from https://ourworldindata.org/co2-emissions, 16 August 2024.

Robert, N. (2019). Gender difference and the consumption of luxury cars in the city of Douala-Cameroon. *International Journal of leisure and Tourism Marketing, 6*(3/4), 238–253.

Salmon, F. (2013). Our Great Master Kent' and the design of Holkham Hall: A reassessment. *Architectural History, 56*, 63–96.

Shen, S. (2012). *The environmental impact of expositions: A study of some contributing factors*. PhD Thesis, Victoria University of Wellington.

Shmerling R.H. (2021). *Showering daily—is it necessary?* Retrieved from Showering daily—is it necessary?—Harvard Health, 31 October 2022.

Sissian, T. (2021). *The Block 2021: The secrets to selling luxury property in a pandemic*. Retrieved from https://www.mcgrath.com.au/advice/articles/the-block-2021-the-secrets-to-selling-luxury-property-in-a-pandemic, 29 November 2022.

Smith, A. (1776). *The Wealth of Nations*. Retrieved from https://www.marxists.org/reference/archive/smith-adam/works/wealth-of-nations/book05/ch02b-4.htm, 31 October 2022.

Sotheby's Institute of Art. (n.d.). *At the Intersection of Art and Luxury*. Retrieved from At the Intersection of Art and Luxury (sothebysinstitute.com), 26 April 2023.

Statista. (2021). *Number of visitors to the Louis Vuitton Foundation in Paris in 2016 and 2017*. Retrieved from https://www.statista.com/statistics/995603/visitors-louis-vuitton-foundation-paris-france/, 21 November 2022.

Statista. (2022). *Average size of homes in selected countries worldwide in 2017*. Retrieved from https://www.statista.com/statistics/1052988/average-home-size-selected-countries-worldwide/, 30 November 2022.

Statista. (2023). *Average primary and secondary school building costs worldwide as of 3rd quarter 2021, by select city*. Retrieved from https://www.statista.com/statistics/758120/global-school-building-costs-in-cities/, 30 March 2023.

Stobart, J. (n.d.). *What were Georgian and Victorian servants' rooms like*. Retrieved from https://www.nationaltrust.org.uk/discover/history/what-were-georgian-and-victorian-servants-rooms-like, 20 April 2023.

Stuff. (2017). *TV ratings: New Zealand's eyeballs on the Block, not the rugby*. Retrieved from https://www.stuff.co.nz/entertainment/tv-radio/97051006/tv-ratings-new-zealands-eyeballs-on-the-block-not-the-rugby#:~:text=306%2C476%20people%20in%20the%2025,television%20event%20of%20the%20-year.&text=Leaders%20debates%20between%20Bill%20English,viewers%20aged%2025%20to%2054, 12 April 2023.

Sutton, B. (2018). *An anti-corruption group accused the Louis Vuitton Foundation of fraud and tax evasion*. Retrieved from https://www.artsy.net/article/artsy-editorial-anti-corruption-group-accused-louis-vuitton-foundation-fraud-tax-evasion, 21 November 2022.

The Ritz London. (2023). *Puttin' on the Ritz*. Retrieved from https://www.theritzlondon.com/suites-and-room-packages/puttin-on-the-ritz/, 13 April 2023.

The Royal Household. (n.d.). *Royal Residence: Buckingham Palace*. Retrieved from https://www.royal.uk/royal-residences-buckingham-palace, 21 April 2023.

Thomas, T. (2022). *TV Ratings November 6, 2022: Huge crowd watch record breaking finale of The Block*. Retrieved from https://www.mediaweek.com.au/tv-ratings-november-6-2022-the-block-finale-makes-history/, 12 April 2023.

Ubel, P. (2017). *How the psychology of income inequality benefits luxury brands*. Retrieved from https://www.forbes.com/sites/peterubel/2017/02/28/do-luxury-brands-benefit-from-income-inequality/?sh=2f997f6727b5, 9 November 2022.

Van Hal, A. (2006). Leveraging the economical benefits of sustainable building. *WIT Transactions on Ecology and the Environment, 98*(35), 41.

Veblen, T. (1994). In C. Ward (Ed.), *The theory of the leisure class*. Dover publications.

Vora, S. (2021). *7 luxury homes with over-the-top bathrooms on the market right now*. Retrieved from https://www.architecturaldigest.com/story/7-luxury-homes-with-over-the-top-bathrooms-on-the-market-right-now, 9 October 2023.

Wainwright, O. (2015). *Beach café, billionaire's retreat, wedding marquee: second lives of the Serpentine pavilions*. Retrieved from https://www.theguardian.com/artanddesign/2015/jun/16/serpentine-pavilion-second-lives-zaha-hadid-toyo-ito-frank-gehry, 25 April 2023.

Waite, R. (2019). *Junya Ishigami Serpentine Pavilion snapped up by tech spa firm*. Retrieved from https://www.architectsjournal.co.uk/news/junya-ishigami-serpentine-pavilion-snapped-up-by-tech-spa-firm, 27 April 2023.

Ward, D. (2004). *Unacceptable verses*. Retrieved from https://www.theguardian.com/world/2004/aug/12/religion.uk, 2 November 2022.

WhichCar. (2019). *Avoid new car remorse: How to predict model updates*. Retrieved from https://www.whichcar.com.au/car-advice/how-to-predict-model-updates-and-avoid-new-car-remorse, 30 March 2023.

Wilkinson, R. G., & Pickett, K. (2018). *The inner level; how more equal societies reduce stress, restore sanity and improve everybody's wellbeing*. Allen Lane.

World Architecture Community. (2008). *Puma City*. Retrieved from https://worldarchitecture.org/architecture-projects/pmhv/puma-city-project-pages.html, 23 November 2022.

Wychavon District Council. (2007). *History of windows and glass*. Retrieved from https://www.wychavon.gov.uk/component/fileman/file/Documents/Planning/History%20of%20windows%20and%20glass%20document.pdf?routed=1&container=fileman-files, 19 April 2023.

Part II

On Architecture and Wealth

The three chapters of Part II have a greater focus on architecture and its relationship to wealth. Taking a historical focus, definitions of architecture are examined to show how it evolved from having a practical focus to being a fine art. The influence of wealth in this change is examined. The effect of this change on social and environmental justice is explored.

7 Defining Architecture

The Architectural Voice

The profession of architecture at one time embraced what are now recognised as the separate professions of planning, urban design and project management. The medieval architect would manage the project, and in the past architects also designed parts of cities, or even whole cities in the case of Walter Burley Griffin, the architect who designed Canberra. Modern architects only tend to design buildings, leaving larger scale work and project management to other professionals. This suggests that what we understand as architecture may have changed over time.

The issue of how we view architecture affects what the profession of architecture is all about, what the profession designs and how members of the profession are educated. Is a building being described as architecture dependent on the wealth of the person who designs it or orders it, or the size of the available budget, as shown in Chaps. 5 and 6? Can there be architecture for the poor by the poor? Given that climate change may well affect poorer nations more than the wealthier ones, as poorer nations have a much higher dependence on agriculture and that: "Increasing climate variability and extremes, linked to climate change, are negatively affecting all dimensions of food security and nutrition" (FAO et al., 2021), should architecture be part of the solution of addressing climate change (SDG 13 Climate Action) through the design of an appropriate built environment (SDG11 Sustainable Cities and Communities) for all (SDG 10 Reduced Inequalities), or should it continue on the path of only applauding weirdly shaped buildings for the relatively few? More importantly, why is architecture not already at the forefront of climate mitigation, especially as Chinese research has shown that architecture is the missing link in implementing green building technologies in China (Wang et al., 2018). This chapter sets out to explore how architecture has been defined both in the present and in the past to see whether its current definition is at the root of its inability to really grapple with the world's current problems of "climate change, poverty, and inequality" (World Bank Group, 2023) and as set out in the SDGs. In other words, to reach the SDG goals, perhaps, we need to redefine architecture.

Library Classifications and Architecture

Although the internet has changed everything in terms of how to access knowledge, most Schools of Architecture still have libraries of real books and journals about every aspect of the discipline. How the access to architectural information through books and journals is viewed begins with how they are catalogued. Two important classification systems form the way most written information is accessed by architectural students and

E. J. García, B. Vale, *Wealth and the Built Environment*, Sustainable Development Goals Series,
https://doi.org/10.1007/978-3-031-81954-4_7

anyone interested in the subject. These are the Dewey and National Library of Congress systems.

The American librarian Melvil Dewey devised the Dewey decimal system in 1873 for the Amherst College Library (McKenna, n.d.) and published it 3 years later. The online history of Amherst College notes that "Melvil Dewey (Class of 1874) Assistant Librarian (1874–76), starts subject catalogue on cards" (Amherst College, n.d.). Under the Dewey system, 500 denotes the sciences, 600 the useful arts (now renamed Technology (Applied sciences)) and 700 the fine arts. This bridging of science and fine arts through the useful arts would seem to apply directly to architecture, since the aim of architecture is to produce a building that functions well and looks good. The idea of the Useful Arts appealed to the engineer and cartoonist Timothy Hunkin (n.d.):

> My favourite part of a public library has always been the large amorphous section between the pure sciences (the 500s) and the fine arts (the 700s), usually simply labelled `the useful arts'. It contains everything from rocketry to crochet, engineering, hobbies, cookery, etc. Mr Dewey, who devised the absurdly ambitious classification of all human knowledge for the libraries, has become a hero of mine. There are many connections between the arts and the sciences, but Mr Dewey's seems particularly strong.

This reminiscence is partly mistaken as crochet is found with knitting under technique in the Textile Arts (746.43 in fine arts) in the Dewey system, while sewing and hat making come under Home Management (640 in useful arts), although a crocheted hat pattern would be under 746.43, so in fine arts (OCLC.org, n.d.-a). Still, the point is that the way of classifying knowledge by its proximity to other kinds of knowledge creates the context for its definition.

The problem for architecture is that the system splits the subject between 700 (fine arts) and 600 (useful arts) with building and engineering housed within useful arts. The system may contain useful reminders "For structural engineering, see 624.1; for engineering design and construction of buildings, see 690; for solar heating, see 697" (OCLC.org, n.d.-b), but the point is that for the education of architects, how to build is eternally split from what to build and what others have built, as the books will be shelved in different places under the different classifications. This gives lie to the claim made for the system by the inventor (Dewey, 1876).

> Thus all the books on any given subject are found standing together, and no additions or changes ever separate them. Not only are all the books on the subject sought, found together, but the most nearly allied subjects precede and follow, they in turn being preceded and followed by other allied subjects as far as practicable.

The implication is that building construction and discussion of the materials to be used is about as important as dressmaking and cooking—something that can be taught in the home at mother's knee rather than in a university or polytechnic. In contrast, real architecture rather than just building sits with painting and sculpture as a fine art. Another way to view Dewey's claim for his system is that knowledge of building in the 600s must precede any knowledge of architecture in the 700s, on the assumption no one would want to access the architecture books in the 700s without first having a very good grounding in how to build as found in the 600s, given the close relationship between these two subjects.

Splitting what architects need to know in this way has other implications by suggesting that architecture as a subject only resides within the fine arts, thereby implying that it is a fine art along with painting and sculpture and exists only to produce a response from the viewer, whether this response be intellectual or emotional based on the creativity of the creator. This is in contrast with craft or applied art which has a functional purpose, although being equally capable of eliciting an emotional response. What makes the art "fine" is that it only has the purpose of producing a response. Patently, most architecture has a functional purpose, but its Dewy classification suggests that the aesthetic response to the creativity of the design is more important than any function (see Chap. 8), such as keeping the rain out or preventing the heat from escaping through the walls and roof. Frank Lloyd Wright is credited with saying, "If the roof doesn't leak, the architect hasn't been creative enough," a sentiment echoed by the 1970s statement emanating from the Oslo School of Architecture that "all good architecture leaks" (Hardarson, 2005). These ideas place creativity over performance and put such architecture firmly in the fine arts category, though the unfortunate owners of leaky buildings might rather wish it had not done so.

There is another interpretation of fine arts, as along with fine wines and fine dining, fine art is a more costly and elitist version of the same product. The origin of fine dining is often attributed to the French Revolution, after which the chefs who formerly cooked elaborate food for the French aristocracy opened restaurants and made such food available to those who could afford it. Spang (2000:141–142) argues that this is a misinterpretation of the writer Mercier, who insisted that "aristocratic delicacies corrupted republican decency and contaminated honest citizens." In other words, the failure of the French Revolution was through the redistribution of wealth and the creation of a new middle class who could afford fine food, thereby losing the "egalite" that had underpinned the revolution. With wine, it seems clearer that a fine wine will often have a small production run and may come from a special section of the vineyard, so its relative rarity ensures it has a higher price. The other issue is whether fine art is more expensive than useful art. As an example, a 1908 English Arts and Crafts extending dining table by Ernest Gimson sold for £17,500 in 2016 (Christie's, 2023), while a 1908 painting by English artist Waterhouse known as The Soul of the Rose in the style of the Pre-Raphaelites sold at Christie's New York for US$4,695,000 (£3.7 million) in 2019 (Art.Salon, n.d.).

The two originators of these works were contemporaries—Gimson 1864–1919 and Waterhouse 1849–1917. Whereas Waterhouse was the son of two artists and devoted his life to painting and teaching, Gimson's father was an engineer and iron founder, and Gimson started his career as an architect, learning on the job as an articled pupil and only later turning to furniture design. Lea Cottage (Fig. 7.1), Grade II, listed and one of three cottages Gimson designed as holiday homes for family members and one of which, Stoneywell, is now owned by the National Trust, was on the market for £1.9 million (Savills, 2023) when this was text written. Although extended to a six-bedroom house and sitting on more than 6.5 acres of gardens and grounds, attributes which undoubtedly add to the very high cost, why should a "useful" house be only half the price of a painting that can only be looked at?

Work on Gimson would appear under architecture in the fine arts section of the library, but the irony is that as an Arts and Crafts architect and designer, for Gimson the building or table would have to be a reflection of how it was made and any creativity was limited to working with a particular set of materials to meet a particular brief, whether for a building or a table, in a particular location. The other two nearby cottages designed by Gimson around the same time—Stoneywell and Rockyfield—use similar materials but differ because of location and brief (Figs. 7.2 and 7.3).

Accepting that these cottages were holiday homes and therefore emanated more from wealth than necessity, even though in the absence of the family Stoneywell Cottage was home to a caretaker, who had a bedroom above the kitchen accessed by a separate stair from that leading to the family bedrooms (Weaver, 1913:102), they are still linked by an architecture, the prime aim of which is not to be overly creative. Given that creativity underpins the idea of fine arts (see Chap. 8), it could be argued that the Gimson cottages would fit equally well in a useful arts category, because their appearance is essentially linked to their materials and how these are assembled. This is not vernacular architecture based on traditional ways of making a building but is a design where materiality in expressing how a thing is constructed is given precedence over other aspects, such as cost, especially since Stoneywell Cottage came in well over the £500 estimate. Alexander (1964:29) makes the point that design can never equally satisfy all parameters: "What is true is that designers do often develop one part of a functional programme at the expense of another. But they do it because the only way they seem able to organize form clearly

Fig. 7.1 Lea Cottage by Gimson. (Source: Creative Commons, n.d.-a)

is to design under the driving force of some comparatively simple concept." This suggests that in design there is not the same unlimited creativity as found in say art or sculpture, where the only parameters are the materials used and skill of the artist/sculpture in transferring the mental image to reality. Creativity for the designer only occurs within the limits of the problem or programme as set, at its simplest the brief, context and budget. This makes architecture as a design based fine art very different from painting and sculpture which normally, though not always, lack a context.

So far, the position of architecture only within the Dewey classification system has been examined. The National Library of Congress library classification system is also a product of the end of the nineteenth century and also splits architecture from building or making, with architecture falling under N for the arts (NA for architecture) but building coming under T for technology (TH for building construction). Weirdly, whereas in Dewey, handicrafts come under decorative arts (745) in the National Library of Congress system, handicrafts and arts and crafts come under technology (TT), although books on Arts and Crafts architects fall under NA. This suggests that separating the search for what to make from how to make it is problematic even for librarians. In the past, many builders learned from those who were more experienced, who in turn had learned their craft in a similar way without frequenting libraries. Additionally, much of the historic built environment has probably been designed at the hands of builders rather than architects. In conversation with the Architecture and Design Library at Victoria University of Wellington, there are more resources under NA with its 23 sub-headings than TH with its 16 "to support the particular courses we run here". It also seems from what the librarian said that students are more likely to borrow TH construction books later in their course as they specialise.

Fig. 7.2 Stoneywell Cottage by Gimson. (Source: Creative Commons, n.d.-b)

Obviously, some NA books will discuss construction as part of architecture but given that the first sub-category under TH is "building construction", this would seem fundamental knowledge for anyone designing a piece of architecture, even at the start of a course of study. Maybe, new students turn to their lectures or the internet for their knowledge on how the process of building might underpin their design ideas, but experience in education suggests that for the new student, the idea of what makes architecture is not necessarily informed by how it might stand up.

In engineering, it might seem that function overrides appearance since an engineering collapse is an engineering failure, whereas as noted above an architect designed building that leaks by failing to keep the weather out is not necessarily viewed as a failure. Alongside that, many bridges designed by engineers have a beautiful elegance that in part comes from the way materials and construction methods are used to create that bridge, since one of the major problems in making a bridge is how to construct it across the span, especially when that span is over water. The bridge is also not a stand-alone object like a building but part of a network, such as that of roads or railways, which will influence the design as it may be one of a series of bridges. An example would be the 1950s concrete M1 motorway bridges of Owen Williams in the UK. "Each bridge is derived from the same blueprint, simply skewing or stretching the same design to meet the needs of the individual locations in terms of length, width, depth, height and angle" (Marshall, 2022). An architectural parallel would be the Georgian terraced houses of London or Bath, where the architectural delight comes more from the whole streetscape than the individual building, but then such houses were built by developers rather than architects. It is apparent that both the concrete motorway bridges and Georgian terraced houses are an integral part of a broader artistic and cultural landscape, since they are aesthetic expressions that can be shared with other people. This echoes the definition of fine arts in Britannica online (Kuiper, 2023), yet bridges as engineering would not be classified under fine arts in either the Dewey or National Library of Congress Systems. This suggests that the design of a building is somewhat different from the design of a bridge.

Fig. 7.3 Rockyfield Cottage by Gimson. (Source: Creative Commons, n.d.-c)

Perhaps it is a difference in the range of creative freedom that puts bridges in engineering and architecture in fine arts. Once the basic material for the bridge has been chosen—wood, stone, steel, concrete—what results comes from the context, the budget and the creativity of the engineer. In contrast, although a building needs structure, the material used for the structure could vary throughout the building, and the choice of the materials for other components, such as windows and walls, is also left to the context, budget and aesthetics of the designer. The architect has more to play with than the engineer. However, when it comes to painting the artist is far more limited in choice as, once the medium is chosen, like the engineer what results is what the artist makes of the chosen medium. Poetry as a fine art is even more restricted, since a word is a word and only the word order is available to the poet. In fact, engineering seems more akin to painting and poetry in the restriction of the medium within which creativity occurs than architecture.

Fundamentally, architecture is a useful art, along with sewing and rocketry, because it has to do something, such as house people, cars or library books. Even a built folly that appears to serve no functional purpose is still designed to be part of a landscape and so must not only stand up but also survive the weather to fulfil its scenic purpose. Arguably, a sculpture might have to do exactly that, and sculpture clearly falls into the fine arts category. Most architecture, however, does have a function, whether to house people, pigs, or statues and as such, it is essentially useful. Its library classification as a fine art may well have been detrimental to the education of many architects, since this appears to place creativity ahead of usefulness or technical skill, whereas, in the great buildings of the past, both are evident to a high degree of resolution. One such example is Westminster Hall (Fig. 7.4), "…one of the most

Fig. 7.4 Westminster Hall. (Source: Creative Commons, n.d.-d)

vital monuments ever built" (Harvey, 1946:50) by Henry Yevele (c1320–1400) with Hugh Herland, who was responsible for the timber roof. Looking back to who paid for such a magnificent structure, the UK Parliament (2024) states: "The money for this project was raised by compelling those who had fled or been banished from the country, and wished to return to their native soil, to purchase a licence for that purpose for a considerable sum."

The Classification of Architecture in the Past

Architecture has not always been classified as a fine art. In their book, Parcell and Winer (2012) discuss in depth four views of architecture ranging from the Greek *techne* to the medieval idea of architecture as a mechanical art, the Renaissance view of architecture as stemming from drawing and finally the emergence of architecture as a fine art around 1750 with its inclusion in the five *beaux arts* (painting, sculpture, poetry, music and architecture) (Parcell & Winer, 2012:178). The history of ideas about beauty and aesthetics and the effect these had on architecture, particularly in the middle of the eighteenth century is the subject of the next chapter. Here, the interest is on how information on architecture can be accessed through how it is classified. Ontology is the study of entities and how they relate to each other to both explain what we know and how we know it. How architecture is classified thus guides those learning as to both what it is and how we know what it is. In this way, an investigation of classification systems should reveal how architecture was viewed in the past and in the present.

Dr. Raphael Sandoz of the University of Geneva has helpfully posted on the web 255 diagrams of trees of knowledge that start with Plato c. 360 BCE and end with Piaget in 1967 (Sandoz, 2023). Of these 255, few knowledge trees mention architecture, whereas many include music, thereby already positioning architecture as less important in the sum of human knowledge. The first entry, that of Plato c360 BCE, does not include architecture. However, Plato divides the arts into the acquisitive and productive, and under the latter are listed agriculture, medicine, construction, shipbuilding, painting and sculpture. This was at a time when painting was used to adorn buildings and other useful objects like shields and vases, just as sculpture would be an adjunct to a building (Steven, 1933). Together, therefore, construction, painting and sculpture could be viewed as making architecture. Aristotle does not mention architecture in his tree of knowledge, though his useful views on art and ethics are discussed in Chap. 9. In 55 BCE, Cicero divides the arts into minor (painting, sculpture, poetry, acting, athletics) and major (medicine, architecture, agriculture, teaching, military arts), the major arts being subjects essential to the functioning of society, whereas the minor arts are useful but perhaps non-essential additions to society. This suggests that for Cicero the understanding of architecture was more like the idea of construction as found in Plato's earlier tree of knowledge. An echo of this idea occurs in the c.1190 table of knowledge by Peter of Poitiers who places architecture under the mechanical arts, along with weaving, military arts, agriculture, manufacturing and navigation. This is not to exclude creativity from any of these branches of knowledge, but again the emphasis is on the production of something. Alberti's schema of knowledge of 1428 keeps architecture in the practical arts along with navigation, painting and sculpture, as opposed to the humanities or liberal arts, and in the various maps of knowledge this is where architecture stays when it is mentioned at all. As an example, the 1507 tree of the English mathematician and astrologer, John Dee, starts with "Sciences and Artes Mathematicall", which he divides into principal and derivative, with architecture under derivative being defined as "a Science garnished with many doctrines, and divers Instructions: by whose judgement, all workes [sic] by other workmen finished, are judged" (Sandoz, 2023), suggesting a situation that is still current in that the people most interested in architecture are architects.

In 1605, Francis Bacon produced a different chart in his *Advancement of Learning.* Bacon divides human learning into history, poesy and philosophy (reason). Architecture now appears under a branch of natural philosophy called "Mixed Mathematics", which include disciplines like music and astronomy that "can never be with sufficient subtilty comprehended, nor demonstrated with sufficient perspicuity, nor accommodated to use with sufficient dexterity and certainty, without the Aid and intervening of the Mathematiks" (Sandoz, 2023). In 1630, Johann Heinrich Alsted places architecture within illiberal knowledge but whereas his branch of the mechanical arts subdivides into fine arts (painting, engraving, sculpture, singing, dance) and agriculture (viniculture, horticulture) among other subdivisions, architecture lies within the compound branch of the illiberal arts along with history and chronology. Alsted, a German born Calvanist minister, simply defines architecture as "Bene aedificandi scientia" (good building knowledge). In 1650, the English philosopher Thomas Hobbes has architecture back in the mechanics along with engineering and navigation. The 1693 knowledge tree of Gottfried Leibniz, the mathematician, philosopher and scientist, has architecture as a branch of mathematics—mechanics—which also includes navigation and military art, but makes no mention of painting or sculpture.

Things begin to change after 1700. In 1711, the Frenchman Gabriel Martin had five branches of knowledge, of which the sciences were divided into philosophy, mathematics (including music), medicine and arts. The arts group consists of mnemonics, painting, sculpture, architecture, military art, pyrotechnics and gymnastics. In 1728, Christian Wolff, the pupil of Leibniz, separated mathematics and its sub-disciplines from what would now be considered "art" subjects, including architecture. In Wolff's tree of knowledge, architecture is a liberal art along with grammar, rhetoric, poetry, argumentation and medicine. However, Wolff acknowledges that knowledge of history, philosophy and mathematics would be necessary for those engaging in such arts. His 1738 treatise on architecture formed part of a volume of his *Elementa Matheseos Universae,* the most common textbook on mathematics in eighteenth-century Germany (Beiser, 2009:46). For Wolff, architecture should be "more like a science, so that each of its precepts are derived from first principles according to the mathematical method" (Beiser, 2009:55). Architecture defined in this way has a chance to play a role within a built environment that tries to acknowledge the SDGs, because its essence is in addressing problems.

The link between architecture and mathematics is continued in Chambers' 1728 *Cyclopaedia; or, An Universal Dictionary of Arts and Sciences.* In this work, poetry was viewed "as the archetypical art, but he [Chambers] did not separate creative arts from the arts in general. Chambers did, however, embrace hierarchy within the arts, viewing manual arts like 'sculpture, architecture, [and] agriculture' as inferior poetry" (Schatzberg, 2012). In Chambers' table of knowledge, architecture is grouped with sculpture and trades and manufactures as a sub-group of mechanics, which in turn is a sub-group of mixed mathematics. Painting also appears under mixed mathematics, via optics (Shiner, 2001:81). In the 1746 publication by Charles Batteux's *Les beaux arts réduits à un même principe,* three classes of arts were outlined: "those that simply minister to our needs (the mechanical arts); those whose aim is pleasure (the beaux-arts par excellence); and those that combine utility and pleasure (eloquence and architecture)" (Shiner, 2001:83). However, when Chambers *Cyclopaedia* was translated into French by Diderot and d'Alembert, a serious attempt was made to raise the status of artisans as they held the knowledge of the arts. In 1751, Diderot wrote "The speculative side consists of nothing more than knowing the rules of an art without using them, while the practical side is but the habitual and unreflective employment of these same rules" (Schatzberg, 2012). Obviously, what is needed is a combination of the two—creativity within the limits of the rules. Diderot implies that a manual art, such as building, should be given the same status as other arts, and hence, since architecture relies on an understanding of building, it can be among the other arts, and hence d'Alembert's listing of the beaux-arts as "poetry, painting, sculpture, music and architecture" (Shiner, 2001:84).

The placement of architecture within the fine arts appears in the 1818 knowledge tree of Samuel Taylor Coleridge. He divided knowledge into pure, mixed and applied, with fine arts (poetry/psychology, painting, music, sculpture and architecture) being a sub-branch of applied knowledge along with the useful arts (agriculture, political economy, commerce and manufacturers). Coleridge defined fine arts as "applied sciences [knowledge], namely those which are applied to the purposes of pleasure, through the medium of the imagination" (Sandoz, 2023). A definition which places science and imagination at the service of a hedonistic purpose makes architecture closer to becoming a luxury rather than a discipline with a role to play in achieving the SDGs.

Views of Architecture

In terms of the classification of human knowledge, architecture is a minor player, for many years being regarded as a part of mechanica, which in turn relate back to mathematics. Only with the rise in interest in aesthetics, which forms the subject of the next chapter, does architecture find itself a fine art, but at the start at least with the implication that all fine arts rest in a basis of skill in manipulating the necessary medium, whether paint or bricks and mortar.

The relationship with mathematics is salutary for architecture today. Although there is no agreed definition of mathematics, Yadav's (2017) simplification is helpful when it comes to considering why architecture would be part of mathematics. Yadav argues that: "Mathematics is the study of assumptions, its related properties and applications". Any piece of architecture similarly starts with an assumption, perhaps about plan or form, the properties of which must be explored to see if they will satisfy the requirements of the client and the client's brief and context, followed by the application of the assumptions through making the design idea into a real building.

Between the arguments of Diderot and d'Alembert in the eighteenth century and Dewey in the nineteenth sits the industrial revolution with its profound effect on craft and those who practised it. In 1835, Andrew Ure, a professor from Glasgow, wrote a treatise on the benefits of industrialisation and how "...the benignant power of steam summons around him his myriads of willing menials, and assigns to each the regulated task, substituting for painful muscular effort on their part, the energies of his own gigantic arm..." (Ure, 1835:18). For Ure the new factory worker was better paid, better fed and led an easier life than the former artisan, but it was now the machines and those who designed them that had the creative knowledge, no longer the workers or former artisans. However, as this creative knowledge drew on science to create the new mechanical devices, this process moved the idea of making away from art towards applied science or technology. Schatzberg (2012), quoting from the eleventh edition of the *Encyclopaedia Britannica*, published in 1911, notes that "the word Art, becoming appropriated to the fine arts", was no longer used for the "large number of industries and their products to which the generic term Art...properly applies". This use of the word art would have surprised Aristotle who although seeing both arts and sciences as important used the word "art" with freedom, as in "even the highest arts, such as economy, and the art of war" (Aristotle, cBC335-322:Book1/II). It should come as no surprise, therefore, that architecture is seen as a fine art in library systems, since architects no longer need building skills in order to practice their craft of the design of buildings on paper (or virtually). The knowledge needed to build is part of technology not of fine art, although that was not the original intention of those who saw architecture as a fine art in the eighteenth century. In an ideal world, architecture is a subject that combines utility and pleasure, and all publications related to this ideal would be found in the same class of knowledge.

The Implications of Classifying Architecture as a Fine Art

Seeing architecture primarily as a fine art also has implications for wealth, since patronage of the fine arts has long been associated with wealth (Zorloni, 2016:ix).

> Throughout history art patronage has played an important part in the wealth of ultra-high-net-worth families. Although the Renaissance is regarded as the golden age of patronage, the patronage concept was actually born during the Roman Empire. Rulers, nobles, and the very wealthy used patronage of the arts to endorse their political ambitions, social, positions, and prestige.

It should be no surprise, therefore, that despite the high numbers of architects working in small practices doing small buildings and parts of buildings, what is shown as architecture in books and magazines tends to favour the large, expensive and extraordinary, sending a false message to those learning about what they might expect in their future careers. Perhaps what is needed is a reconsideration of the nature of architecture and what architecture schools should be teaching, given the need to deal with climate change, a growing human population and the ever-increasing demands being made on the resources of the planet.

A Definition of Architecture

In their book probing the nature of architecture and how it has changed over the centuries, Parcell and Winer (2012):10) state: "Another academic context for the present study is the history of architectural practice. In the modern era, three characters (designer, builder, dweller) and three things (material, drawing, building) are basic elements that mark out a conventional field of practice in architecture." They compare this to music with its "three characters (composer, performer, listener) and three things (sound, score, performance)" (Parcell and Winer 2012:11). They suggests that music can be perceived in different ways. In the conventional concert, the listener is isolated from the performers who in turn, if the composer is dead, will have no link with his or her intentions except for what is written in a notated score or what is communicated through the conductor's interpretation of that score. There can only be one interpretation if the musicians are to perform in "concert". This is very different from carol singing on the beach or in the park in a southern hemisphere Christmas celebration, where the listeners are the performers and what is heard is a collective interpretation of the words with no score and the tunes coming from memory. Dancing to music is another way in which listeners become performers, here working with the musicians who may speed up or slow down the music according to the ability of the dancers.

There are obvious parallels with architecture. Most illustrations of new buildings in the journals are like the formal concert with no dwellers and certainly no builders in sight. The architect will be named but not seen. In a self-built slum house, the designer is the builder and the dweller; although there may not be drawings, there will be materials and building skills, with these determining what can be constructed. The final building may also not be final, being adjusted over time as resources permit. However, it is the pristine, new building in the journal that will be perceived as architecture not the slum dwelling. Parcell and Winer (2012):11) comment on how the two concepts of designer, builder, dweller and material, drawing, building find themselves treated in education.

> Descriptions of architecture in current publications and lectures typically emphasize two elements of practice and let the rest disappear into blind spots. Consequently, we are led to believe that designers produce buildings. This oversimplifies and distorts the field of practice. Framing the field in this narrow way promotes two beliefs: that the designer is a creator, and that the building is an aesthetic object. This has several consequences: It marginalizes the builders and the process of construction; it neutralizes the dwellers and the social program for the building; it disregards the building's materiality and earthly context; and it skips over the representations with which the design is developed.

To this observation, which so reflects what is currently taught by way of architectural design in most schools of architecture, could be added the perception of the client in this process. Using the musical analogy, the client (dweller) is not just a listener in the concert audience, but because of wealth often appears to be the only listener having bought the whole orchestra and concert hall, with everyone else standing outside hoping to catch a faint echo of the performance inside. Ultimately, unless architecture as a profession abjures the idea that good or creative architecture leaks there is no way that architecture will have any role in dealing with climate change, poverty and inequality and hence any role in achieving the SDGs. This is a pity as out of the fine arts architecture would appear to have more to offer climate change in terms of creating buildings that dramatically reduce emissions, or neighbourhoods based on walking and cycling as a contribution to climate change mitigation, as well as designing buildings to cope with the effects of climate change such as flooding and unusual temperatures. In WWII, architects were involved in the design of wartime buildings, whether air raid shelters, barracks, temporary housing or whole new neighbourhoods such as created in the USA for the Manhattan Project at Los Alamos, Oak Ridge and Hanford (Overstreet, 2020). Ouroussoff (2011) argues that the creative architectural energy that went into such wartime projects left a post-war legacy. "The drive toward standardization was echoed in the conformity of cold war-era planning strategies. And the 'decentralization' of cities proposed by planners worried that they were easy targets for bombers continued, on a much larger scale, as suburban sprawl." Meanwhile, some painters were employed as war artists as "a Committee was appointed to choose artists who could make what was described as 'an artistic record of the war in all its aspects'" (Anon, 1944:1). This led to such productions as the Oxford University Press series of *War Pictures by British Artists,* first published in 1942. Series one consisted of *War at Sea, Blitz, R.A.F.,* and *Army,* while series two published in 1943 was made up of *Women, Production, Soldiers* and *Air Raids.* However, not all artists could be employed in this way and *Recording Britain* was set up in 1938 "primarily as an unemployment relief scheme for artists" (Mellor, 1990:9), using a grant from the USA Pilgrim Trust to paint the changing aspects of Britain. As such it echoes an earlier 1910 book *Vanishing England* that set out to record what was left of an England before industrialisation and the car and before "the air will be conquered, and aeroplanes, balloons, flying-machines and air-ships, will drop down upon us from the skies and add a new terror to life" (Ditchling & Roe, 1910:2). Perhaps, we shall know when the battle against climate change has been lost, when artists are commissioned to record the changing state of the country as the winds and heavy rain do their damage. Meanwhile, it is to be hoped that architects realise that they have a serious role to play in climate change mitigation and start acting accordingly, including in the education of future architects.

To do this, there needs to be a return to the original idea of a fine art, given this is where architecture now sits, as expounded by Diderot. Such a definition accepts that the art in question needs to have a deep understanding of the craft that underpins it. It may be no accident that in the Dewey system building and engineering come before fine arts, so that knowledge of creative design builds on knowledge of how to build. What is a pity is that much architectural education fails to grasp this simple idea. Probably, Alsted came nearest to a definition of architecture that could deal with current problems—Bene aedificandi scientia—good building knowledge.

References

Alexander, C. (1964). *Notes on the synthesis of form*. Harvard University Press.

Amherst College. (n.d.). *Library history & facts*. Retrieved from https://www.amherst.edu/library/about/policies/history, 3 June 2021.

Anon. (1944). *War pictures at the National Gallery*. Curwen Press.

Aristotle. (cBC335-322). *Nicomachean ethics*. Retrieved from https://standardebooks.org/ebooks/aristotle/nicomachean-ethics/f-h-peters/text/titlepage, 8 February 2023.

Art.Salon. (n.d.). *John William Waterhouse, R.A.: The soul of the Rose*. Retrieved from https://www.art.salon/artwork/john-william-waterhouse-r-a_the-soul-of-the-rose_AID57752#:~:text=In%20October%20 2019%20Christies%20in,range%20of%20USD%20 3%2C000%2C000%20%2D%205%2C000%2C000, 15 March 2023.

Beiser, F. C. (2009). *Diotima's children: German aesthetic rationalism from Leibniz to Lessing*. Oxford University Press.

Christie's. (2023). *Ernest Gimson 1864–1919*. Retrieved from https://www.christies.com/en/lot/lot-6029747, 15 March 2023.

Creative Commons. (n.d.-a). "File:Lea Cottage, Ulverscroft, Leicestershire from Lea Lane.JPG" by RobinLeicester is licensed under CC BY-SA 3.0.

Creative Commons. (n.d.-b). "Stoneywell, Ulverscroft" by Martinvl is licensed under CC BY-SA 4.0.

Creative Commons. (n.d.-c). "Rockyfield, a cottage designed by Ernest Gimson - geograph.org.uk – 2131886" by Stefan Czapski is licensed under CC BY-SA 2.0.

Creative Commons. (n.d.-d). "Westminster Hall" by George Hawkins, is licensed under CCO 1.0

Ditchling, P. H., & Roe, F. (1910). *Vanishing England*. Methuen and Co Ltd.

FAO, IFAD, UNICEF, WFP, & WHO. (2021). *The State of Food Security and Nutrition in the World 2021. Transforming food systems for food security, improved nutrition and affordable healthy diets for all*. Food and Agriculture Organization of the United Nations.

Hardarson, A. (2005). *All good architecture leaks: Witticism or word of wisdom?* Retrieved from https://www.ntnu.no/documents/20658136/1241076122/All_good_architecture_leaks_Hardarson_2005.pdf/37ccf00a-dab7-4c5c-8aa5-55595fa79677#:~:text=The%20 phrase%20%E2%80%9Call%20good%20 architecture,on%20design%2Drelated%20construction%20defects, 15 March 2023.

Harvey, J. H. (1946). *Henry Yevele: The Life of an English architect*. B.T. Batsford Ltd.

Hunkin, T. (n.d.). *The useful arts*. Retrieved from https://www.timhunkin.com/90_useful_arts.htm, 3 June 2022.

Kuiper, K. (2023). *The arts*. Retrieved from The arts | Britannica, 20 March 2023.

Marshall, C. (2022). *M1 Architecture*. Retrieved from M1 Architecture | Roads.org.uk, 20 March 2023.

McKenna, A. (n.d.). *Dewey decimal classification*. Retrieved from https://www.britannica.com/science/Dewey-Decimal-Classification, 3 June 2022.

Mellor, D. (1990). 'Recording Britain': A history and outline. In D. Mellor, G. Saunders, & P. Wright (Eds.), *Recording Britain: A pictorial domesday of pre-war Britain*. David and Charles.

OCLC.org (Online Computer Library Center). (n.d.-a). *600 Technology (applied sciences)*. Retrieved from https://www.oclc.org/content/dam/oclc/webdewey/help/600.pdf, 6 June 2022.

OCLC.org. (n.d.-b). *700 the arts*. Retrieved from https://www.oclc.org/content/dam/oclc/webdewey/help/700.pdf, 6 June 2022.

Ouroussoff, N. (2011). *Wartime Architects: Creating Amid Chaos*. Retrieved from https://www.nytimes.com/2011/06/02/arts/design/architecture-in-uniform-in-montreal-review.html, 27 March 2023.

Overstreet, K. (2020). *Unraveling the urban planning mysteries behind the Manhattan Project*. Retrieved from https://www.archdaily.com/948013/unraveling-the-urban-planning-mysteries-behind-the-manhattan-project?ad_campaign=normal-tag, 27 March 2023.

Parcell, S., & Winer, S. (2012). *Four historical definitions of architecture*. McGill-Queen's University Press.

Sandoz, R. (2023). *Interactive historical atlas of the disciplines*. Retrieved from https://atlas-disciplines.unige.ch/, 21 March 2023.

Savills. (2023). *Polly Botts lane*. Retrieved from https://search.savills.com/property-detail/gbntrsnts210224, 15 March 2023.

Schatzberg, E. (2012). From art to applied science. *Isis, 103*(3), 555–563.

Shiner, L. E. (2001). *The invention of art: Cultural history*. University of Chicago Press.

Spang, R. L. (2000). *The invention of the restaurant: Paris and modern gastronomic culture* (pp. 119–145). Harvard University Press.

Steven, R. G. (1933). Plato and the art of his time. *The Classical Quarterly, 27*(3/4), 149–155.

UK Parliament. (2024). *The hammer-beam roof*. Retrieved from The hammer-beam roof – UK Parliament, 16 June 2024.

Ure, A. (1835). *The philosophy of manufactures*. Charles Knight.

Wang, W., Zhang, S., Su, Y., & Deng, X. (2018). Key factors to green building technologies adoption in developing countries: The perspective of Chinese designers. *Sustainability, 10*(11), 22.

Weaver, L. (1913). *The "Country Life" book of cottages*. Country Life Ltd.

World Bank. (2023). *Climate change*. Retrieved from https://www.worldbank.org/en/topic/climatechange/overview, 13 March 2023.

Yadav, D. K. (2017). Exact definition of mathematics. *International Research Journal of Mathematics, Engineering and IT, 4*(1), 34–42.

Zorloni, A. (2016). About the book: Introduction. In A. Zorloni (Ed.), *Art wealth management. Management for professionals* (pp. ix–xi). Springer.

8 Aesthetics and Wealth

Introduction

This chapter continues the discussion of the links between wealth and the built environment. More precisely, it hypothesises the existence of an aesthetic of wealth that has had a much more far-reaching impact on society and the environment than the XVIII century development of aesthetics as a field of study. The assumption is that through aesthetics, the design of the built environment, under the influence of the ideals of wealth, tightens production and consumption into a never-ending cycle of demand and use of resources that is neither responsible nor sustainable. This relates directly to SDG 12 (Responsible Production and Consumption). Would consumerism even be possible without aesthetic manipulation for wealth accumulation, and unless this link is broken can SDG 12 be achieved?

Currently, the SDGs do not acknowledge the transformative power of aesthetics or its social, economic, and environmental impacts as felt through the built environment. Therefore, it is difficult to trace the line between what is promoted as "sustainable consumption" and the ordinary consumerism of products, albeit with a green label. An understanding of the links between wealth and aesthetics raises awareness of a system of ideas that justifies and motivates irresponsible production and consumption, something that cannot be fixed until the ensuing negative effects are exposed. For those involved in the design of the built environment, this process must start with an education that aims to teach a path to sustainability rather than one of becoming, or serving, the wealthy.

The Boom in Aesthetics

The German expert in aesthetics, Wolfgang Welsch (1997: 1), states that we are experiencing an "aesthetic boom" that affects the economy and ranges from individual styling to urban planning, producing an "aestheticisation of everyday life" (Featherston, 2007: 64). This occurs when aesthetics is inextricably linked to the production of commodities and becomes indistinguishable from them. As a commodity, aesthetics can be bought. This is a common experience for architects, urban designers and planners. Clients hire designers expecting they will produce something aesthetically pleasant and also liveable. Through aesthetics and design skills, architects are expected to do "something functional, elegant, beautiful and timeless. Good, poetic, intelligent design is the most efficient way of ensuring the best value for money. This is the fairy dust, the magic—the stuff we pay them for" (Waite, 2017). This quote highlights the value of aesthetics in

E. J. García, B. Vale, *Wealth and the Built Environment*, Sustainable Development Goals Series,
https://doi.org/10.1007/978-3-031-81954-4_8

making architecture and consumer goods more saleable, thus securing a good deal for both producer and retailer. In the modern market, aesthetics is used to make something look expensive and thus increase revenue. It can also help to raise an experience by investing in "high-quality pieces that exude sophistication" (De Castro, 2023), or simply be part of luxury items that tend to have adverse environmental impacts (see Chap. 6).

In the built environment, aesthetics, at all scales, from the domestic scale of houses to the beautification of neighbourhoods, can increase rents and land value and, consequently, produce gentrification. Gentrified spaces are "in part, filled, and driven, by the aesthetic, fantasies and desires of the dream-world—a relationship that must be explored through further research" (Carroll, 2021: 35). To understand why this happens, it seems best to start asking where the dreams and desires implied in aesthetics have come from.

In 2022, Architects Can! (ACAN), a network of individuals concerned with the built environment and the ecological crisis, released a public letter accusing the Royal Institute of British Architects (RIBA) Stirling Prize of greenwashing and calling for the RIBA to "stop celebrating an architecture that is bad for the planet." The trigger for this concern was The Elephant Park regeneration scheme. The project was shortlisted by the RIBA, but according to ACAN, 1194 social housing units were being replaced by 2700 new units, with only 92 remaining socially rented. The project also showcases a glass façade that can only contribute to higher energy consumption (ACAN, 2022). The gentrification has meant previous owners have not been able to relocate with the money given to them in compensation. The aesthetic ideas of "regeneration" and "place making", terms that no one really knows the meaning of, were used as an excuse to evict everyone and make a great business for the developer. After all, is it possible to create something beautiful if it is beautiful only for the satisfaction and enjoyment of a few? The goal of SDG 11 is to make cities "inclusive, safe, resilient and sustainable" (United Nations, n.d.), inclusive being the important word in this context.

The Boom in Aesthetics and the Power of Aesthetic Judgements

Everyday life is communicated, shared and inundated with images. In 2023, two-thirds of the world spent 2.5 h (151 min) per day looking at images on social media (Dixon, 2023), most of them situated in the built environment. People share what they eat, where they go and the book pages they like on social media. All businesses have their websites plus a presence on social media in order to stay virtually close to their clients. This continuous flow of images is produced, exposed and uploaded. People also get feedback from virtual receivers. In this way, they can assess the impact and, therefore, the popularity of the images and content uploaded. However, all the images must be judged before being shared, resent and forwarded, which implies that people exercise many aesthetic judgements per second. The capacity of aesthetic judgements to create qualities and add value to an object is music to the ears of businesses, firms and brands. "The entrepreneur sees in X the potential for Y. But that is also what artists do. The aesthetic faculty is a cognitive faculty for seeing one thing as another thing. The 'tricky' economic actor or economic system is one that harnesses the imagination" (de la Fuente & Murphy, 2014: 6). In architecture, it is quite common for designers, media or both to create a narrative, fall in love with the potential of a building and be oblivious about what the building is for, or what its consequences are.

Architects and students are equally influenced by anything posted in the media, just because it has been published. This can be enhanced, if the projects are shortlisted or win prizes, in which case, images are not even judged; they are just swallowed and reproduced. The aesthetic boom, the overflow of images in the age of social media and the influence of wealth are tightly linked in a vicious circle where the judgement of the aesthetic of images adds value out of nowhere. Therefore, with the boom in aesthetics also came a boom in aesthetic judgements. These judgements imply a power for aesthetics, a quality realised by the German philosopher Immanuel Kant during the Enlightenment period.

The *Critique of Aesthetic Judgment,* the first part of Kant's 1790 book *Critique of Judgment*, is undoubtedly an essential work for understanding the intellectual foundation of aesthetics. Kant proposed that the judgement of taste is aesthetic, which refers to the feeling of pleasure or displeasure experienced by a subject (Kant & Guyer, 2000: 89). For this judgement to be fair, Kant set a condition: "the judgement of taste is without any interest" (Kant & Guyer, 2000: 90). Judging the beauty of an object or picture, in the case of social media, should be done impartially. "Taste is the faculty for judging an object or a kind of representation through a satisfaction or dissatisfaction without any interest. The object of such a satisfaction is called beautiful" (Kant & Guyer, 2000: 96). The definition of the beautiful "is that which, without concepts, is represented as the object of universal satisfaction," which means that if someone says that something is beautiful, it implies that it is beautiful to everyone not only for the individual, otherwise it would be just agreeable. Therefore, Kant insists on defining beauty as something that "pleases universally without a concept" (Kant & Guyer, 2000: 104) or without a purpose, rationale or anything to gain from it. This is very different from posting something to get more followers or using aesthetic judgements to divide opinion when, in Kant's understanding, beauty exists only if it is universally agreed.

The Intellectual and Humble Origins of Aesthetics

Kant's parents were not wealthy. His father was a saddler from Tilsit in East Prussia, now Sovetsk in West Russia, who moved to Königsberg, a Prussian city in Germany before Germany was a country (the German Empire was founded in 1871). He grew up within the morality of the guild structure of self-sufficiency, honour and the doctrine of Pietism, which was part of the Protestant church with a focus on religious feelings and emotions (Rohlf, 2023). Perhaps, Kant's background played a role in his understanding of aesthetics as the world of the senses and rationality within a religious or transcendental meaning. His work on aesthetics argued for a faculty of judgement that allows us to perceive beauty and, through this, justify the purposiveness of nature, the existence of God and the possibility of a universal moral order.

Kant was also not wealthy. Many years after, he graduated from college and went to university, his only monetary source remained his job as a private tutor, which compromised his capacity to make ends meet. His economic situation changed when the Russians invaded Prussia, more money was circulated, and Kant was made Dean of the university. However, his ideas were not always welcomed, since they were considered too soft for the rationalists and too liberal for the church. He became a problem for the Prussian government, since he did not advocate for the absolutism of the monarchy and justification of divine revelations (Rohlf, 2023).

At the time, Kant was one of many in a complicated situation due to his beliefs and ideas. One of the philosophers who developed German metaphysics before Kant, Christian Wolff, lost his university position and was exiled due to his take on theology. Like Kant and Leibniz before him, Wolff (see also Chap. 7) was a philosopher with a strong background in physics and mathematics that posed troubles for an absolutist monarchy. Like Kant, Wolff came from a non-privileged background. Wolff's opinions about art were classical, linked to the Greek "ars" in "techne", being the aptitude and skills to perform a craft. Therefore, his definition of art included various activities, like woodcutting, medicine, agriculture and architecture. Wolff defined his philosophy of art as "technology" because to make art, the body (especially hands) uses tools to produce a work, in this case, a work of art. Wolff paid particular attention to architecture and even wrote a book *Elementa architecturae civilis* where "he treats architecture as a series of physical problems which are to be solved by the application of general physical principles" (Beiser, 2010: 52). The rationalism of Wolff is found precisely in this subjugation of all arts to the principles of physics through the concept of technology.

Curiously, one of Wolff's disciples, Baumgarten, was the first author to use the word aesthetics to study taste and its link to beauty. This speculative philosophy was a different way of using the concept of aesthetics, a word that in ancient Greece was used to define the doctrine of sensibility. Part of the effort of those of the time dealing with aesthetics was put into proving its legitimacy as a field. However, Kant criticised Baumgarten for trying to rationalise aesthetics using empirical principles, namely, by pointing out that something subjective, like taste, cannot obey a priori pragmatic rules as this means it would no longer be subjective. In contrast, Kant's point was that aesthetics was more complex. It could be subjective and still obey pragmatic rules. Tolstoy (n.d.: 64), biased by religious beliefs and morals, would later criticise Baumgarten for instigating a view of art that "justifies the false position in which a certain part of society is living" and that, therefore, became an "article of faith" for that sector of society.

The most influential philosophers who cemented the foundations of aesthetics, like Kant and Wolff, were from a non-privileged socio-economic background. They were not eccentric artists working for an elite that came out with the foundations of aesthetics but humble teachers with a strong background in mathematics, physics, philosophy and technology. Curiously, an architect of the same period, Sir Christopher Wren, also had a background in astronomy, mathematics and physics (see Chap. 10). Most importantly, philosophers like Kant and Wolff had high moral standards driven by strong ideas that challenged the political and religious status quo of their time. They were definitively not social media influencers.

Art, Architecture and Usefulness in Everyday Life

As explained in Chap. 7, understanding and classifying architecture helped define the profession. Kant stated that architecture is an art but also that it must work, which did not add much to what was already known since Plato, Vitruvius and Alberti, who also felt that architecture was an art providing the building was built to last and could accommodate people's necessities. However, Kant proposed that: "Architecture is the art of exhibiting concepts of things that are possible only through art, things whose form does not have nature as its determining basis but instead has a chosen purpose, and of doing so in order to carry out that aim and yet also with aesthetic purposiveness" (Kant & Pluhar, 1987: 191).

Kant implies that having aesthetics as a purpose, instead of a rational goal of accommodating people's necessities, made architecture an art. This echoes Pevsner's quote: "A bicycle shed is a building; Lincoln Cathedral is a piece of architecture. Nearly everything that encloses space on a scale sufficient for a human being to move in is a building; the term architecture applies only to buildings designed with a view to aesthetic appeal" (Ratcliffe, 2016). Kant also states that architecture is an art that must work. "In architecture, the main concern is what use is to be made of the artistic object, and this use is a condition to which the aesthetic ideas are confined" (Kant & Pluhar, 1987: 192). However, in Kant's view, aesthetics remains the purpose of architecture, and use only plays the role of being a limitation.

Kant perceived that "temples, magnificent buildings for public gatherings, or again residences, triumphal arches, columns, cenotaphs, and so on, erected as honorary memorials…we may even add to this all household furnishings belong to the architecture realm" (Kant & Pluhar, 1987: 191–192). Within the examples mentioned, cenotaphs, memorials and triumphal arches belong to the idea of architecture as pure art, while temples, magnificent buildings and residences lean towards the second statement of architecture as art that must work, buildings which people must use. The wealth ideology emerges from the examples; architecture is only those exceptional buildings in a city, while the most frequent small and humble houses are probably not considered in the residence category. However, tourists' experiences in the most visited cities are often related to the delight of the ordinary dwellings, whether the terraces and squares of Georgian London or the old merchants' houses of Amsterdam.

The SDGs imply that the built environment has a role in encouraging the development of sustainable communities, achieving inclusiveness and doing something towards ending poverty. However, the SDGs do not consider the implications of the aesthetic bias of those who have a role in building sustainable communities. If aesthetics is the final purpose of architecture and what makes it an art, and this only applies to certain buildings as proposed by Kant, how can architects contribute to achieving the SDGs for all? In a climate change era, can global concerns help question aesthetics and redefine architecture to make it more relevant and inclusive?

Aesthetics and Climate Change Action

One problem, especially in an era of climate change, is that the aesthetic and artistic value of images prevails over the function or necessity of an object or product. "Aesthetics invest objects with fascination and interest. People want to buy those things" (de la Fuente & Murphy, 2014: 8), even if they do not need them. Wealth and aesthetics intersect, when this fascination is paired with an aspiration to consume unique, expensive products. The aesthetics of wealth elevates the perception of an object. Whether a lemon squeezer, a computer or a house, if it looks expensive, it should be more aesthetically agreeable than anything that looks cheap and humble. The Juicy Salif citrus squeezer designed by Philippe Starck in 1990 has a form not usually associated with a lemon squeezer as, according to the product details its "long, spider-like legs are tall enough to fit your glass underneath—ready to collect your lemon, orange or lime juice". While held in different museums as an art object, it is still a lemon squeezer which, for US$87, should do an exceptional job to beat the regular handheld lemon squeezer that can be bought for only US$6. The problem is that it does not do a better job, because it normally makes a mess of the countertop, and the design does not facilitate the juice flow. Moreover, it is made of polished aluminium with larger embodied energy than lemon squeezers of glass or ceramics. According to the National Museums of Scotland (Raup, 2019):

> Many people buy the Juicy Salif just for the sake of its design, with no intention of ever using it to squeeze any citrus. Perhaps that's the point. In a publication released by Alessi the 25th anniversary of the design, Philippe Starck states that the original intent of the product was not to actually be functional but to spark conversation.
>
> And it certainly achieves that purpose. It is a popular culture icon, and still captures the imagination of the masses.

The aesthetic of wealth is the medium for consuming images and buying objects more related to wants than needs. Starck's lemon squeezer, was designed "to demonstrate that you appreciated design, a signifier of taste" (Heathcote, 2015). Only rich people can afford an overpriced kitchen gadget without the intention of using it. What is bought defines the consumer's taste and status, which is made possible because of the aesthetic value attributed to it by part of the community. In a society dominated by an aesthetic of wealth, the role of design is to seduce and fascinate consumers, so they want to pay more for things they do not need. There is a close link between the aesthetics of wealth and design; they nurture each other. The aesthetic imposed by the wealthy defines what will be wanted by consumers, and designers must deliver expensive and useless objects damaging to the environment to create such a need.

Geniuses and Architects

In Kant's aesthetics, fine art combines taste with genius. To clarify what this means, everybody has a taste that helps them perceive beauty, but not everyone can produce fine art. The production of fine art is reserved for genius, and Kant stated accordingly: "Fine art is the art of genius" (Kant & Pluhar, 1987: 307). Following Kant's statement, if architecture is considered a fine art designed by architects, then architects should be geniuses. The implications of this statement are relevant to the SDGs, because they are linked to the education and role of architects and designers within the built environment (see Chap. 10). However, whether genius in architecture is contributing to the SDGs is debatable. Are wealthy and well-known architects designing to reduce inequalities and create sustainable communities?

> The pursuit of genius in architecture is what has most contributed to the unstitching of our urban fabric, giving us those buildings in outlandish shapes and unsightly materials that take a chunk of the city and make it into somewhere else, as Morphosis did with New York's Cooper Square, or Zaha Hadid with the Port Authority Building in Antwerp. (Scruton, 2020)

Kant defines a genius as someone gifted by nature and through whom nature dictates the rules of art (Kant & Pluhar, 1987: 307). In his explanation, all arts have rules as a foundation. Without these rules, they would not be artistic products because they could be made randomly. However, within the world of artistic rules, Kant explains that the judgement of beauty in fine arts does not depend on rules or concepts, even though they have them, another strange contradiction. Whether unquestioned or misunderstood, the idea that fine art cannot explain its rules and that the judgement of beauty remains purely subjective has been inherited from Kant. Geniuses are part of an elite who set new rules for others to follow when, in fact, they could not have followed any of their own rules. All these ideas go against the classical understanding of the art of producing architecture, which used to be the outcome of following clear and rational rules, something that was useful in the past and could be extremely helpful in a climate change era with global concerns like ending poverty.

The myth of genius is deeply rooted in the life of architects, from their education to their professional careers and, with it, the myth of originality, uniqueness, obscure processes and arcane language. Considering that SDG 4 concerns having a quality education, what does it take to teach someone how to be a genius? Kant proposed that a genius has a unique talent that cannot be taught or learned and that "the foremost property of genius must be originality" (Kant & Pluhar, 1987: 175). Given this idea, in an architecture class, there are possibly a few geniuses to whom nothing can be taught, and nothing is worth saying to the rest of the class because genius cannot be taught. The myth of genius is deeply embedded in the practice and education of architects, which affects the profession and could lead the built environment to a genius environmental collapse. Since nature runs through their veins and gives the rules to the art they produce, they do not have to be accountable for it. It is important to highlight that for Kant, scientists cannot be geniuses because they can clearly show and explain rationally how discoveries were made (Kant & Pluhar, 1987: 175). With the excuse of genius, a tutor or a firm leader can encourage a total disinterest in logical and scientific explanations and a lack of empathy towards people and their contextual problems (Dickinson, 2020: 3–5). However, architecture can be taught to all students using ideas that can be explained and followed rationally, with clear rules, little by little, even with a humble pen and tracing paper.

In the climate change era, the egocentrism of geniuses has led to the production of luxurious buildings that consume more resources than necessary and do less for society. The planet is getting smaller by the minute due to the increasing human ecological footprint. There is no more room for architectural geniuses detached from people's problems and needs. However, there is ample room for humble, local architects to help those who need them most.

From Geniuses to Entrepreneurs

In a competitive market, a poor genius is not a genius. Would Steve Jobs or Elon Musk be considered gifted with visionary minds if Apple or Tesla had not succeeded commercially? After all, they did not invent anything. Electric cars have existed since the 1900s (see Chap. 4), and Musk simply financed the pre-existing Tesla car company (Edelstein, 2021). Computers existed before Steve Jobs was born, mostly thanks to Alan Turing and his team. However, Musk and Jobs behave like Kant's description of geniuses and are venerated as such simply because they made a lot of money. Therefore, you can buy the qualities and the demigod's power of a genius, as described by Kant, and demand to be treated as such, as long as your brand, labels and logos inundate the market. To do this, aesthetics and often ridiculously expensive prices are essential.

The narratives and aspirations of brands are aesthetically consumed through their labels, logos and images as displayed in their advertisements. "Labels matter", assert Eduardo de la Fuente and Peter Murphy (2014: 128), because they have stylistic value in a world classified according to aesthetic preferences. The perception of the stylistic value of electronic gadgets or clothing is influenced by the image projected by the marketing of brands, and their logos become the symbol of a community. People buy items from clothes to electronic gadgets following images, packaging design and logos. When it comes to clothing, there is a difference between buying the finished product and buying the raw material to make up for yourself. Cloth is not branded, and the consumer judges by cost and type suitable for the job in hand, feel (quality) and appearance (aesthetics). In contrast, the logo of a brand adds value to the product with the complicity of the consumer. Wearing a brand's logo on your clothes or electronic gadgets can create a sense of belonging to a community and even identify or show socio-economic status (the value of a Hermès scarf is discussed in Chap. 6). Curiously, most of the sports brands that are expensive in the developed world are produced in developing countries in working conditions that are hardly human, while consumers and influencers fight on the internet to defend their favourite brands.

Architecture is not an exception in the marketing of brands. Architectural firms that developed a style have become brands that can eventually add "value" to a building. In 2022, Zaha Hadid Architects (ZHA) paid £3.4 million in licensing fees to the Zaha Hadid Foundation to keep using the trademark name. In the last five years, ZHA has paid £11.9million (US$16 million) to the foundation (Hopkirk, 2022) to keep using the name of the deceased architect. This is a considerable amount of money. However, when a name becomes a trademark, it can boost real estate sales and open opportunities to work internationally in big developments.

L'Avenue Libertador, in Buenos Aires, Argentina, is the first project of Zaha Hadid in Latin America. The project and its construction are a collaboration between ZHA, Grupo Portland (the client), lacuzzi Colombo Arquitectos (project and building manager), BMA and Lopatin Arquitectos and Grupo Go Desarrollos. What is most discussed is the aesthetic of the façade, highlighting stylistic recurrences found in other ZHA buildings (BMA, 2024). By promoting the brand ZHA, the development can attract wealthy buyers and prepare them to pay more for the same. Wealthy firms can hire agencies to reinforce their images at a cost that someone else will have to pay. In L'Avenue Libertador, Conway+ was in charge of the brand identity, strategy, website design and development, brochure, coffee table book, interviews and a film (Conway & Partners, 2024). The aesthetic and branding in this type of architecture are so important that the façade of the building is the main element that nurtures it. The design of the rest of the building, what makes it useful, liveable and domestic, could be done or finished by associated firms if necessary. The idea that the uniqueness and talent of an architect can be judged by their capacity to design complicated facades is not new, but it has become essential for the image of architectural brands and their capacity to attract the wealthy.

When the name of an architectural brand is attached to images, projects are more expensive. The penthouse of the luxury condo building at 520 West 28th in New York, designed by ZHA, was listed in 2018 at US$50 million (Fig. 8.1). The rest of the apartments have been sold at an average of US$9.7 million.

The apartments are promoted as having a great view towards the High Line, a linear public park built on the disused freight rail line elevated above the streets in Manhattan West's Side. However, they are just apartments and nothing special from a design point of view, given that at least 15% of the total area in each apartment is

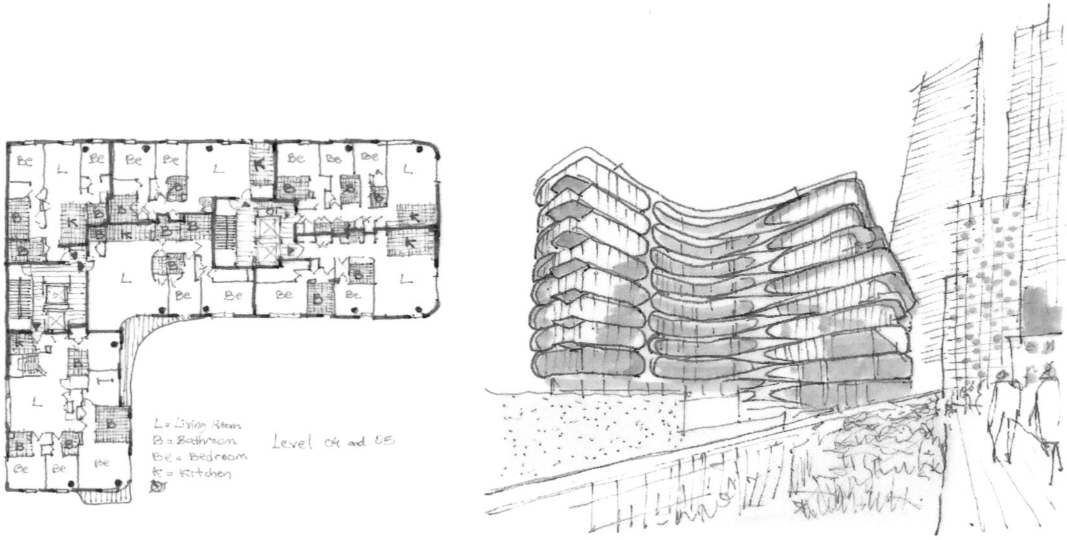

Fig. 8.1 ZHA 520 West 28th apartments: View and plan. (Source: Gonzalez (2018))

dedicated to circulation, floor area that could have been welcomed in the kitchen, dining room and living room. The east wing of the building has two apartments with very long corridors. Guests will have to go across the private and service areas of the house before arriving at the kitchen, dining room and living room. The service and corridor areas of the bedrooms are as big as the bedrooms, possibly inspired by hotel rooms. Half the apartments face South and have a favourable orientation, but the other half face North-East and North-West, creating a designed inequality. However, that will not be a problem for someone who can pay high electricity bills without caring about the environment.

Wealth and Taste

The impression is that taste is as vast as the colour palette, and everyone is entitled to have different tastes. However, taste can be influenced by socio-economic contexts. For the French philosopher Bourdieu, taste is typically bourgeois and limited by social status, income and education. In a survey on ideal homes and their interiors, participants were asked about the characteristics that describe ideal homes, like clean, tidy, practical or harmonious, and Bordieu weighted these according to their relevancy for different socio-economic groups, from manual workers to professionals. Aesthetic properties, like "harmonious", "imaginative" and "studied", were more important for high socio-economic groups, while functionalist characteristics like "clean", "easy to maintain", "practical" and "comfortable" were much more important for lower socio-economic groups. The survey confirms Bourdieu's hypothesis that taste is a socio-economic product (Bourdieu, 1984: 254–255). While lower socio-economic groups will spend more money on food, higher groups will spend it on how they present themselves to the world. For this reason, it is understandable that people from a privileged background cannot see a clean, comfortable or practical house as having any special value because, for them, this goes without saying. They are moved by something else. Therefore, it is not surprising that the wealth view of the role of architects, many of whom are from a privileged background, is to provide something else. The more you have, the more eccentric the taste for rarer things that are not necessarily practical or useful. The real luxury is having the capital to choose the useless over the practical.

> Tastes thus obey a sort of generalized Engel's law. At each level of the distribution, what is rare and constitutes an inaccessible luxury or an absurd fantasy for those at an earlier or lower level becomes banal and common, and is relegated to the order of the taken-for-granted by the appearance of new, rarer and more distinctive goods. (Bourdieu, 1984: 247)

Arches and columns were elements of classical architecture, frequently associated with big, expensive buildings, until they became accessories for the entrances of houses of wealthy families, making non-wealthy people more eager to include them in their houses, increasing their popularity, and decreasing their perception as a luxury only for the wealthy. These trends are no more than aesthetic impositions designed to please the insatiable taste for distinction of a minority. The search for the distinction of taste is a form of domination that establishes vertical differences within social classes to keep the wheels of consumerism turning (Glevarec, 2020: 40). Contrary to Kant, Bordieau proposed that there is nothing pure about the judgements of taste, since they are "preserved by, and preserve, the structure of capitalist society" (Young, 2006: 25).

The Tastes of Clients and Architects

Taste can be a conflict zone for architects when dealing with their clients. The look of the building is important for the image and brand of the architect and the client's expectations, and both may have different tastes. A factor that can level or tip the conversation is money. The architect is employed and paid for by a client, therefore, there is an implied responsibility for delivering a building that the client likes.

A colleague from Argentina once told the story of a discussion about taste with a client. The architect made a point about what he thought, as a professional, was the appropriate design for a detail in a house, while the client disagreed and had a different opinion. These classic conversations between clients and architects about taste can continue for a long time until one of the sides gives up. The client said: "Well, Architect, in the end, it is all just a matter of taste, and there is nothing written about taste", implying that there is nothing dogmatic about it. The tired and frustrated architect answered: "Excuse me Madam, but there is a lot that has been written about taste, even if you have read nothing about it". The architect had to apologise later. You cannot lose a client like that, especially when they pay your rent.

On one hand, the client had a good point since they have taste and are paying. Does this mean that the work of an architect should be to satisfy every demand from the client? In the example above, the architect felt it was their responsibility to define the aesthetics of the proposed house because they assumed they had seen, studied, and knew more about beauty and aesthetics in buildings than the client. In other words, architects feel they have a better or more refined taste than the client unless they are the client. In an interview, Victor Legorreta, the son, leader and heir of Ricardo Legorreta's firm, declared that "educating clients is nonsense because it sounds arrogant. It sounds like the architect knows more than the client does" (Belogolovsky, 2017). The fact that the client hired the architect in the first place should prove that they were looking for a professional, an expert in something they cannot do for themselves. They are also willing to pay for that service. However, where ego and money come together, everything is complicated.

In architecture, a challenging thing about taste is producing something that the client finds beautiful, and that other people do too, since everyone lives in the same place. Architects have social responsibilities other than a duty to themselves. The RIBA Ethical Practice Knowledge Schedule (RIBA, 2021), which tries to support ethical reasoning and decision-making, highlights that architects have a "duty to society and the end user" as well as a "duty to the wider world" (which includes the climate, biodiversity and sustainability), suggesting that what the architect designs needs to satisfy not just the client and the architect but everyone and the environment.

The Education of Taste in Architecture

In architecture, the taste developed by students is influenced by social media, design websites, magazines, books and the tastes and preferences of tutors and lecturers, especially if the latter are grading the work. The possibility of attending classes in traditional universities in developed countries, which means having the money to afford expensive fees in exclusive universities, will put these students in a privileged situation from the start.

One argument is that a student of architecture, urban design or planning at Harvard or other top-ranked universities will have a better education than a student at the University of Buenos Aires, Universidad Nacional Autónoma de Mexico or Sao Paolo (all of them public and free), simply because universities in developed countries tend to be better ranked than universities in developing countries (QS, 2024). Rankings only measure which universities are the wealthiest, those with the largest funding and donations, which are the most entertaining and have the largest campuses, and where research is written up in English. This has nothing to do with the quality of education but a lot to do with the wealth gap between institutions. Unfortunately, education based on a wealth gap across countries is reproduced within countries. Universities in capital cities or the largest city centres within countries usually tend to attract more students, funding and privileges and have wealthier students than smaller or poorer universities in the provinces or states. The aesthetic component is that graduates from the wealthiest universities will be seen, heard and perceived differently from the rest. The best thesis projects from the wealthiest universities are published in architectural magazines and websites and exposed worldwide; all eyes are on them, and they impose what all eyes should see. What is not always perceived is that the promoted work is often not producing something new but reproducing an aesthetic with a uniform taste.

Architectural Competitions, Aesthetics and Wealth

A good example of imposed uniformity is found in the winning projects of the competitions organised by the magazine *eVolo,* which describes itself as a "design journal focused on technological advances, sustainability, and innovative design for the 21st Century. Our objective is to promote and discuss the most avant-garde ideas generated in schools and professional studios around the world" (EVOLO, 2006a, b). It is important to highlight the language used, with the grouping of keywords "technological advance", "sustainability" and "innovation" without explanation of what this is supposed to mean except that it has to be "avant-garde". The magazine started its competitions in 2006, and records of the winning projects are available from 2009 to 2023 on its website.

The passion for the projects is an encouraging example of the work ethic of the students and tutors who support them. The competition entries were probably done after class, at night, or between work shifts. Such a workload and lifestyle are not healthy but are traditionally accepted by architects as necessary for any successful project. All the projects have good intentions or "aesthetic purposiveness", in the words of Kant. However, are intentions enough in architectural competitions? Should ideas be feasible or merely possible in the context of climate change?

In the eVolo competition entries from 2009 to 2023, 27 out of 45 (60%) first, second and third places were given to projects whose topics were concerned with environmental issues (EVOLO, 2006b). Regardless of the nationality of the winners, some projects are very difficult to imagine in any urban context, especially the flying buildings. However, it seems entrants are trying to save the world with skyscrapers. This starting point is a sophism since tall buildings have a higher environmental impact than high-density low-rise buildings (Pomponi et al., 2021). A building alone, even when presented as a giant and advanced gadget, cannot fix a city or solve the climate change problems of a civilisation. It is like thinking that buying the most expensive kitchen appliances will give your family better meals, make you a better cook and solve the food supply of your entire neighbourhood. In this context, the SDGs and their targets can be used to tick boxes and validate individual and isolated efforts to solve collective issues without questioning if the solution is feasible or could create more problems than it solves.

Would the entry projects look the same if they were houses for non-wealthy people drawn on paper by hand, with no colour? Is the imposed aesthetic of the competition giving an advantage to wealthy people? To produce images that correspond to the aesthetic of previous winners, competitors will need expensive computers with expensive software licenses, among other resources, without which the expected aesthetic cannot be produced. Competitors from a non-privileged background normally need a part-time job to study in universities (even when they are public and free). As such, they will have to work ten times harder to satisfy the aesthetic demands of the competition. In the case of the eVolo competitions from 2009 to 2023, almost two-thirds, 31 out of 45, or 68% of the first, second and third prizes were won by entries identified with developed countries. Projects from the USA alone won 10 prizes (22%) and it is the only country to win the first prize 4 times (26%) (China and Poland 3 times) (see Fig. 8.2). Projects from developing countries were placed first, second or third 14 times (29%), but half were from China (EVOLO, 2006b). The success of Poland is impressive for a relatively small country compared to Mexico, Brazil, Argentina, or the entire African continent, which have many more people and hundreds of schools of architecture but not a single winner in any category. Are they not good enough or simply not interested in participating in competitions that cannot do anything relevant for their people?

Money, human resources and time availability give an advantage to students and young professionals with a privileged background who live in developed countries where accessibility to electricity, the Internet and computers is not a problem or who attend universities whose facilities provide them with the resources needed. As in Formula One racing, the money invested in the car tends to overshadow the skills of the driver.

The problem with these competitions is that they limit competitors to reproducing an aesthetic instead of solving real problems, in real places by feasible means. Having a budget is a taboo idea, of course. Most of the projects are not avant-garde enough to deal with climate change. They

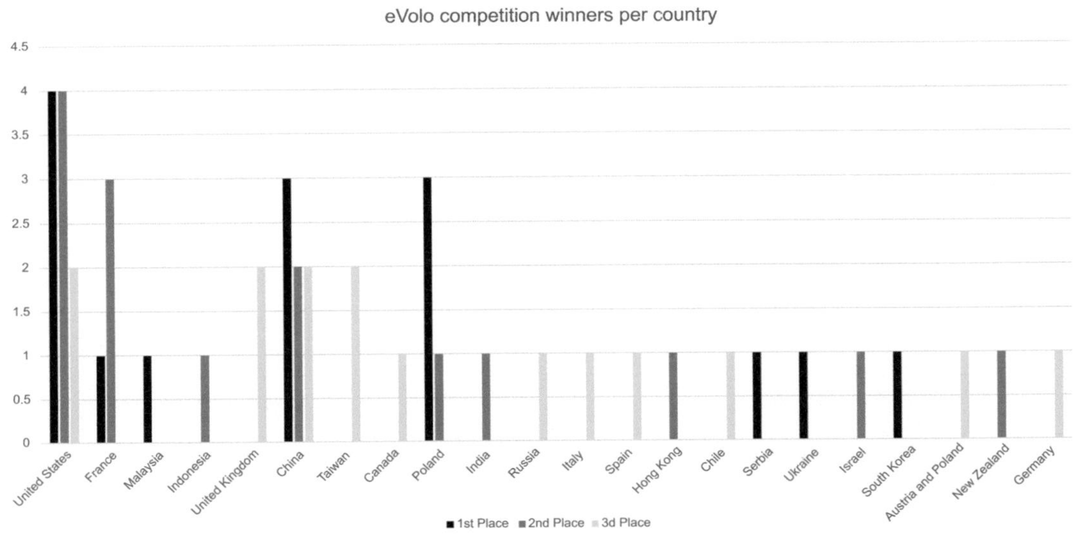

Fig. 8.2 eVolo competition winners per country. (Based on EVOLO (2006b))

have too much design and too little common sense. They repeat ideas from post-WWII architects, especially the English group Archigram and the Italian collective Superstudio. The aesthetic and influence of Archigram's Instant City can be found in the Elevated Plastic Skyscraper, to be situated somewhere in Africa. The Elevated Plastic Skyscraper is supposed to recycle plastic by turning it into building blocks that can be inflated once attached to the walls and later released as balloons. It is unclear how the balloons are supposed to be used or by whom after the towers decompose into a rain of plastic garbage. The winner from 2023 is a city in the air to avoid sea level rise, which their authors describe as "metabolism agglomeration". It is inspired by projects like Yona Friedman's Spatial City, Constant's Babylon City and the Metabolists' Tokyo Bay plan after the bombing of WWII. As young designers, there is nothing wrong with learning from precedents, like Archigram's Walking City (Fig. 8.3), but these designers need to understand that they were ideas from a different context without climate change mitigation urgency.

The post-war architects were optimistic about the technological possibilities in construction, communications and mass production, and they explored the new aesthetics of everyday life. The aesthetic was fundamental, because it allowed them to use diagrams, collages and colourful illustrations to propose utopian ideas that could stimulate the fantasies of a burgeoning middle class and, more specifically, other architects. All winning projects proposed megastructures, a term coined by Fumiko Maki (1964: 8) referring to "a large frame in which all the functions of a city or part of a city are housed", pretty much in the same fashion as Soleri's ideas of designing a building that is a city (Garcia & Vale, 2017: 76). However, the nature of the eVolo competitions does not seem to encourage simpler ideas that do more with less, using local materials and traditional construction methods, in other words humbler buildings.

Designers are subtly pushed to make their projects resemble scenes from classic science-fiction movies like Metropolis, Blade Runner, Star Wars or Aeon Flux (Fig. 8.4).

Every illustration must be a knockout punch for the jury members looking for projects with an

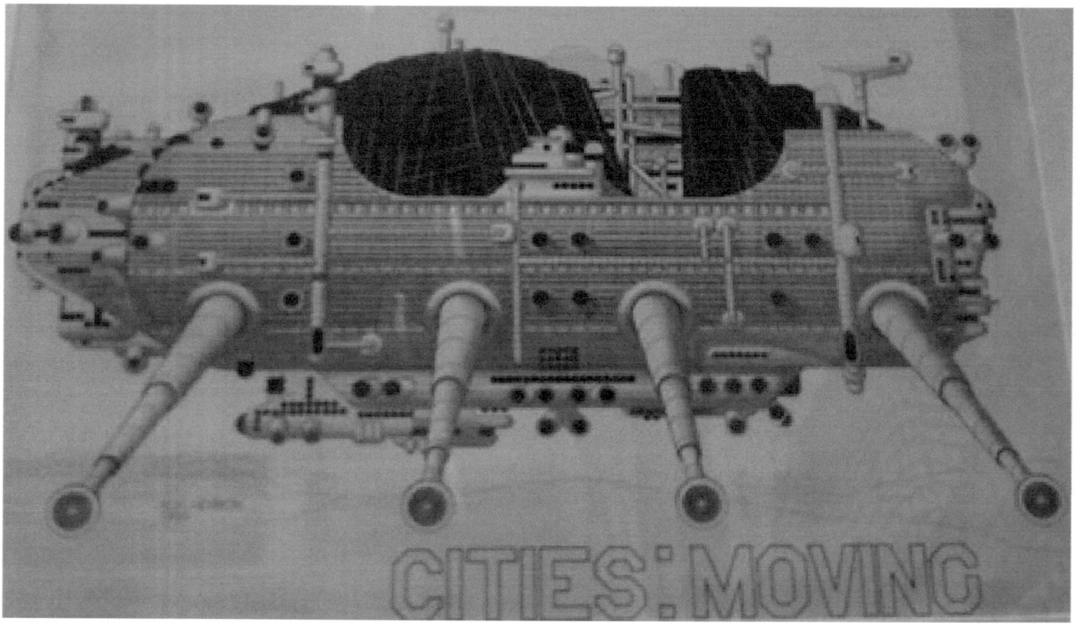

Fig. 8.3 Archigram walking city. (Source: Creative Commons (n.d.-a))

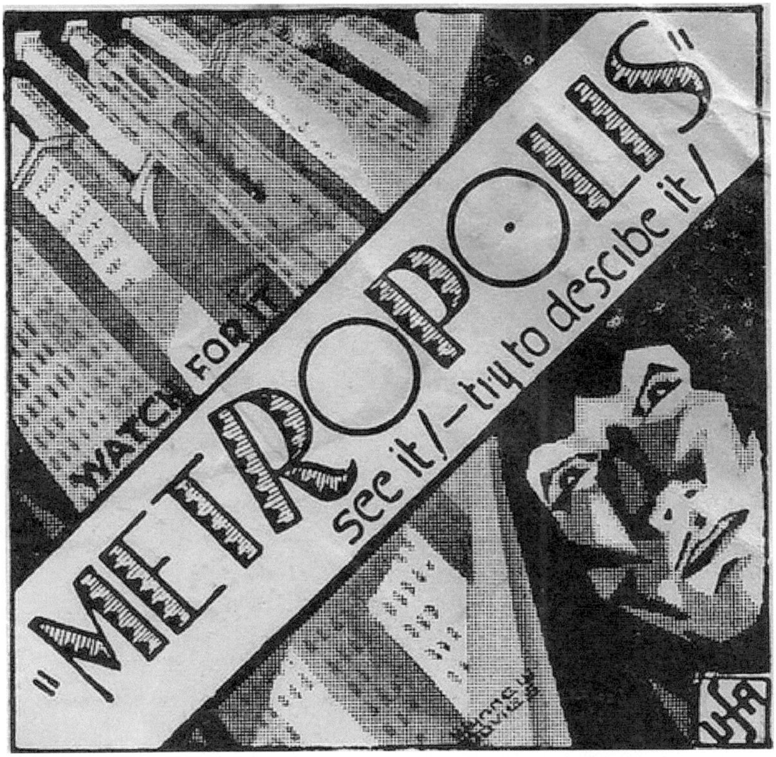

Fig. 8.4 Poster for 1927 film Metropolis. (Source: Creative Commons (n.d.-b))

"edge". There is no way to prove whether the projects are technologically advanced or sustainable, as claimed. The participants do not have enough space to create impressive illustrations and draw decent, clear, complete, logical floor plans and sections. Given this, they choose illustrations of the building over its architecture, because it must satisfy the jury's taste and influence their aesthetic experience and judgement, using Kant's terms. The only evidence of performance is in the aesthetic of the building, as to whether it looks sustainable or has the appeal of a technologically sound building. In this context, a technologically advanced look is largely based on the aesthetics created by Piranesi's study of prison interiors, Viollet-le-Duc's technical drawings of possible renovations of churches, and Ledoux's and Boulle's purism. None of these ideas, drawings and projects were ever built, nor did they help any poor families. However, their narratives and illustrations are similar to the conditions expressed by eVolo and can be easily aligned verbally with the intentions of the SDGs. How is it responsible, using SDG terms, to encourage future architects to waste their efforts, time and resources on producing and consuming these images instead of fixing one house for a real family in need? Part of the answer is another question: who can make a business out of simply fixing houses for refugee families or poor people?

Institutions, professionals and academics should make young designers aware that reproducing images that do not help to humanise architecture only contributes to commodifying it. Architectural competitions of visionary ideas use aesthetics as a medium to impose the belief that through technology and wealth, we can fix all problems, including climate change issues, without acknowledging that these are part of the problem. The way wealth uses aesthetics as a means of imposing these ideals affects and alters any possibility of responsible consumption and production, the goal of SDG 12 and, therefore, works against SDG 11 (Sustainable Communities and

Cities). All of these consequences are rooted in the ideological origins of aesthetics.

Ideology, Aesthetics and Responsible Consumption

Holub (1978) proposed the emergence of aesthetics can be linked with the beginning of the commodification of art. In this way, wealth and the commodification of art and architecture were made plausible through aesthetics, a field aligned with the ideology of an elite. Piketty proposes that an ideology is "a set of a priori plausible ideas and discourses describing how society should be structured. An ideology has social, economic and political dimensions. It is an attempt to respond to a broad set of questions concerning the desirable or ideal organisation of society" (Piketty, 2020: 3). This implies that a group of people have a pre-existing concept of an ideal society, usually shared among only some, which can be imposed on others. From this point of view, an ideology has no ethics and can be used to support noble ideals, like equity in a society, or to justify unreasonable inequalities based on imposed perceptions. To a certain extent, ideologies can be used as a criterion to judge plans, actions and decisions, and this is why they matter for the SDGs, which look ideologically neutral when they are not (see Chap. 1).

Contrary to Kant's proposal that aesthetic judgement should be done in a vacuum, aesthetics emerge from a context with unavoidable ideological content. Aesthetics have an ideological component, and ideologies use aesthetics to showcase their content. Any ideology that privileges wealth and the wealthy uses aesthetic means to do it. In the case of luxury (Chap. 6), the price of products influences their aesthetic perception. This reinforces the aesthetic value of the product, as in Starck's lemon squeezer that can only be bought by people with money, who already own a working lemon squeezer. When the acquisition of expensive objects provides status, other people may also want to buy them. For Veblen (1994: 75–77), through aesthetics, the taste of an elite is imposed on the rest of the society to define a sense of beauty, or confirm the expensiveness of the articles valued, which is largely responsible for nurturing the habit of conspicuous consumption and waste. What is seen and imposed as beautiful is driven by the taste and needs of an elite. In this way, the taste of an elite does not feel imposed, since it has been accepted and consented to from the moment something is found desirable and pleasurable.

Wealth is key, especially when the consumption of goods is linked to how people at higher socio-economic levels look or what they do and the desire to belong to that group. Without understanding the links between ideology, aesthetics and wealth, it is difficult to promote any responsible consumption independent from producing wealth for a few. The ideology of aesthetics explores what is in a taste, what people agree to and support when something is found agreeable, and why there is irresponsible consumption and production of goods, leading to the need for SDG 12 (Responsible Consumption and Production).

In the built environment, as ideologies take shape, they become buildings, streets, parks and gardens. The system of parks and boulevards still present in the urban framework of Paris was executed by Georges-Eugène Haussmann between 1854 and 1870 and had the purpose of social and political control, by allowing a fast deployment of military troops, circulation of goods and, therefore, more money for a few. The aesthetic urban design of Haussmann, perhaps the father of urban beautification and gentrification, hid the fact that thousands of people were displaced to make room for the growing new middle class and their businesses (Nelson, 2013: 103–119).

Aesthetics Means Money (But Only for a Few)

Currently, aesthetics means money, and money can buy an aspirational image and more aesthetics as long as it is economically successful. Sales can define the perception of aesthetics. An expensive product like those from luxury brands, or one that has suddenly sold many items, like Crocs footwear, must be considered aesthetically pleasant, because it has become an economic success that temporarily defines a trend or fashion. Featherstone (2007: 64) proposed that material goods are consumed as signifiers of taste and distinctive lifestyles. He continues (2007: 66) by noting that the "aestheticization of everyday life refers to the rapid flow of signs and images which saturate the fabric of everyday life in contemporary society." The aestheticisation process erases the boundary between reality and the image, leading to production and consumption that affect everyday life, including the production of the built environment. Images used to signify a lifestyle are usually based on what is supposed to look luxurious and what is used by an elite (see Chap. 6). The problem with inundating everyday life with aspirational images of wealth (or other drivers) is that the real thing gets lost in a hyperreality where the real cannot be distinguished from the image that simulates a reality (Baudrillard, 1981: 10).

The reality is that most people are not wealthy, even if some pretend to be by wearing an expensive brand or a fake version of it. However, this promotes the idea that the only way to be someone or part of a community is by pretending to be wealthy. This can lead to expenditure on unnecessary items and unaffordable goods. It is difficult to advocate for responsible consumption and production or any sustainable goal within a hyperreality dominated by the simulation of wealth, because consumers and producers are detached from the reality of the finitude of resources. The CEO of the LVMH group in North America, Anish Melwani, stated that their average customers do not care about sustainability. "They say they do, but when you actually get down to the brass tacks, we don't have people walking into our stores today and asking us, 'Hey, is this done sustainably?'" (Ravenscroft, 2023). The eldest son of billionaire Bernard Arnault, Antoine Arnault, who is also the head of environmental initiatives in the LVMH group, explicitly stated that LVMH would not sacrifice growth for environmental concerns (Boland, 2023). The group has made efforts to decrease its carbon footprint in its shops, but it is the supply chain where the highest impact resides. The promoted architectural efforts like the new Hermès maroquinerie (leather-goods shop) in Louviers, France, which got the E + C label, a certification given to buildings that make efforts to be energy efficient (Pagliacolo, 2023), helped to create a veneer of sustainability while the largest impact happens in the background. Projects, promises, intentions and investments in new buildings create an illusion of being sustainable that architecture makes possible.

Many practitioners and educators in architecture behave exactly like clients of luxury shops; they will say that they care about the environment, but when choosing between buildings that privilege aesthetics and buildings that are good for the environment, they always pick the aesthetic of wealth. While wealthy consumers only care about looking wealthy, wealthy businessmen only care about growing their businesses, and some designers only care about the aesthetics of wealth, it is very clear who does not care about the environment.

The Aesthetic of Wealth as Influencer

Although everyone with a phone and access to the internet can enjoy equal opportunities to exercise their aesthetic judgements, they do not necessarily have equal opportunities and means to consume what is now desired. Architecture and the design qualities of the built environment play a key role in any aspirational picture to be uploaded. From the architectural and interior design of spaces arranged to be photographed and look appealing on vacation rental websites to the design of Instagrammable buildings, aesthetic manipulation through the design of the built environment pushes up the value of the image consumed. On Youtube, the most viewed channels of children, as well as most of the toy gun battle channels, are all filmed in and showcase luxury houses.

The Nerf Blasters battle videos on YouTube have grown in popularity in the last six years. Nerf blasters are the marketing name for plastic guns that shoot bullets made of foam and are sold by the toy firm Hasbro. They can range from US$10 to more than US$100. The most popular Nerf War battles-related YouTube channel is PDK Films, with over 17 million subscribers and over 3 billion views (PDK Films, 2023). The built environment plays a key role in these. The size and appeal of the houses used in the videos are essential, because they are the set where the action happens. The wealth these houses exhibit can be showcased in the title of the video, like "Mansion Battle" (PDK Films, 2020), where the action happens in a house within an extremely large plot. This is not an exception, since most videos showcase houses with a modern style, always with a large pool, long corridors and a never-ending number of rooms. After renting and evaluating so many houses for their videos, it would not be crazy to speculate that in the future, they will be doing a real estate programme. In the meantime, children and teenagers receive only one message: a big wealthy house, full of excess, is the only means of having a decent playtime.

In 2015, Ryan Kaji's parents started a YouTube channel recording Ryan's playtime. He was three years old. Since then, it has become the most popular channel for children. In 2023, he became a 12-year-old millionaire with 35.5 million subscribers (Ceci, 2023), one of the highest-paid YouTubers. When they moved to Honolulu, the house included several bedrooms, two floors and a big pool. The size of the kitchen is very generous by traditional European standards, considering that it also has an additional space with a sink and storage, but his mom found it disappointing not to have an extra space for a pantry (Kaji Family, 2021a). The built environment plays such an important role that in the same year, they temporarily moved to another house in Los Angeles with more features. This time, the kitchen was bigger and the pool had an artificial waterfall. The master bedroom could be the envy of any four-star hotel. The ensuite bathroom was the size of a small bedroom with a shower for two people plus a bath (Kaji Family, 2021b). This is the background of Ryan's YouTube life, and what children will perceive as the standard of a normal house. Anything less than this would be strange. Do not be surprised if, out of nowhere, children ask: are we poor?

The Aesthetic of Wealth Turns Designers into Social Media Influencers (Against the Environment)

In the social media era, the power of deciding by aesthetics means what is beautiful, desirable, arousing and likeable becomes a commodity since likes, subscriptions and followers have a price. In the case of influencers and social media aficionados, the only condition necessary for monetising the production of social media content is to grab people's attention constantly. Why and how do people influence other people and persuade them to do or buy things?

The need for validation, appreciation and a sense of belonging to a community are key drivers for photo-sharing content on social media. There is a compulsion to share what is personally appreciated, satisfaction in finding others equally interested and relief in their approval. Gaining followers is the way of having a concrete reward and developing a reputation within a community (Serafinelli, 2018: 59). This is precisely the side that can be commodified and manipulated and has an aesthetic linked to wealth that makes all this possible.

Kevin Systrom, a co-founder of Instagram, studied at Stanford University, which has an annual tuition fee of US$61,731 plus expenses for accommodation and living, making up to US$87,833 per year (Stanford University, 2024). He graduated from a private secondary school whose tuition is US$72,050 per year, plus US$1200 for book spending (no additional fees) (Middlesex, 2024). Before starting the start-up, Kevin Systrom received almost a million dollars from his parents just for education, hardly a humble beginning. The fantasy stories of start-ups are built on money and privilege that help find wealthy investors. When it was bought and included in the smartphone Apple iPhone 4, which had new upgrades in its camera, Instagram was used and reached by millions of people

(Serafinelli, 2018: 56). Wealthy people developed Instagram to seduce wealthy entrepreneurs from Silicon Valley using expensive smartphones. Evidently, wealth can allow a disconnection from reality to produce things nobody needs.

The designers of the wealthy have been acting as influencers long before the creation of social media. The architects of the wealthy are used to competing for people's attention since most architects work for wealthy clients, and to do this, they need to build a reputation, a name and a successful business; in other words, to catch the attention of wealthy people. Through social media, designers now have a way of crying out for attention that can reach a larger virtual community than formerly, which sounds like a win–win situation for those looking for attention. This can also mean accepting and producing the same images for the same target group without solving any social or environmental problem. Basically, it is a branding and marketing exercise using buildings. What makes a building Instagrammable is its ability to stand out aesthetically and cry for attention, so people stop and take a picture of it. "The best way to set yourself aside from your competitors on social media is, therefore, by providing your guests with a visual sense of amazement, creativity and fun the moment they step into your space" (Valé, 2018: 4).

The media platform is not the true problem, but how people, and in this case designers, use such platforms and for what purpose. It is good to hear the voices and see the work of talented people from different countries that, in the past, could not be known unless published. There is more diversity in media resources, but with them, there is more confusion. The Internet and social media are excellent environments for spreading lies and confusing and dividing people. Conspiracy theories against COVID-19 vaccines or promoting buildings as if they were sustainable find fertile soil in the connectivity given by the internet. There is a lot of content, but it means very little without a sense of social and environmental justice.

In the long term, constantly crying for attention is unhealthy and can negatively impact the environment. It is very common behaviour in children between three and seven years old, who can be loud or annoying to grab their parents' attention. This attention-seeking behaviour manifests because children do not yet have the maturity to distinguish and accept the difference between needs and wants. While designers who can make a change remain self-centred, crying for attention like children and focused on what they want for their designs, they forget the planet and people's needs. Hopefully, the new generation is coming and has already been striking on the streets and participating in protests to wake us up. We have to question the aesthetics of wealth and its ideology to show the new generation that making pretentious buildings is not all architecture can offer.

The Dollar-Green Aesthetic of "Sustainable" Buildings

Even though just a wall with colourful patterns, greenery or graffiti can do the trick of grabbing attention (flytographer, 2023). It is aspirational and pretentious scenarios that make images more Instagrammable. Buildings with exotic views in places where flight tickets are expensive or, simply, pictures of private pools tend to be identified with living at a higher socio-economic level. People take more pictures when they travel, possibly because they feel in a good mood, which increases the need to share that feeling and post images on their preferred social media. This is especially true, when tourists travel to places of scenic beauty and with stunning views, colourful streets and unique architectural styles, these being factors that are found in the most Instagrammed places in the world (normally in the background of a selfie), from Notting Hill in London to the Opera House in Sydney and the Marina Bay Sands hotel in Singapore (Pradeep, 2023). However, being a tourist and posting inspirational images is costly both economically and environmentally. There are cheaper ways of travelling, like applying for a work holiday visa and working for a year in a country while living in a van. However, there is still the need to pay for flight tickets, which are very expensive, especially for young people living in developing countries who have to work to pay their rent, study or contribute to the economy of their households. Hence, taking a sabbatical year before starting work, something that has existed since the Grand Tour, is a luxury and privilege only for those who can afford it.

> The rigid form of it [the Grand Tour] that we associate with the eighteenth century was actually the popularizing of a travel phenomenon in which the British went to the Continent for travel and study to prepare for careers as diplomats and public servants, but the Tour quickly evolved, taking various forms though the places visited remained largely similar. A tour of the Continent was seen as the ideal means of imparting culture, taste, knowledge, self-assurance, and polished manners. (Colletta, 2015: ix)

Travel today may still be associated with such high ideals but making sure your friends instantly know where you are and what you are looking at seems equally, if not more, important.

Most people who travel do not sleep in used vans. They go to Instagrammed places and sleep in hotels. Tourism, hotels and Instagram were meant to be together. As an example, the Marina Bay Sands hotel and casino in Singapore has it all—imageability, pools and views. On its Instagram account, the hotel promotes itself as having the world's largest infinity pool (Fig. 8.5) (marinabaysands, n.d); however, it does not mention the casino, which is hidden in the menu under "fun & games", even though it is very big at 160,000 sq. ft. (a little under 15,000m^2) and is the largest revenue contributor to the complex, producing 55% (US$756 million) of the total revenue (US$1.37 billion) (Zhu, 2023). Perhaps the word casino is unpopular among influencers, but the Marina Bay Sands is definitely a casino with, pool, shops and pricey bedrooms. Room prices range between US$840 and US$1619 per night (Marina Bay Sands, 2023). Compared to the average Singapore salary of US$4315.62 per month, each night in the hotel represents between 19% and 37% of the average month's salary. In Singapore, it is possible to rent a one-bedroom apartment with an in-house maid for a month or

Fig. 8.5 Marina Bay Sands pool. (Source: Creative Commons (n.d.-c))

an apartment with two bedrooms and two bathrooms for less than one night at the Marina Bay Sands (PropertyGuru). People cannot even have free access to the SkyPark at the top of the complex, so how can it be an example of achieving any Sustainable Design Goal?

One does not have to be an engineer to guess that adding the weight of that pool at the top of three very tall buildings makes the structure more expensive and uses more resources. Considering that the building is a hotel with a casino, a luxury that many cannot afford and many other facilities to maintain, including the largest infinity pool in the world, it is perhaps surprising that the building is considered an example of sustainable tourism. In 2023, Marina Bay Sands (MBS) was accredited by the Global Sustainable Tourism Council (GSTC, 2024), an organisation formed in 2007 in collaboration with 32 partners initiated by the Rainforest Alliance and several programmes from the United Nations, including United Nations World Tourism Organization, all of them attempting to create "a common understanding of sustainable tourism" (GSTC, 2022). Some of the key points of the press release were that MBS has reduced its carbon footprint by 50% since 2012 due to integrated systems that help to use less energy, and they are wasting less food as well as sponsoring events to raise money for charity (Marina Bay Sands, 2023b). The narrative is that if you visit MBS, you are part of a selected group of "environmentally conscious travellers", as explained on the Forbes website (Indra, 2023), and that wealth and technology can make luxury tourism look sustainable.

However, there is no evidence that MBS is sustainable, since there are no available numbers and figures that discuss how the carbon footprint was measured, nor the data used, only reports with hundreds of pages with an aesthetic narrative of responsibility. The GSTC Industry Criteria with Performance Indicators for Hotels and Accommodations (GSTC, 2016) has two sections: demonstrate sustainable management and maximise social and economic benefits while minimising negative impacts. Many criteria have indicators, every one of them aligned with SDGs, but there is not a single concrete way of measuring anything; methods and data are not available. The demonstration of sustainable management is done with words and by buying technological gadgets. The social aspects are covered by highlighting how many people they employ and by sponsoring events. Since the SDGs have parameters that tend to measure economic factors, targets may be more easily achieved when a building can be a good business without the certitude of achieving any sustainability.

During the COVID-19 period, the world did well regarding emissions, because everybody travelled less domestically and internationally, which proves that tourism is intrinsically not sustainable. Having a building promoting gambling is not the most socially responsible thing to do for a community. There is no need for complex certification with their boards and councils to understand what is fundamental. The rest is just aesthetics.

Concluding Thoughts: The Aesthetic Purpose of Wealth

Before Kant, it was very clear that the beauty of buildings was dependent on their capacity to last and their utility for people. The more abstract aspects of aesthetics were linked to mathematical proportions, which could be explained rationally. After Kant, aesthetics became so complicated and inaccessible that it could only be approached through interpretations. The criticism of the ideological aspects of aesthetics found it vulnerable to manipulation. Aesthetics stimulate modes of production and consumption that consolidate elitist knowledge and capital accumulation in fewer hands. It frequently uses an obscure understanding of philosophy as opposed to Vitruvius, who wrote the following (Pollio, 2009: 9):

> Philosophy, in fact, makes the architect high-minded and ensures he will not be arrogant, but rather flexible, fair and trustworthy without greed, which is the most important quality, since no work can be carried out satisfactorily without loyalty or integrity; and [philosophy also ensures] that he will not be avaricious or preoccupied with receiving awards.

Young professionals are being pushed to sacrifice their family time to participate in competitions that have no sense to make a name in social media and, eventually, to make a living. Tutors preaching and encouraging their students to keep drawing visionary architecture are fomenting and reproducing the ideology of wealth. Wealth imposes the idea that innovation in architecture is only attainable through implementing the latest technological gadgets by attracting and driving other designers towards aesthetic mannerisms considered a new trend, a new fashion or both. Such architectural innovation is an obstacle to climate change mitigation even when the themes used are inspired by or allegedly intended to fix climate change issues. Because the aesthetic of wealth always asks for more, never less, projects must be more structurally complicated and sacrifice the environment, functionality and cost to have an edge over other projects. Feasibility is less important than the illusory potential of an architectural idea. It is this illusion that nurtures the aspirational quality of the aesthetic of wealth. The aesthetic of wealth has this expanding effect of cascading down and seducing people, so they can all passively like the same thing. Consumerism has made possible Kant's proposal of a universal agreement on aesthetic judgements: taste means expensive, and beauty means wealth.

In a free market, the aesthetic experience is the experience of buying someone else's taste. Poorer Schools of Architecture buy the taste of wealthier Schools of Architecture. Small Schools of Architecture buy the taste of bigger Schools of Architecture, just as Schools of Architecture from colonised countries buy the taste of their colonisers. They buy their books, magazines, ideologies and rules. They pay for accreditations to belong to things imposed; they click on the advertisements on their websites. There is an entire periphery of non-wealthy consumers of a wealthy aesthetic, forming a big chunk of the world that has been aesthetically excluded. Perhaps, the American writer Bukowski (2009) had a point when he suggested that "bad taste creates many more millionaires than good taste". Whether the aesthetic of wealth can be ethically justified is explored in the next chapter. Does something have to look different to be good for people?.

References

ACAN. (2022). *ACAN responds to RIBA stirling prize—Stop celebrating architecture that is bad for the planet.* Retrieved from https://www.architectscan.org/post/acan-responds-to-riba-stirling-prize-stop-celebrating-architecture-that-is-bad-for-the-planet, 19 January 2024.

Baudrillard, J. (1981). *Simulacres et simulation*. Editions Galilée.

Beiser, F. C. (2010). *Diotima's children: German aesthetic rationalism from Leibniz to Lessing*. Oxford University Press.

Belogolovsky, V. (2017). *Victor Legorreta: "Sometimes, Architects Take Themselves Too Seriously"*. ArchDaily. Retrieved from https://www.archdaily.com/883766/victor-legorreta-sometimes-architects-take-themselves-too-seriously, 6 February 2024.

BMA. (2024). *L'Avenue.* Retrieved from https://bmaestudio.com.ar/proyectos/lavenue/, 17 June 2024.

Boland, H. (2023). *We won't sacrifice growth for environmental concerns, says billionaire Louis Vuitton heir.* The Telegraph. Retrieved from https://www.telegraph.co.uk/business/2023/12/20/wont-sacrifice-growth-for-environment-louis-vuitton/, 16 January 2024.

Bourdieu, P. (1984). *Distinction: A social critique of the judgement of taste*. Harvard University Press.

Bukowski, C. (2009). *Hollywood.* Canongate Books.

Carroll, L. (2021). Gentrification and the critique of the contemporary urban dream-world. In T. Tunali (Ed.), *Art and gentrification in the changing neoliberal landscape* (pp. 21–38). Routledge.

Ceci, L. (2023). *YouTube: Most subscribed kids content channels 2023*. Statista. Retrieved from https://www.statista.com/statistics/785626/most-popular-youtube-children-channels-ranked-by-subscribers/, 1 December 2023.

Colletta, L. (2015). *The legacy of the grand tour: New essays on travel, literature, and culture*. Fairleigh Dickinson University Press.

Conway and Partners. (2024). *L' Avenue Libertador/Portland Group*. Retrieved from https://conwayandpartners.com/projects/l-avenue-libertador/, 17 June 2024

Creative Commons (n.d.-a). "Archigram walking cities" by luisvilla is licensed under CC BY 2.0.

Creative Commons (n.d.-b). "Metropolis (1927 film)" by Archives New Zealand is licensed under CC BY 2.0.

Creative Commons (n.d.-c). "Infinity Pool, Marina Bay Sands, Singapore" by silaskhua is licensed under CC BY 2.0.

De Castro, O. (2023). *Dress To Impress: Using Personal Style To Support Your Career.* Retrieved from https://www.forbes.com/sites/forbesbusinesscouncil/2023/08/25/dress-to-impress-using-personal-style-to-support-your-career/?sh=7702847e4c64, 14 February, 2024.

de la Fuente, E., & Murphy, P. (2014). *Aesthetic capitalism*. Brill.

Dickinson, D. (2020). Christopher Alexander and the inadequacy of genius in the architecture of the coming age. *Urban Science, 4*(2), 17.

Dixon, S. (2023). *Average daily time spent on social media worldwide 2012–2023*. Statista. Retrieved from https://www.statista.com/statistics/433871/daily-social-media-usage-worldwide/, 30 November 2023.

Edelstein, S. (2021). *Tesla existed before Elon Musk: Founders on how they pitched the idea*. Retrieved from https://www.greencarreports.com/news/1131215_tesla-existed-before-elon-musk-founders-on-how-they-pitched-the-idea, 2 September 2024.

EVOLO. (2006a). *About*. Retrieved from https://www.evolo.us/about/, 8 February 2024.

EVOLO (2006b). *2024 Skyscraper competition*. Retrieved from https://www.evolo.us/category/competition/, 8 February 2024.

Featherstone, M. (2007). *Consumer culture and postmodernism*. SAGE Publications.

Flytographer. (2023). *Top 10 Instagrammable Walls Around the World*. Retrieved from https://www.flytographer.com/blog/instagrammable-walls/, 24 January 2024.

Garcia, E., & Vale, B. (2017). *Unravelling sustainability and resilience in the built environment*. Routledge.

Glevarec, H. (2020). La distinction n'est pas une différenciation: Caractéristiques de la distinction sociale selon P. Bourdieu et critères d'une différenciation contemporaine. *Recherches Sociologiques Et Anthropologiques: RS&A, 51*(1), 39.

Gonzalez. (2018). *520 West 28th/Zaha Hadid Architects*. Retrieved from https://www.archdaily.com/891620/520-west-28th-zaha-hadid-architects, 15 February 2024.

GSTC (Global Sustainable Tourism Council) (2016). *GSTC Industry Criteria. Performance Indicators for Hotels and Accommodations and corresponding SDGs*. Retrieved from https://www.gstcouncil.org/wp-content/uploads/GSTC-Industry-Criteria-for-Hotels-with-SDGs.pdf, 8 February 2024.

GSTC. (2022). *History of GSTC*. Retrieved from https://www.gstcouncil.org/about/gstc-history/, 8 February 2024.

GSTC (2024). *Marina Bay Sands becomes certified to the GSTC Industry Criteria*. Retrieved from https://www.gstcouncil.org/marina-bay-sands-becomes-certified-gstc-industry-criteria/, 8 February 2024.

Heathcote, E. (2015). *Design horrors: The bad, the ugly and the dysfunctional*. Retrieved from https://www.ft.com/content/ff047176-7805-11e5-a95a-27d368e1ddf7, 14 September 2023.

Holub, R. C. (1978). The rise of aesthetics in the eighteenth century. *Comparative Literature Studies (Urbana), 15*(3), 271–283.

Hopkirk, E. (2022). *Zaha Hadid Architects' £12m bill to use founder's name*. Retrieved from https://www.bdonline.co.uk/news/zaha-hadid-architects-12m-bill-to-use-founders-name/5115600.article, 5 February 2024.

Indra, C. (2023). *The Art Of Responsible Tourism: Travelers Can Be A Force For Good*. Retrieved from https://www.forbes.com/sites/forbesbusinesscouncil/2023/08/25/the-art-of-responsible-tourism-travelers-can-be-a-force-for-good/?sh=3eb9b58826a5, 8 February 2023

Kaji Family. (2021a). *Ryan's New Hawaii House Tour*. Youtube. Retrieved from https://www.youtube.com/watch?v=Pha-zU3tB4Y, 1 December 2023.

Kaji Family. (2021b). *Ryan's New LA House Tour*. Youtube. Retrieved from https://www.youtube.com/watch?app=desktop&v=PLOOx3cGuhk, 1 December 2023.

Kant, I., & Guyer, P. (2000). *Critique of the power of judgment*. Cambridge University Press.

Kant, I., & Pluhar, W. S. (1987). *Critique of judgment*. Hackett Pub. Co.

Maki, F. (1964). *Investigations in collective form*. Washington University.

Marina Bay Sands. (2023). *Book your Stay*. Retrieved from https://www.marinabaysands.com/booking/rooms.html?Rooms=1&CheckinDate=2024-02-05&LOS=1&Adults_1=1&Children_1=0&ChildrenAge_1=&Currency=USD&flow=tf&multi=false&offerCode=, 24 January 2024.

Marina Bay Sands. (2023b). *Marina Bay Sands, Singapore's largest hotel, earns global sustainability certification*. Retrieved from https://www.marinabaysands.com/content/dam/revamp/company-information/media-centre/2023/feb/largest-hotel-earns-global-sustainability-certification.pdf, 8 February 2024.

Middlesex. (2024). *Facts and figures*. Retrieved from https://www.mxschool.edu/about-mx/facts-and-figures/, 24 January 2024.

Nelson, B. (2013). *The remaking of Paris: Zola and Haussmann*. Cambridge University Press.

Pagliacolo, E. (2023). *Lina Ghotmeh Unveils a Bioclimatic Workshop for Hermès*. Azure. Retrieved from https://www.azuremagazine.com/article/lina-ghotmeh-hermes-workshops/, 31 January 2024.

PDK Films. (2020). *Nerf War: Mansion Battle*. YouTube. Retrieved from https://www.youtube.com/watch?v=lYW4I6QsZuc, 1 December 2023.

Piketty, T. (2020). *Capital and ideology* (A. Goldhammer, Trans.). Harvard University Press.

Pollio, V. (2009). *On architecture* (R. Schofield, Trans.). Penguin UK.

Pomponi, F., Saint, R., Arehart, J. H., Gharavi, N., & D'Amico, B. (2021). Decoupling density from tallness in analysing the life cycle greenhouse gas emissions of cities. *Urban Sustainability, 1*. article 33. 10pp.

Pradeep, S. (2023) *The 22+ Most Instagram Travel Destinations in the World 2024*. Retrieved from https://amazingworld.travel/top-instagram-travel-destinations/, 24 January 2024.

Ratcliffe, S. (2016). *Nikolaus Pevsner 1902-83*. Oxford essential quotations: Oxford University Press. Retrieved from https://www.oxfordreference.com/view/10.1093/acref/9780191826719.001.0001/q-oro-ed4-00016323. 5 February 2024.

Raup, E. (2019). *The most controversial lemon-squeezer in history?* Retrieved from https://blog.nms.ac.uk/2019/08/14/lemon-squeezer/#:~:text=The%20design%20simply%20is%20not,Perhaps%20that's%20the%20point. 14 September 2023.

Ravenscroft, T. (2023). *"Average customer doesn't care terribly much about sustainability" says LVMH US head.* Retrieved from https://www.dezeen.com/2023/12/14/anish-melwani-interview-lvmh-us-head/, 16 January 2024.

RIBA (2021). *The six duties of care underpinning the RIBA Ethical Practice competence.* Retrieved from https://www.architecture.com/knowledge-and-resources/knowledge-landing-page/the-six-duties-of-care-underpinning-the-riba-ethical-practice-competence, 5 February 2024.

Rohlf, M. (2023). *Immanuel Kant*, the Stanford encyclopedia of philosophy (fall 2023 edition), Edward N. Zalta & Uri Nodelman (eds.). Retrieved from https://plato.stanford.edu/cgi-bin/encyclopedia/archinfo.cgi?entry=kant, 5 February 2024.

Scruton, R. (2020). *The problem with architectural "Genius"*. Retrieved from https://theimaginativeconservative.org/2020/09/problem-with-architectural-genius-timeless-roger-scruton.html, 3 September 2024.

Serafinelli, E. (2018). *Digital life on Instagram: New social communication of photography.* Emerald Publishing Limited.

Stanford University. (2024). *The student budget*. Retrieved from https://financialaid.stanford.edu/undergrad/budget/index.html, 24 January 2024.

Tolstoy, L. (n.d.). *What is art?* Walter Scott, Ltd.

United Nations. (n.d.). *Goal 11: Make cities inclusive, safe, resilient and sustainable.* Retrieved from https://www.un.org/sustainabledevelopment/cities/, 20 June 2024.

Valé. (2018). *Instagram Design Guide.* Valé Architects. Retrieved from https://valearc.com/shop-1/instagram-design-guide, 24 January 2024.

Veblen, T. (1994). *The theory of the leisure class*. Dover Publications.

Waite, R. (2017). *Feature: Seven clients reveal what they really want from architects.* Architects Journal. Retrieved from https://www.architectsjournal.co.uk/news/feature-seven-clients-reveal-what-they-really-want-from-architects, 16 January 2024.

Welsch, W. (1997). *Undoing aesthetics*. Sage Publications.

Young, D. A. (2006). The fine arts of distinction: Bourdieu on society and art. *Australian and New Zealand Journal of Art, 7*(2), 23–44.

Zhu, M. (2023). *Marina Bay Sands Q3 revenue jumps 34.3% to $1.37 billion as tourism recovers*. The Strait Times. Retrieved from https://www.straitstimes.com/business/marina-bay-sands-q3-revenue-jumps-343-to-137-billion-as-tourism-recovers#:~:text=Its%20casino%20segment%20remained%20the,the%20same%20quarter%20in%202022. 24 January 2024.

9 Ethical Dilemmas of Wealth in the Built Environment

What Is the Link Between Wealth and Ethics?

In his *Politics,* Aristotle distinguished between two notions of wealth: the natural and the non-natural. Natural wealth is related to managing the needs of a household through acquisitions that provide families with the "necessaries of life to exist", such as the provision of food (Aristotle, 1885: 14). This is the true wealth that Aristotle considers limited to the domestic realm, making it necessary and honourable (Aristotle, 1885: 19). For Aristotle, true wealth, from the "right sources" (Beresford, 2020: 76) produced within the boundaries of your property, has limitations because it is linked to finite natural sources and whether a fair amount can be given to all. These honourable conditions are relevant in the current context of environmental crises. The natural wealth of Aristotle could be linked with sustainability ideas associated with "living with the means at hand" (Gracia & Vale, 2017) and even with the production of an autonomous architecture (Vale & Vale, 1975) which not only consumes but tries to help families to produce energy and food (see Chap. 10 for an example). Where wealth comes from, how it is used and by whom, meaning by what type of person, matters for the ethical use of wealth if the sustainability of natural resources is to be taken seriously. However, there is no possible sustainable design discussion if there is a belief in wealth as a life purpose or life's only measure of success. For this reason, the ethical approach to wealth is of great concern for thinking about, teaching and practicing architecture.

Non-natural wealth is the "art of making money" (Aristotle, 1885: 15) through retail trade or exchanges that involve "coins". As you can perceive, Aristotle would not have been a great admirer of cryptocurrencies, since they are still coins. Aristotle considers non-natural wealth unnecessary because it is based on always desiring more money without limits. This form of wealth, where "men gain from one another" is censurable and resembles usury (Aristotle, 1885: 19). The main difference between natural and non-natural wealth is in their ends. Natural wealth serves families to live well; however, the only pursuit of non-natural wealth is to make more money, or as Aristotle expressed it, "the birth of money from money" (Aristotle, 1885: 19). Therefore, for Aristotle, economics that only considers wealth creation and ethics were irreconcilable (Meikle, 1994: 41). To some extent, concealing money with politically correct ends is what the SDGs are trying to accomplish. However, as stated by Aristotle, a critical misunderstanding is assuming wealth as an end or, in the case of the SDGs, a goal.

> As for the money-making life, it is something quite contrary to nature; and wealth evidently is not the good of which we are in search, for it is merely useful as a means to something else. So we might

E. J. García, B. Vale, *Wealth and the Built Environment*, Sustainable Development Goals Series,
https://doi.org/10.1007/978-3-031-81954-4_9

> rather take pleasure and virtue or excellence to be ends than wealth; for they are chosen on their own account. But it seems that not even they are the end, though much breath has been wasted in attempts to show that they are. (Aristotle, 1885: 8)

From this point of view, wealth is just a means to an end that can be positive and useful or unnecessary and negative, depending on the intentions and aims of the wealthy person. In Aristotelian logic, virtue and the virtuous use of wealth are tightly bound to the point that actions, along with their ends, define people's ethics and vice versa. The implication is that a reasonable person does good things with wealth, but the improper use of wealth is done by people who are not virtuous. From Aristotle's reflections, it is possible to derive a series of categories for analysing the impact of wealth. There are three essential elements: means, ends and people. These elements are susceptible to being explored through a series of questions. Why pursue wealth (ends)? Who owns it (people)? Where does wealth come from (means)? How is wealth used and spent (means) and for whom (ends)?

How wealth is produced, according to Aristotle, sheds some light on its ethics and perception by differentiating ways of using and acquiring it. Using money means spending money or giving it away, but acquiring wealth is about making money and keeping it and is about possession more than use (Beresford, 2020: 76). Greed is not a virtue. It attacks the two basic characteristics of fairness in the Aristotelian sense, which are to act following laws and social norms and to take an equal share. That is why Aristotle defines an unfair person as someone who "takes too much" (Beresford, 2020: 105).

The virtue, for Aristotle, is in the moderation of acquiring and managing wealth to do good to others. Wealth can encourage "right giving", which aims at helping people without self-interest. From this point of view, giving is more important than "taking", since "taking" is only justified as an instrument for "giving". Generosity is the way of finding moderation to use wealth positively. However, ethical virtues are not mentioned in the SDGs, which makes it difficult to find a mid-point between taking and giving, especially if giving comes at the expense of taking.

To place these ideas in context, it is important to start asking why ethics matter now. What do ethics mean for the current global political context? What is the role of ethics in reducing inequality and fighting climate change? And more specifically, how do ethics relate to the built environment and architectural design?

Ethics and the Built Environment

Ethics in the built environment are important for mediating between personal desires and client expectations, private and public interests, individual and community needs and people's demands on resources and the limitations of the environment that provides them. What architects care about and prioritise in their teaching, design and building work is based on trade-offs between their understanding of right and wrong and their client's versions of the same thing. The built environment is the arena and the background for a culture where a community and even a civilisation can develop and where actions taken to satisfy a client can affect someone else beyond the client's interest. The architecture of wealth tends to compound individual needs with community group values, forgetting that the concept of group, the general, already contains its parts. Here is where morality, ethical dilemmas and social justice all get mixed in the same bag.

Fisher (2018: xiv) organised ways of approaching ethical dilemmas in the built environment by covering different aspects and ethical scales, like problems that deal with individual or group ethics. On an individual scale, the individual character traits of a professional, like honesty, temper, and empathy towards others are called "virtue ethics". Here, virtue refers to the Greek concept of "character virtues", those human qualities for doing good things that are developed by "habit" (ethos). They express all the "ways of being good", as Aristotle defined it (Beresford, 2020: 61). The "habit" is an important concept because it implies that ethics is not a theoretical field but a practical one about understanding right and wrong but most importantly applying this to being a good person, otherwise as Aristotle added, "what would be the point of it?" (Beresford, 2020: 30). Character virtues are the product of a habit that develops through experience and education. The practice of these virtues should be reflected in the duties of the professional.

For architects, this would involve the present and future inhabitants of a building and the habitat of any species affected by the new building. Unfortunately, not all the codes of ethics recognise taking care of the environment as part of the ethical responsibilities of architects. Fisher (2018: xxv) proposes a couple of questions for evaluating the merits of a building according to the virtues and duty at the individual scale. "Does the design have the right character given its context? (virtue ethics), does it do the right thing despite potentially negative consequences? (duty ethics)." On a group scale, social contract ethics and utilitarian ethics deal with the social responsibilities of architects. "Does it enhance the good of a place and of those who use it? (contract ethics), does it achieve the greatest good for the greatest number? (utilitarian ethics)" (Fisher, 2018: xxv).

Wealth can mess with the types of ethics at all scales because it essentially separates the individual from the group; it disconnects and pulls apart buildings from their owners, forms from their needs and destroys traditions through fashion. This disconnection across levels and scales produced by the impact of wealth on ethics also has a temporal dimension, since wealth can be inherited. Inheritance can perpetuate privileges which can undermine some of the most fundamental values of a society, like merit and equality. This is another point where the personal contribution to the economic growth of a country (SDG 8—Decent Work and Economic Growth) can go against reducing inequalities (SDG 10—Reduced Inequalities).

The External Forces: Why Do Ethics Matter Now?

Ethics matter because they are about doing the right thing. This is at the core of institutions and their laws, professional practices, everything taught at schools and universities and any sense of justice. In the context of the SDGs, ethics matter. Without them, it is difficult to achieve the goals of one SDG such as SDG 8 (Decent Work and Economic Growth) while also delaying progress in others, like Life Below Water (SDG 14) and Climate Action (SDG 13). The life of many people depends on other people's choices to do the right thing. This is an ethical problem on a global scale and depends on many people caring for others with the hope that if they do the right thing in reducing emissions, this will be the best chance of protecting each other. What ethics share with sustainability is that both rely and depend on making decisions related to the present and future of other people and the fair use and distribution of resources. However, the post-modern understanding of ethics is not related to actions taken in a vacuum but in muddy waters. "Ethics assumes that, based on the facts of a situation, reasonable people can come to some agreement as to what is right or wrong" (Fisher, 2018:xvii). If ethics depend on an agreement, then whoever exercises more power and has more wealth could selfishly manipulate the terms of the "agreement" and impose a version of right and wrong to justify inequalities and keep undeserved power for the sake of increasing and protecting their wealth. When they do that, they push everyone to play by their rules or remain afraid of being left behind.

Ethics matter if leaders and institutions fail to do their job, which is to be fair and to defend people who do the right thing. What should happen with current institutions that defend the privilege of a wealthy few, or remain attached to values that are irrelevant to the fight against climate change? Since "laws and institutions no matter how efficient and well-arranged must be reformed or abolished if they are unjust" (Rawls, 1999: 3), this poses questions about what justice means.

Problems with ethics can be derived from a lack of consensus or agreement about what is just. Rawls (1999: 9) defines justice simply as "fairness" and highlights how the distributive principle is essential for understanding it, namely what fairness is for everyone. Everyone has an idea of justice that can be based on fair or unfair justifications. What is fair for one person or a group might not be shared by everyone. The perception and criteria to define fairness are vulnerable to the influence of wealth. Some people might find it sensitive to reduce speed limits, especially those who are used to walking and cycling, but wealthy people might find it unfair to not be able to enjoy the speed of their very expensive and fast cars. What is the point of driving a Lamborghini at 50kph?

In contrast, why do some people care for other people beyond their individual interests? "That brings us to another way of thinking about ethics: Immanuel Kant's categorical imperative, that we judge the rightness of an action according to whether or not we would want it to be universal." (Fisher, 2018: 30). The idea of a "universal good" was mentioned in the work of Aristotle and is a starting point for the explorations of ethics (Beresford, 2020: 7). However, rich people have consumption-based carbon footprints a hundred times bigger than poor people (Hurth & Wells, 2007), and they keep on getting wealthier at the expense of the latter. What happens when the actions of an elite can endanger many people?

Ethics in the Context of Unethical Leadership

Writing about ethics would be less urgent in a fair and utopian world without inequality, climate change and some current political leaders. Unfortunately, ethics matter because many leaders, politicians and wealthy people behave unethically and get away with it. Therefore, it is important to acknowledge that the education and subsequent professional practice of architects and other built environment professionals often happen in a context governed by leaders with questionable ethics, which impacts what is designed and how this is done. This creates an ethical dilemma. Respecting the laws and not taking advantage of others are the two pillars of Aristotelian ethics (Beresford, 2020: 104), but what happens if the laws are unfair or the leaders that promote them have questionable ethics?

Donald Trump, the ex-president of the USA, made the headlines for his contradictory and uninformed messages on social media and his complete lack of understanding of the environmental crisis, which he called a "hoax invented by China" (Worland, 2019). Trump also reduced federal taxes for the richest in American society as well as inheritance tax. From an ethical point of view, Trump has defined the portrait of what a leader should not be. Wealth and greed played an important role in his unethical behaviour. On May 30, 2024, he was found guilty of "falsifying business records" (Sisak, 2024), though this will not legally stop him seeking to become president again, which now has happened (2025). In the meantime, his skyscrapers stand tall in the urban landscape of Chicago, Honolulu and New York. He will also probably find architects keen to build new towers for him in the future.

Trump is not alone. Emmanuel Macron, whose politics have been compared with Trump's (McArdle, 2018), has not been less effective in providing examples of questionable leadership by abolishing the wealth tax in France, a situation that caused social unrest and division (Piketty & Couper, 2021: 159). Most importantly, his tax reforms did not encourage the wealthiest French to invest more nor benefit all the people as promised (Tonnelier, 2022). This revealed a total lack of interest in building a fair society, since it exposed vulnerable people by widening the inequality gap, with a rich minority that has grown disproportionally (Oxfam, 2019). This also makes it impossible to address climate change, since the richest lifestyles pollute much more than the poor (see Chap. 2), who then need to accept and pay for rises in carbon tax for carbon emissions for which they are not responsible (Piketty, 2017). Macron, to use one example among many other politicians, represents the politically correct leader who promotes a green, apparently ethical agenda, but only commits to it if it does not touch the economic interest of the elite, which is unethical because it is not fair.

Recently, the French president has committed to reducing carbon emissions by 2030 through the *planification écologique* package. The French president was supposed to be doing this anyway, since France hosted and signed the Paris Agreement in 2015. The intentions of investing 7 billion euros in the budget to make buildings more efficient, aid farmers, motivate people to use fewer cars and develop new technology are all good. However, the same package promotes the production of more electric cars (which pollute less but are still cars whose batteries rely on lithium, among other metals that need mining for their extraction) without specifying if this will help to make them more accessible or just create a bigger business for a few. While the budget stimulates new economic investments in technology, the impact of this development is not acknowledged. Most importantly, it supports pollutant industries without the certainty that they will reduce their emissions. Since previous commitments and promises regarding the environment have not been delivered, scepticism still prevails. "Macron wants to show that he is committed to emissions-reduction goals while not burdening citizens through costly obligations — a perfect example of the centrist president's signature *en même temps* ("at the same time") rationale where he frequently argues for and then against a case" (Leali, 2023).

The "productive ecology" of Emmanuel Macron represents the politically correct discourse of the twenty-first century, which is partially based on science as long as it does not compromise the economic growth of a few (Gatinois, 2023). This so-called "competitive ecology" has a dubious ethical commitment to the environment because it is an "ecology that creates an economic value" above any other value (Meslet, 2023). The problem is that the environment is bigger than the economy, and without the environment, there is no economy, and indeed, no humanity. Macron represents the politician who wants to bridge and balance the critique of the left (or green, depending on the country) and the climate change denial from the right, but does neither (Gatinois, 2023). The transition proposed by Emmanuel Macron has been described as "cornucopian", since it is based on the faith that economic growth and investment in technology can make a painless transition to a prosperous economic future as if resources to make the transition will be unending (Legrand, 2023). This shows how the so-called transition has an ideology and political agenda that can score very well in the SDG targets but achieve very little for the environment and people.

The case of Macron in France matters because France and Germany are the two largest economies and polluters of the European Union, and both have fallen behind the targets to reduce their emissions by 2030 (Lafrance & Wehrmann, 2023). Decisions at a country scale, regarding what money goes where and to benefit whom, can change things for better or worse, affecting everyone's future in the case of climate change issues. If there are no concrete laws to limit and redistribute the disproportional accumulation of the wealth of a minority, countries can host and sign as many "sustainable" agreements as they want. These are all doomed to fail due to their poor environmental and ethical values.

Using climate change as an excuse to impose a political agenda and gain access to power is not an ethical end. One of the consequences is that people start getting annoyed and confused with the topic because it is suddenly associated with people with a political agenda or interests unrelated to the environment but more to personal business and wealth accumulation for a few. The sustainable transition to a fair future, instead of being a process of reflection on what is necessary to keep and how to save resources, is now promoted as an opportunity to start an economic race between countries to compete for green investments in car manufacturing and energy businesses (see Chap. 2). Therefore, the transition is perceived as developing "green infrastructure" for more consumption and economic growth for the same elite.

There is also the possibility of increasing taxes for the same end, even though they might not be invested most effectively, in technology that does not lead to less but more consumption in the long term. Some politicians may use the perception that taxes to fight climate change are not correctly used and campaign against doing anything at all for the environment with the excuse of protecting household economies. In 2023, the then Prime Minister of the UK, Rishi Sunak "defended defying the recommendations of the Climate Change Committee (CCC) and said he had "absolute confidence and belief" the country was on track to meet its end goal" (Allegretti, 2023). Therefore, he promoted not banning petrol cars or gas boilers, among other measures, presumably to gain the sympathy of sceptical voters and detach himself from Labour, the opposition party, whose leader had climate change as the heart of his political agenda. The concern, in terms of environmental ethics, is that even in a democratic nation, doing the right thing for a country can become less important than being different from the political party in opposition. This is how the chances of a society dealing with climate change are flushed down the (political) toilet.

People Who Say "No"

Hopefully, not all important changes depend on political leaders. As a rule of thumb, projects with dubious ethics, usually potentially harmful to the environment and only beneficial for a few people, try to advertise themselves aggressively. The more dubious the business, the bigger the promise. Promising to create plenty of new jobs is a classic manipulative marketing tool. However, not all people are greedy and selfish, as conceived in the Hobbesian world of business and wealth accumulation. Recently, Ecuador had a referendum to decide if permission would be granted to the largest oil company in the country to drill in one section of the Ecuadorian Amazonian Forest. The initiative resulted from two decades of rising concern about the impact of the oil industry on the forest and its indigenous community. This was a long fight between a small group of humble people claiming rights to their land and resources and a big oil company. Defying all odds, 6 out of 10 people in Ecuador voted No, and the company will not be able to drill any more in that section (Collins, 2023). Once again, this proves that most people have their hearts in the right place. It is the wealthy minority who need to learn from and follow them.

Should Architects Work for Rich Leaders with Questionable Ethics?

If a country's elected leaders cannot provide examples of ethical behaviour through their actions and decisions, how can they set ethical standards for the rest? If they do not comply with the laws of their countries, how can they take effective actions to reduce emissions and lead people to behave sustainably?

However, Trump's and Macron's examples of poor leadership in developed countries fall short when compared to Latin American and Middle Eastern leaders, many of whom were convicted after their periods of office, were elected with a criminal record or were accused of crimes against humanity. After finishing her mandate in Argentina, ex-president Cristina Fernandez de Kirchner went on trial, accused of corruption and charged with illicit association and fraudulent administration (Alcoba, 2022). She not only got away with these accusations but also avoided going to trial for money laundering, all of which did not stop her from becoming and remaining in power as the vice president of Argentina (Alves, 2023). Pablo Grippo was the preferred architect of Cristina Kirchner, even before she was president. He designed her house, and the collaboration has not stopped since. Grippo was involved in constructing the Hotel Los Sauces, the Kirchner residence in El Calafate, a town in the Patagonian province of Santa Cruz, Argentina. He was also the architect for the redesigned hotel Las Dunas, among other buildings (Pizzi, 2019). However, he came into the public eye when one of Grippo's companies received the concession (after nobody else presented an offer) to work on two yachts on the glacial lakes in the National Parks Los Glaciares. He was later accused of being a man of straw using Kirchner's money, an accusation he denied (Manzanares, 2021). The architect was interviewed about the origin of the money to buy the yachts, especially because one was so large with its 20 rooms and a restaurant. He was then called to justice in 2018 after not paying the fees to use the National Park. Apparently, in Argentina, being the preferred architect of a corrupt political leader can be a very lucrative business, especially when architects are not interested in knowing where the money to pay them comes from.

In Brazil in 2022, Luiz Inacio Lula da Silva went from being in prison, accused of corruption, to becoming the president of Brazil (Meyerfeld, 2022). How these people are elected usually obeys the law of being the least worst candidate among the options. In Brazil, for example, Lula won because the other option was Jair Bolsonaro, a person who has proven on more than one occasion to have ethical values that make sense only to him. From declaring himself proudly homophobic, to disbelief in COVID-19 that cost the lives of many people (Downie, 2022), to the lack of protection of Amazonas (Brown, 2023), the actions of Bolsonaro speak for themselves. Then why would Bjarke Ingels, the leader of the architectural firm BIG, be interested in having a business meeting with him? The architect was on a mission to study the possibility of developing a so-called 'sustainable tourism' project (if this is even possible) with the developer Nômade Group, who also did a project in Tulum, Mexico, and is supposed to have a way of building in the rainforest without creating any impact. Ingles provided many reasons to explain the intention to work in Brazil under Bolsonaro's interest. He stated that the government of Brazil was not his client and he would not make a blacklist of countries where he would not work since that would be an "oversimplification of a complex world", so anything goes. According to Ingels, it is "ignorant" and unreasonable to be "dividing everything into two categories" if architects want to "positively impact the world" (Ravenscroft, 2020). The whole point of ethics is understanding and defining criteria to discern right and wrong. Never mind, Aristotle did not live in the complex world of Ingels.

Since powerful and rich people can still get away with behaving unethically, why should architects avoid working for them? Table 9.1 provides some of the reasons given by star architects, and what these imply for their ethical positions.

Architects that serve the wealthy have already picked a side in their role as enablers of their clients' wealth.

The French architect Jean Nouvel's justification for human rights being old is not correct. The Universal Declaration of Human Rights dates from 1948, while Amnesty International started in 1961 and Human Rights Watch in 1978. They are all less than 100 years old, which means many people still alive probably lived in a world without human rights. Even if he considered it old and not fashionable enough, the important question is always about relevancy, not age. His justification is an act of complete denial. However, ethics does not disappear when it does not suit you or the law. The supply of drugs in France is a criminal activity that could lead to a prison sentence, while working for a government that has violated many human rights, like Saudi Arabia, is not. Perhaps empowering human rights agencies could help architects make better choices.

Koolhaas' statement is cynical and hypocritical, but it admits that architects sell their work to the highest bidder. The cynical part is that he frames architects as victims of an unfair system but affirms that wealth is a precondition for architectural innovation. Are the strong sponsors not also the people making the system unfair? The hypocritical side is to try to gain credibility by victimizing architects while supporting the forces he criticised. Bjarke Ingels complements Koolhaas' idea of selective innovation by wealth. Bjarke Ingels infers that his only ethical value, an implied sense of loyalty, is to his creativity, echoing the description of genius by Kant (see Chap. 8), and whatever it takes to implement it, when in fact, a great deal about ethics is to be fair and do good to others. Jacques Herzog put less emphasis on the creative process but was more direct in defending money as an end. However, money has

Table 9.1 Ethics in the words of architects

Architect/ firm	Justification	Ethical position
Bjarke Ingels	"I'm happy to share the ideas and ideals that I hold with any government willing to listen...especially if they hold different beliefs. If there's a responsibility that comes from the creative platform we've created, it is to use that platform to change the world for the better." (Wainwright, 2020)	Any unethical action in the present is justified on the promise of a potential change that has no chance of happening
Jean Nouvel	Alluded to human rights as an "old question" after being commissioned to design the Louvre Abu Dhabi, the National Museum in Qatar and a luxury resort in Saudi Arabia. (Wainwright, 2020)	Denial
Zaha Hadid	"I have nothing to do with the workers," the response given after questions about the hundreds of migrant workers that died in the stadium in Qatar (Wainwright, 2020)	Passive. The Pontius Pilate approach
Rem Koolhaas	"On our own, we can at most have good intentions. But we cannot represent the public good without the larger entity, such as the state." "To make matters worse, the more radical, innovative and brotherly our sentiments, the more we architects need a strong sponsor." (Wainwright, 2020)	Victim. Justify his action as a victim of an unfair system. Adds that architecture with big ideas needs strong sponsors
Jacques Herzog	"Only An Idiot" would have turned down the project on moral grounds", referring to the design of the Birds Nest for the Beijing Olympics stadium (Wainwright, 2020)	No ethics, only money
Wolf Prix	"I have no bad feelings. I design what I want to build and take responsibility for the building." (Ravenscroft, 2023)	Machiavelli. The end justifies the means

no ethics, as clearly explained by Aristotle, since it can be used to support or destroy people.

In the case of the architects of the wealthy, the ethical dilemma emerges too late, only appearing when the media has unveiled what the architects are doing for their wealthy clients and when they must find a way to justify it. They are not passive victims, and the project size does not oblige them to say yes to criminals. What is more important—to do what designers want for their ego and greed or what is right? It can't be both.

This is also a problem of scale. The bigger the project and the money invested, the more exposed and vulnerable architects are to external forces, from corrupt leaders to numerous ramifications of the effect of their design in unknown social and cultural contexts. When the commission is so big and economically tempting, designing to satisfy a few rich people is the best way of not doing what people need, what is fair for everyone. The alleged complexities faced by architects of the super-wealthy and their ethical dilemmas have a common factor: it is all about money and vanity.

Money Corrupts Architects and Architectural Design

At the World Architecture Festival in Amsterdam in 2018, David Adjaye told the audience, "It's all about money now. Architecture in the west is absurdly expensive – projects cost billions and it's crazy." Adjaye added that "money has totally corrupted the ability to make a meaningful form" and that projects are driven by "elitism that is to do with hyper-commercial liberalism and who controls money" (Block, 2018). Adjaye lectured the audience on the social responsibility architects should have for their cities even though they work for clients. This is also implicit in the ethical code of the RIBA mentioned earlier, with its six ethical practice aspects of: duty to the wider world; duty to society and the end user; duty to those commissioning services; duty to those in the workplace; duty to the profession; and duty to oneself (Dixon, 2021). However, there is no sense that you can pick and choose which aspect to embrace but rather it has to be all. Similarly, all professional planners in New Zealand have to sign up to a code of ethics that encompasses responsibilities to the public, the profession, clients and self (New Zealand Planning Institute, 2024).

In his address Adjaye added that architects "are always in debt to ensuring the city is giving the best back to its constituents" (Block, 2018). He used one of his projects, the Smithsonian National Museum of African American History and Culture in Washington, as an example of giving back to the city (Fig. 9.1). From the points of view of artists and politicians, this is a way to encourage social equity. The entrance to the general public is free, which is not always the case for museums. However, the building cost US$540 million, with half of the money from taxpayers and the rest from rich people and their companies (Cotter, 2016). It is not a modest building, but at least, on paper, it has good intentions. It would be interesting to know what the non-wealthy sector of the Afro-American community has to say about the museum's contribution to creating equity in their everyday lives. Unfortunately, their voices are not published in art and architecture magazines.

The form of the building was intended to allude to West African architecture, although it is hard to find a building in the West African Region whose vernacular architecture could remotely resemble the museum. The choice of a double façade is rather intriguing, considering that it is non-existent in the vernacular architecture of West African countries. In times of climate emergency, the architect designed a glazed façade for a museum whose core needs to be enclosed for functional reasons. The problem is that in architectural design, decisions tend to have a domino effect, especially the wrong ones. With the unnecessary glazed skin, a double façade was needed to protect common spaces from overheating and the exhibited work from direct sunlight. The original idea for the double façade was to use bronze perforated panels, but it was downsized to 1200 ornate cast aluminium panels painted in a bronze colour. It is no surprise that the building was so expensive and had mediocre scores in the LEED assessment's categories "energy and atmosphere" (14/35) and "material and resources" (6/14) and very poor in the category "indoor environmental quality" (9/23) (USGBC, 2018). Regardless of the low scores, the building was awarded a gold LEED certification (see Chap. 3). It seems that for Adjaye Associates and the Smithsonian, LEED does not care about the amount of material

Fig. 9.1 National Museum of African American History and Culture. (Source: Creative Commons (n.d.))

used or its justification. Designers can waste as much material as possible if certified as recyclable. Adding photovoltaic panels to the roof will do the rest of the job. Would the remaining humble residences in the West Africa region, made by hand, with scarce resources, intelligently used, that still work, also receive gold LEED certifications now? Probably not because they could not afford the cost of being LEED-certified.

> It is a building with many narratives—relating to the context, the history and the program," David Adjaye, founder of Adjaye Associates stated: "This narrative is articulated immediately by the silhouette—borrowing from the form of a Yoruba sculpture—while also resonating with the angle of the Washington Monument. (Cornachio, n.d.)

As part of the architecture of wealth, the building is fully injected with aesthetic aspirations to manipulate people. The instrument of manipulation is the "narrative", which is the perfect excuse to justify any caprice from the designers, clients or both, since the narrative can be adjusted to justify any excess or nonsense. Would these "narratives", in Adjaye's words, have been possible without a rich budget? The Smithsonian National Museum of African American History and Culture in Washington has catapulted upwards Adjaye Associates' career. It was named Cultural Event of the Year by The New York Times. Adjaye and his firm both received innumerable awards. Adjaye was knighted by Queen Elizabeth II in 2017 and recognised as one of TIME magazine's most influential people of the year. He was the recipient of the 2021 RIBA Royal Gold Medal and then, in 2022, appointed to the Order of Merit. With more recognition come more prizes and involvement in larger projects, leading to even more prizes. This is the vicious circle of wealth.

Apart from the large commissions won in recent years, the portfolio of Adjaye Associates would seem alien to regular architects who pay their bills by mostly working on tiny or medium-sized houses and bathroom renovations. Adjaye Associates has done more exhibitions (39) and pavilions (13) than residential (5) and interior designs (33) together. This is a characteristic of the architecture of the privileged, who can make a living by building pavilions and exhibitions for the media. In the past, pavilions and exhibitions were used to learn, though a small sample, what was happening in other countries (see Chap. 6). Not everyone was invited to participate. However, after the internet was made more accessible, building pavilions for the multiple biennale and other elitist events seems outdated, expensive and irrelevant. A larger sample of architectural examples from around the world is available online, and weekly updates can be received by email for free. The only reason for the existence of biennales is as a business for the elite who can afford to travel to them. When privileged architects have the chance to design for these events, sometimes it pays off. Adjaye Associates has been involved in larger and larger projects since the Smithsonian Museum opened. They even joined the team of architects working in NEOM, the linear city in the Saudi Arabian desert (see Chap. 2). However, the government of Saudi Arabia does not have a good record when it comes to human rights. "Peter Frankental from Amnesty International UK said architecture studios designing Neom in Saudi Arabia are facing a 'moral dilemma' and should 'think twice' about their involvement in the project" (Bah, 2022). This has not stopped the firm from joining the project even when Sir David Adjaye received the 2021 Crystal award from the World Economic Forum "for his leadership in serving communities, cities and the environment" (Schwab, 2012). However, in 2023, after an investigation carried out by the Financial Times (Spero & Raval, 2023), Adjaye exited from project NEOM "following the publication of allegations of sexual misconduct, including sexual harassment and sexual assault" (Spocchia, 2023) allegations that the architect has denied.

These design practices pose unresolved moral dilemmas that do not seem to bother either people giving awards or those who read about them. Is it ethical to work for criminals? Is having a good intention with a poor execution ethical? Is it morally correct to design against the environment in an environmental crisis? Do buildings represent and honour the content they exhibit (in the case of museums) in terms of their historical meaning or solely the aspirations of their designers? It seems big money shapes big architecture and corrupts architects. This must impact the creation of sustainable communities and cities (SDG 11).

Ethics at Multiple Scales in the Built Environment

To analyse the internal process related to the design of the built environment from an ethical point of view, it is essential to acknowledge that the built environment is neither a big, flat, homogeneous entity nor a grey stain on a two-dimensional map. Unlike a flat panel of medium-density fibreboard, the built environment is like a sinuous puff pastry concoction with multiple layers of people, buildings and landscapes, simultaneously interrelated and independent. From the domestic environment of a family to the whole metropolitan area, the built environment has multiple scales, each with its contexts and ethical dilemmas, and all are potentially different. Yet, ethics makes possible finding the common ground that bridges the individual and the collective, the big and the small, the general and the specific. Through ethics, it is also possible to identify the dilemmas that belong to one scale and influence another or that are specific to the one scale. Scales can be the starting point for integrating ethics into the design process.

There is a mismatch between the scale of work done by architects and their impact on the different scales of the built environment. The impact is much wider than a building, simply because the built environment is made of buildings, people and an environment that supports them. Scales pose different challenges to ethical behaviour that range from the individual space that one body occupies, to inhabiting a small room, to the collective scale of a city, a region or the entire planet.

> Unethical behavior arises from a spatial and temporal disconnect between our actions and their consequences, a disconnect that can blind us to the potentially negative effect of what we do because its impacts frequently occur in some far-off place or unknown future. Likewise, ethical behavior stems from an understanding of its consequences not only on others in our immediate sphere, but also on distant populations, future generations, and other species. (Fisher, 2018: 213)

This argument can be used to make manifest the responsibilities and limitations that designers have or that are used as an excuse to care only about self-interest justified by the incommensurability of actions as seen in the analysis of the external forces. This complete disconnection between the actions and consequences of unethical behaviour can be seen in the environmental impact and the inequality of resource consumption due to the unconsidered behaviour of a few towards the rest. Inequality in the use and consumption of resources in the rich northern hemisphere, with its consequent impact on the increase in emissions, increases the vulnerability of those in the southern hemisphere to the impacts of climate change. The same disconnection occurs at multiple scales and can be seen in the built environment. Ethical dilemmas emerging from a family and a community tend to be opposed, even though they share a continuum in the urban landscape. One cannot exist without the other. However, if wealth is the only way of mediating, the continuum is broken, because wealth acquisition encourages everyone to be against everyone else as winner takes all. Identifying what ethical problems are caused by the impact of wealth and at which scale is important both for academics wanting to include ethical principles in their teaching and for practitioners working on larger projects. These ethical dilemmas are place-specific and deserve to be studied in detail. Some examples are provided below.

Scales have a hierarchical logic in the built environment. Big scales contain smaller scales, and smaller scales are embedded within bigger scales. A house is contained within a plot, and multiple plots are embedded within a street, which defines blocks that cluster several plots. Many blocks make a neighbourhood, and several neighbourhoods a city. Wealth tends to break that logic by separating and imposing an individual will over the group interest. Examples of this are when a big building becomes more important than an entire neighbourhood, or when a building tries to be more important than an entire city. The example of the Amazon warehouses and their effect on the neighbourhood given in Chap. 2 illustrate the first point. For the second point, arguably Gehry's Bilbao Museum is more well known than the former industrial city of Bilbao, given its effect on tourism (Chicca et al., 2023: 232–3).

In the built environment, all scales interact, sharing big problems like climate change that affect all scales while keeping others more contained within the boundaries of a specific scale, like complying with building code specifications that are sometimes unfair or demand lower standards than what might benefit the wider society. This is a problem because obeying laws becomes somewhat optional when they are not trusted. A lack of prescribed minimum qualities for a bedroom could open the door for opportunistic interpretations to save money. However, when a project gets bigger because the client has more money, the minimum is not a problem but rather the lack of a maximum. All bedrooms in the luxury condo building at 520 West 28th in New York, designed by ZHA (see Chap. 8), have ensuite bathrooms as if they were mandatory or unavoidable. In the single-bedroom apartment of the same building, almost half the apartment's floor area is dedicated to the bedroom, ensuite bathroom and walk-in closet. The ensuite bathrooms usually occupy almost half of the area of the main bedrooms and will only ever be used by one or two people. Most social housing apartments have only one bathroom for a whole family. Ethics and fairness start at home.

Ethics and Wealth in the Sustainable Design of the Built Environment

One reason ethics have less importance in the curriculum of architectural studies and architectural firms is that they need to be consistently implemented through all the stages of the design process. As highlighted by Aristotle, ethics is a practice, not a theory.

Trying to propose (or impose) a sustainable design can be a futile exercise without implementing ethics in the design process. A good example is found in the Sustainable Development Goals (SDGs), which have a heroic set of targets and good intentions, but which can only be practised with a more straightforward definition of ethics. Even though SDG 10 aims to "Reduce Inequality within and among countries", and SDG 16 aims to "Promote peaceful and inclusive societies for sustainable development, provide access to justice for all and build effective, accountable and inclusive institutions at all levels", there are no explicit actions linked to limiting wealthy people from taking too much. The first goal of the SDGs is to "End poverty in all its forms everywhere", but there is no single target or indicator that limits the greed of the wealthy, and no mention of fiscal instruments for inducing a fairer distribution of wealth. It is impossible to end poverty if the wealthy few keep widening the inequality gap, especially at the expense of the demise of the middle class. Much of the information presented in the previous chapters shows that without limiting and redistributing wealth fairly, there is no chance of decreasing emissions.

The magic words across all the SDGs are: "adopting and implementing policies". If a country's government can write something on paper, it counts as progress to achieving the target of the goal, even though nothing changes in real life. As Rawls (1999) proposed, it is difficult to build any sense of fairness or justice without ethics. This idea translated to the built environment means that a building must be fair to the environment and the people in it. No matter how it looks, a fair building will help its inhabitants, community and environment to flourish.

A triumph of the architecture of wealth has been to isolate the design process from its ethics and, as a result, to impose the idea of sustainable design as a shopping list of technological gadgets. Ethics remains important for understanding that fairness is essential in any design because it is tangible and affects everyday life. However, it becomes less perceivable in the architecture of wealth, because it is assumed that unfairness is the norm. Unfortunately, architects have been historically trained to design for the wealthy, and fairness has become an option.

From professional practice to the academic exercises of students of architecture and planning, there is an inclination to follow the rules of the market, which inevitably leads to uneven qualities and unequal prices. It is much easier to design a house that is half good and half bad, rather than one that is all good. Half good, half bad houses are often symmetrical, with circulation in the middle and rooms on each side. The problem is that symmetry and equity are different things in architecture. A house with circulation in the middle, such as the typical early New Zealand house plan (Fig. 9.2) creates two different room qualities due to opposite orientations.

Some houses seem to lack a sense of proportion and purpose, such as those with gigantic living rooms and tiny kitchens, or houses with two kitchens, even though one may be labelled Butler's Pantry. With its sink and cooker, this has the same purpose as the kitchen also equipped with sink and cooker (Ray White, 2024). Half-and-half buildings are also found in medium-density apartments. These buildings tend to contain a few good apartments, usually the ones in the flyer or website, and all the rest have a lower quality, often because of poor orientation or lack of a view. The logic of the market is that this predesigned inequality offers the possibility

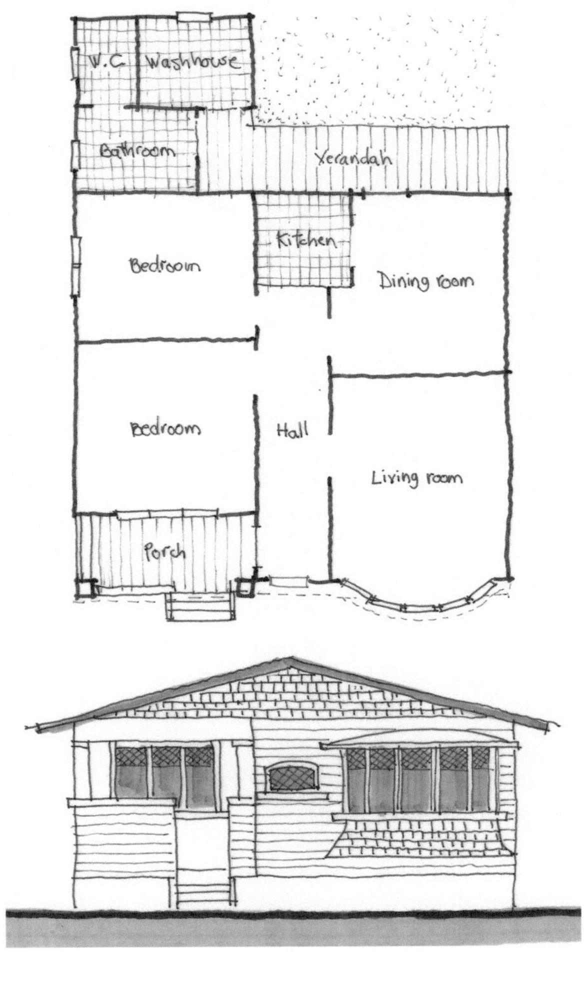

Fig. 9.2 Typical early New Zealand house plan. (Source: Ashford (1994: 29))

to sell apartments at different prices. The best will be very expensive, others only expensive, and the worst apartments will be slightly less expensive. This is how the market, according to the wealth of buyers, fixes problems that emerge from unethical, poor design. Why should the most expensive apartments exist at all if all could have been designed to be equally good?

For these reasons, it is important to detach architectural design from the ideology of wealth to start exercising and teaching fairness. All the rooms in a house, all the buildings in a plot, all the plots in a block, and all the neighbourhoods in a city must be equally good for their purpose and for their communities. This is a very challenging ethical principle related to fairness and equity that demands the skills of a designer because landscapes are heterogeneous, and people have different needs. From the design of a house or multiple units within a plot to the design of mixed-used medium-density buildings, the principle of a fair design is rarely approached because it is difficult to get all the rooms, all the buildings and all the plots or sections of a project to have the same opportunities and qualities.

However, it is possible to design for equality, and there are many examples if you look for them. The works of the architects Candilis, Jossic and Woods exhibit a few principles that can be linked with ethical principles from Kant and Aristotle. A very short note in Candilis' (1973: 20) book for the design of mass tourism, almost hidden among an immense amount of hand drawings, architectural plans and diagrams, reads: "A single-family isolated house, even though excellent, has no interest whatsoever if it does not contemplate the possibility of being integrated within an urban framework or if it does not provoke the creation of a new urban framework". The idea that the value of a building depends on the possibility of it being integrated into its environment is both a clear ethical statement and a design principle. However, this principle needs to be taught or sufficiently emphasised in schools of architecture which, influenced by the architecture of wealth and its ideology, encourage students to design individual works of art without testing if they allow others to do the same.

When Proposing to Build Nothing Means Everything

From an environmental point of view, the primal ethical problem for architects and designers is not the style of a new building, nor what its concept or meaning should be, but instead, why a new building should exist. This is not an easy challenge, but it should emerge whenever a new client arrives at an architectural firm. This is the zero point of any sustainable practice because it is environmentally ethical, which means trying to do something good for the planet, your client and, therefore, your firm. Logically, the most sustainable project would be the one that can solve the client's problem or needs without building anything because it uses no resources other than the designer's ingenuity and client openness. However, this is not how the business of architects works. Questioning clients about the need to build, design or create could be a very effective way of losing them. This might compromise the persistence of the practice and the jobs of its employees and their families. What is ethically correct for the planet could be perceived as detrimental to architects' business and socio-economic stability. This is a problem where wealth and the environment clash since wealth can freely be imposed, and the environment has to provide. Justifying the need for a building beyond a clients' perception is not a standard checkpoint in the design process, taught in academia, nor expected by clients. However, to build or not to build is the primal ethical dilemma for architects because it is a problem that requires design. This chapter proposes that an essential ethical work of an architect is to be able to justify the existence of a new building on a planet that is undergoing a climate change crisis.

In 1996, the Bourdeaux City Council initiated a project to "embellish" local town squares, including the Place Léon Aucoc, whose project was developed by the architectural firm Lacaton and Vassal. On their website, Lacaton & Vassal explained that from their first approach to the site, they felt that "this square is already beautiful because of its authenticity, lacking in sophistication. It possesses the beauty of what is obvious, necessary, right" (Lacaton and Vassal, 1996). In this statement, it is important to highlight that contrary to the imperatives of the aesthetic of wealth, where originality and sophistication imply luxury and beauty, Lacaton and Vassal, associate authenticity with a lack of sophistication.

The square did not need to be embellished, since it was already beautiful. The firm's belief in this idea and their careful research and interviews of neighbours helped them to make a bold decision: propose building nothing. "As a project, we've proposed doing nothing apart from some simple and rapid maintenance works—replacing the gravel, cleaning the square more often, treating the lime trees, slightly modifying the traffic—of a kind to improve [the] use of the square and to satisfy the locals" (Lacaton & Vassal, 1996). However, that did not stop the council changing things later (Fig. 9.3).

At first, city officials rejected the idea proposed by Lacaton and Vassal, since they were probably expecting something else from the architects, something to be built, something that gave evidence of the intention of the city council to embellish the city (Huber, 2015: 343). Small local firms of architects are no strangers to the practice of building nothing and being rejected. Countless times, architects are informally asked by family members and friends for advice, usually over a coffee or a barbecue, to get an opinion about what would be suitable to do here and there in the house, of course, for free. Sometimes, the best option is to do nothing. Instead of receiving this economically friendly advice as good news, some clients are disappointed with the answer and seek advice from someone with potentially more imagination.

The project of Lacaton and Vassal was executed, although invisible to photographers look-

Fig. 9.3 Place Léon Aucoc in its present state. The fence and the playground have been added, since the architects told the council the square was perfect. (Source Lionel Chatras)

ing for architectural forms, and became an important work in the portfolio of the firm and an example of architects challenging aesthetic expectations (Huber, 2015). In the square, there is no design as a spectacle only "a project that has no object other than the original site itself" (Wilson, 2013: 48). Since then the square has been changed (Fig. 9.3). What is challenged in Place Léon Aucoc is the ethics of the aesthetics of wealth, the commodification of forms and the necessity of always linking architecture with form and aesthetics.

Anne Lacaton made it more explicit in an interview: "we don't believe much in form" (Niksa et al., 2003: 110). She explained that the form comes at the end of the process when things need to take shape to have dimensions and a structure to materialise a space to live, whatever the function might be, but it is never the beginning of the process. The first aim is to produce spaces to live in, which always happens in a pre-existent built environment. In the case of the square, they considered that everything needed was already there. Understanding architecture as a part of something bigger, the built environment and the urban landscape, within which it has to serve the community should be a key ethical premise for architects. Anne Lacaton mentioned that the idea of appreciating what is already there on the site came from their experience of working and living in Africa after they graduated from the School of Architecture in Bourdeaux, giving a school for nomadic children in the Sahara as an example. The school is just a roof of branches with nine supports to create shade. There are no teachers, just a television with educational programmes, powered by solar energy. In that situation, Lacaton said she learned that "What you need to do, you do it very carefully, you do it very well", since resources are so scarce and what is at stake is very important (Niksa et al., 2003: 112).

Several lessons can be learnt from Anne Lacaton's anecdote. One is that it is relevant for the academic and professional growth of students of architecture in developed countries to understand and internalise their positions of privilege. The lack of contact with people, families and communities that have less is a disadvantage when exercising architecture as a social service. The education and practice of the architecture of wealth in a wealthy environment for wealthy people can offer very little to the remaining 90% of the global population. In words produced for the German Pavilion at the 2012 Venice Biennale, Petzet and Heilmeyer (2012: 13) stated "We feel it is our duty to start from scratch with each new project. That can also mean fundamentally questioning our own profession—and, with that, the way architecture is practiced." This is an ethical position.

Ethics, Wealth and the Environment

In cases where something has to be built, the second ethical dilemma concerns what to build, how much and how to do it. It is an old dilemma already discussed first by Vitruvius, and then by Alberti (influenced by the work of Vitruvius) through the concept of "giusta misura", the right measure, the harmony between the whole and its parts. Beyond this classical aesthetic understanding of beauty lies the dilemma for architects between overdoing or falling short (Enia & Martella, 2019: 154), which concerns how much a project or building needs to be designed and built. Clients have some wealth that entitles them to have demands and aspirations, and architects have the skills and vision, but they are also running a business to provide projects and buildings to people who can pay their fees. This situation can frequently create tension between clients' and architects' wants and needs and their potential implications for the environment, from which specific references of how much is too much can be inferred. This is the point where ethics, the environment and sustainability meet.

The RIBA has recently acknowledged the link between ethics and the environment (RIBA, 2019). In 2017, the RIBA Council created a commission to make a report on how to integrate sustainability as a "core value within the profession" and "to consider how the profession can best reflect its core values and engage with the UN's 2030 Sustainable Development Goals" (RIBA, 2018: 3). The concerns were confirmed after consultations in 2018. After public consultation, the results were clear: "83% believe the Institute should engage more effectively" with sustainable development and "only 11% consider RIBA is engaging with the UN Sustainable Development Goals at present" (RIBA:2018: 31). All the sustainable goals, the diagrams of concentric and interlinked circles and the green labels available in the market have done little to reduce emissions significantly, and few understand them. In the context of the environmental and inequality crises, without ethical practice, sustainability ideas are useless. Therefore, the first step towards sustainability is an ethical practice that should change policy and education.

> Designing in a sustainable manner should be in the code of conduct. Sustainable design should be considered better design. It is a design that is exemplary in every dimension. Sustainable design is that which is better for people, better for the environment and better for society. Architects who do not consider the constraints of the planet and the needs of future generations in their design are not effective designers. (RIBA, 2018: 32)

The RIBA is trying to change directions towards a grounded approach to establishing institutional relevancy based on engaging with the most urgent and globally concerning issues.

> RIBA and the Awards Committee have heralded a new direction for the profession, encouraging all architects to focus not only on the privileged but also humanity at large that suffers from disparities, conflicts and climate change. There are innumerable opportunities to implement principles of circular economy, de-growth, transition design, eco urbanism, and what we call Barefoot Social Architecture (BASA) to achieve climate resilience, sustainability and eco justice in the world. (RIBA, 2023)

The call to shift focus from the privileged to the rest is the first step towards doing something good for people and the environment. There is no need to develop 17 goals and countless indicators; just work ethically, and innumerable opportunities will open up from doing this. Time will tell if the RIBA statements are just words towards an institutional aesthetic of looking politically correct or a new phase. In the meantime, the RIBA has taken action to reward people who can provide good examples. The good and ethically correct architecture, the only one that works for people, still matters.

The 2023 winner of the RIBA Stirling Prize is The John Morden Centre in London by UK studio Mae Architects. It is a daycare centre for residents of Morden College, which was founded in 1695 by Sir John Morden and built by Sir Christopher Wren's master builder and successor, Edward Strong. The new building clusters facilities that are currently spread throughout the college. It develops from a colonnade that is the main circulation running across the plot. The multiple-use spaces attach to the colonnade, creating different outdoor areas. The building aims to tackle isolation among older residents by designing spaces where they can interact (Mae, n.d). The RIBA Stirling Prize Jury commented that the building "is a skilfully designed package that minimises the building's impact on our planet's delicate ecology while also harnessing the therapeutic value of the surrounding nature". Elle Van Loon, from the office OMA, stated that the building "sensitively and seamlessly integrates medical facilities and social spaces, delivering a bold and hopeful model for the design of health and care centres for the elderly" (RIBA, 2023). In the short film of the building captured by Jim Stephenson and available on the Mae website, Lori Morley, one of the residents, commented: "It is a fascinating building. It is a very tactile building. So it's touchy-feely...It is an extension of our homes" (Mae, n.d) (Figs. 9.4 and 9.5).

For critics and editors of architecture, this new turn taken by the RIBA lacks bling and spectacle. Journalist Catherine Slessor (2023) thinks that the RIBA's shift "to consistent earnestness is welcome", but she also mentions that these low-key projects are part of "dutifully dull shortlists" that exemplified a phase where "[the] architectural awards programme has become curiously introverted." The *Architects' Journal* headline was that the RIBA shortlist missed the best buildings again (Waite, 2023), arguing that few big buildings with multi-million budgets were nominated. Wainwright (2023), from *The Guardian*, has called the Stirling shortlist "a polite beige consensus" even though he described the buildings

Fig. 9.4 Exterior facade of the circulation and courtyard. (Based on Mae (n.d.))

shortlisted using words like "impeccably detailed", "elegant", "meticulous work" and "tasteful and well made". Still, they seem to lack a "sense of fun" that "quirky buildings" are supposed to have. In the eyes of the critics, there is the impression that they are talking about very good buildings that committed the capital sin of not being funny enough. The architecture that can do good is always judged as not spectacular enough for shows, awards, magazine covers and websites. In the eyes of experts, it lacks the spark and aggressiveness of the architecture of wealth that has numbed the taste of architects. The same unfairness towards good, humble buildings is replicated at all levels. It is replicated in universities, across design crits and student marking moderations, in the design process of smaller architectural firms that want to get an award to be well-known, all looking for distinctions at the price of sacrificing what matters the most, to do well-made buildings that do good for people with a limited budget.

To make things worse for the wealthy rhetoric and aestheticians of architecture, but probably much better for the remaining 99% of people, the RIBA's Royal Gold Medal, a recognition of a lifetime's work for a person or group "who have had a significant influence on the advancement of architecture" (RIBA, 2023), went to Pakistani architect Yasmeen Lari. It is important to highlight that in the past, architects with totally different backgrounds and architecture have won the

Fig. 9.5 Interior courtyard and tree. (Based on Crook (2023))

same recognition, like Dame Zaha Hadid in 2016 or Sir David Adjaye in 2021. The focus of an institution's award and its ethical approach seems very dependent on the chair and committee's ideology and the head of the institution. What is politically correct or susceptible to making the institution more popular or relevant also plays an important factor. Therefore, as Aristotle proposed, awards and prizes cannot necessarily be a way to happiness because awards are not an end since their importance relies on the institutions and people who provide them rather than the person who receives them, and institutional interests can be volatile.

Awards and prizes could help promote good and ethical architecture. This would encourage younger generations of professionals and students to keep trying to do good architecture for people, even when it is hard and rarely recognised. Is it a coincidence that the ethical examples are set by architects from developing countries working for the poor? In an interview that Yasmeen Lari gave to the Danish YouTube channel "Louisiana Channel" (2023), produced by the Louisiana Museum of Modern Art, she started by saying that the role of architects has been defined by the wealthy and for the wealthy:

> ...whether it was the Medicis of Florence or whether they were...what I call the robber barons of the East India Company or the merchant princes of the Industrial Revolution or even the tech giants, you know, the corporate sector today, architects have always believed that they are there to serve the most wealthy and the most exclusive clientele. (Louisiana Channel, 2023)

Lari emphasised that the iconic architecture of the wealthy has only served a minority, and the world today needs architects who use their design skills to help alleviate the disparities of people living with nothing and the scarcity of resources. The design skills referred to by Lari are the ability of architects to "think of the context and design for the context" as well as dealing with multiple disciplines and still being able to serve a client. Lari is convinced that if clients are poor, architects are in a good position to serve humanity. However, this implies changing the definition of architecture, how we see architects and what is expected from them. These are the key ideas behind what she considers the true "decolonisation and democratisation" of architecture, which is to use architecture "to give agency to those who don't have a voice" (Louisiana Channel, 2023).

In times of climate and economic crises, an ethical approach to architecture can be useful to push the profession to redefine what architects do, how they do it, for whom they do it and how much is too much. Since the generation of architects who decide to serve the poor will work with what has been left by the wealthy, they will need to be prepared to excel in scarcity, relinquish control to people and treat them as partners. Cutting the ties with wealth will not be easy, and universities should be prepared to teach an overdose of humility. After all, is there any other road to sustainability?

Ethics and Sustainability to Avoid Unfairness by Design

As argued by Chicca et al. (2023), political and economic systems have environmental impacts that shape the built environment, especially if they are unfair. Ethics matter more than ever, since they are the tool people can use to identify unfairness and injustice, and without them it is impossible to transition to any sustainable future. Discussions about sustainable goals are redundant without institutions that can uphold ethical values.

Ethical dilemmas on the design, management, execution, budget and impact of projects and buildings scale up with wealth. Big buildings plus big budgets and wealthy clients equal numerous ethically compromised decisions. The fast delivery of big buildings and projects with a large budget usually incurs unethical actions. The bigger the project, the more financial interests are being compromised, and the bigger the expectations of a quick positive return on investment, which means less time to develop a design and less involvement of future inhabitants in the design process. The less transparent the process gets, the easier it becomes to compromise ethical values. The reason is very simple: the architecture of wealth takes too much.

If ethics are not directly linked to the design process, there will be no path to sustainability and social justice. More designers need to write about ethics, how they teach it and most importantly how they include it in the design process. A lesson to pass on to young professionals and students at university is to avoid the creation of buildings that promote unfairness by design. It is a humbling experience because when designing for fairness, the goal is not for the architect to shine with their buildings but to make people flourish within their habitats. The shortest way to accomplish it is by respecting and serving the interests of others.

References

Alcoba, N. (2022). *"The corruption case against Argentina's VP Kirchner, explained"*. Aljazeera. Retrieved from https://www.aljazeera.com/news/2022/8/26/the-corruption-case-against-argentina-vp-kirchner-explained, 28 September 2023

Allegretti, A. (2023). *"Rishi Sunak defies critics and presses on with net zero U-turn"*. The Guardian. Retrieved from https://www.theguardian.com/politics/2023/sep/21/rishi-sunak-defends-climate-policy-u-turn-despite-criticism, 17 October 2023.

Alves, J. (2023). *"Cristina sobreseída: cuáles son las 5 causas que todavía debe resolver y qué estrategia planean sus abogados"*. El Cronista. Retrieved from https://www.cronista.com/economia-politica/cuales-son-las-5-causas-que-debe-resolver-y-cual-es-la-estrategia-que-buscan-desplegar-sus-abogados/, 28 September 2023.

Aristotle. (1885). *The politics of Aristotle* (B. Jowett, Trans.). Clarendon Press.

Ashford, J. (1994). *The bungalow in New Zealand*. Penguin.

Bah, S. (2022). *Denzeen Agenda features "moral dilemma" facing architects designing Neom in Saudi Arabia*. Retrieved from https://www.dezeen.com/2022/12/20/peter-frankental-interview-saudi-arabia-neom-project-dezeen-agenda/, 8 February 2023.

Beresford, A. (2020). *The Nicomachean ethics* (A. Beresford, Trans.). Penguin Books Limited.

Block, I. (2018). *Crazy money has "corrupted: architecture says David Adjaye*. Retrieved from https://www.dezeen.com/2018/11/30/david-adjaye-world-architecture-festival-crazy-money-corrupted-architecture/, 7 February 2023.

Brown, S. (2023) *"Report sums up Bolsonaro's destruction legacy and Amazon's next critical steps"*. Mongabay. Retrieved from https://news.mongabay.com/2023/04/report-sums-up-bolsonaros-destruction-legacy-and-amazons-next-critical-steps/, 19 October 2023.

Candilis, G. (1973). *Arquitectura y urbanismo del turismo de masas* (J. Beramendi, Trans.). Gustavo Gili.

Chicca, F., Vale, B., & Vale, R. (2023). *The environmental impact of cities: Death by democracy and capitalism.* Routledge.

Collins, D. (2023) *"Ecuadorians vote to halt oil drilling in biodiverse Amazonian National Park"*. The Guardian. Retrieved from https://www.theguardian.com/world/2023/aug/21/ecuador-votes-to-halt-oil-drilling-in-amazonian-biodiversity-hotspot, 30 October 2023.

Cornachio, J. (n.d.) *Architectural details: David Adjaye's Museum of African American History.* Retrieved from https://architizer.com/blog/practice/details/architectural-details-david-adjaye-smithsonian/, 8 February 2023.

Cotter, H. (2016). *Review: The Smithsonian African American museum is here at last. And it uplifts and upsets.* Retrieved February 8, 2023, from https://www.nytimes.com/2016/09/22/arts/design/smithsonian-african-american-museum-review.html

Creative Commons. (n.d.). "National Museum of African American History and Culture" by John Brighenti is licensed under CC BY 2.0.

Crook, L. (2023). *RIBA unveils 2023 Striling Prize shortlist.* Retrieved from https://www.dezeen.com/2023/09/06/riba-unveils-2023-stirling-prize-shortlist/, 30 October 2024.

Dixon, A.B. (2021). *The six duties of care underpinning the RIBA ethical practice competence.* Retrieved from https://www.architecture.com/knowledge-and-resources/knowledge-landing-page/the-six-duties-of-care-underpinning-the-riba-ethical-practice-competence, 22 November 2023.

Downie, A. (2022). *"Outrage in Brazil as Jair Bolsonaro avoids five charges related to Covid response"*. The Guardian. Retrieved from https://www.theguardian.com/world/2022/jul/27/outrage-brazil-jair-bolsonaro-avoids-five-charges-related-to-covid-response, 19 October 2023.

Enia, M., & Martella, F. (2019). Reducing architecture: Doing almost nothing as a city-making strategy in 21st century architecture. *Frontiers of Architectural Research, 8*(2), 154–163.

Fisher, T. (2018). *The architecture of ethics*. Routledge.

Gatinois, C. (2023). *"Emmanuel Macron incarne une vision de l'écologie productiviste"*. Le Monde. Retrieved from https://www.lemonde.fr/politique/article/2023/09/26/planification-ecologique-emmanuel-macron-endosse-les-habits-d-un-pompidou-vert_6191018_823448.html, 17 October 2023.

Gracia, E.J. and Vale B. (2017). *Unravelling Sustainability and Resilience in the Built Environment*. Routledge.

Huber, D. (2015). Beyond belief. *Artforum International, 53*(9), 342–347.

Hurth, V., & Wells, P. (2007). Averting catastrophic climate change: Confronting wealth. *International Journal of Innovation and Sustainable Development, 2*(1), 63–78.

Lacaton & Vassal. (1996). *"Place Léon Aucoc, Bordeaux"*. Retrieved from https://www.lacatonvassal.com/index.php?idp=37#, 20 September 2023.

Lafrance, C., Wehrmann, B. (2023). *"How Germany's and France's climate policies and greenhouse gas emissions compare"*. Clean Energy Wire. Retrieved from https://www.cleanenergywire.org/factsheets/how-germanys-and-frances-climate-policies-and-greenhouse-gas-emissions-compare, 17 October2023.

Leali, G. (2023). *"Macron pitches non-punitive green transition with new package"*. Politico. Retrieved from https://www.politico.eu/article/france-emmanuel-macron-pitches-non-punitive-green-transition-with-new-package/#:~:text=French%20President%20Emmanuel%20Macron%20on,so%2Dcalled%20planification%20écologique%20package., 17 October 2023.

Legrand, T. (2023). *"La planification écologique selon Macron, le piège du moindre effort"*. Libération. Retrieved from https://www.liberation.fr/politique/la-planification-ecologique-selon-macron-le-piege-du-moindre-effort-20230927_QE5IZ7MK3FCWTEV2YVUWHEGHPM/, 17 October 2023.

Louisiana Channel. (2023). *"Architects Should Serve Humanity. Architect Yasmeen Lari"*. Retrieved from https://www.youtube.com/watch?v=pETSRm001nw, 30 October 2023.

Mae. (n.d.). *John Morden Centre.* Retrieved from https://www.mae.co.uk/projects/morden-college, 30 October 2024.

Manzanares, V. (2021). *"El arquitecto de Cristina Kirchner negó ser su testaferro y aseguró que él es el dueño de un crucero de lujo"*. A24. Retrieved from https://www.a24.com/politica/arquitecto-cristina-kirchner-nego-testaferro-aseguro-dueno-crucero-lujo-05052019_SkABpvniV, 19 October 2023.

McArdle, M. (2018). *Trump & Macron: The all-too-telling tale of look-alike opposites.* Retrieved from https://nypost.com/2018/12/01/trump-macron-the-all-too-telling-tale-of-look-alike-opposites/, 17 September 2024.

Meikle, S. (1994). Aristotle on money. In *Phronesis* (Vol. 39(1), pp. 26–44). PHRO.

Meslet, E. (2023.) *"Planification écologique : Macron a un double discours incohérent"*. L'Humanité. Retrieved from https://www.humanite.fr/environnement/ecologie/planification-ecologique-macron-a-un-double-discours-incoherent, 17 October 2023.

Meyerfeld, B. (2022). *"From prison to Brazilian president: Lula's spectacular comeback but underwhelming victory"*. Le Monde. Retrieved from https://www.lemonde.fr/en/international/article/2022/10/31/from-prison-to-brazilian-president-lula-s-spectacular-comeback-but-underwhelming-victory_6002359_4.html, 28 September 2023.

New Zealand Planning Institute. (2024). *Code of Ethics.* Retrieved from https://planning.org.nz/about-nzpi/code-of-ethics, 25 April 2024.

Niksa, A., Bradic, S., Grimmer, V., & Croatia. (2003). We don't much believe in form. *Oris, 24*, 108–131.

Oxfam. (2019). *The G7's Deadly Sins.* Retrieved from https://oxfamilibrary.openrepository.com/bitstream/

handle/10546/620853/mb-g7-inequality-crisis-220819-en.pdf, 23 October 2024.

Petzet, M., & Heilmeyer, F. (2012). *Reduce, reuse, recycle: Architecture as resource: German Pavillion, 13th international architecture exhibition: La Biennale di Venezia 2012*. Hatje Cantz.

Piketty, T. (2017). "Trump, Macron: Same fight". Le blog de Thomas Piketty; Le Monde. Retrieved from https://www.lemonde.fr/blog/piketty/2017/12/12/trump-macron-same-fight/ 28 September 2023.

Piketty, T., & Couper, K. (2021). *Time for socialism: Dispatches from a world on fire, 2016–2021 [Vivement le socialisme!]*. Yale University Press.

Pizzi, N. (2019). *"El arquitecto de Cristina rompe el silencio: "El crucero es mío, lo compré con préstamos"*. Infobae. Retrieved from https://www.infobae.com/politica/2019/05/05/el-arquitecto-de-cristina-rompe-el-silencio-el-crucero-es-mio-lo-compre-con-prestamos/, 19 October 2023.

Ravenscroft, T. (2020). *"Criticism of Jair Bolsonaro meeting is "an oversimplification of a complex world" says Bjarke Ingels"*. Dezeen. Retrieved from https://www.dezeen.com/2020/01/23/jair-bolsonaro-bjarke-ingels/, 19 October 2023.

Ravenscroft, T. (2023). *"Working on The Line a "tricky argument" for studios boycotting Russia says Wolf Prix"*. Dezeen. Retrieved from https://www.dezeen.com/2023/08/24/wolf-prix-the-line-neom-interview/, 19 October 2023.

Rawls, J. (1999). *A theory of justice* (Revised ed.). Harvard University Press.

Ray White. (2024). *The Block NZ/ House 2*. Retrieved from https://raywhite.co.nz/auckland/auckland-city/point-chevalier/KGL23328/, 24 June 2024.

RIBA. (2018). *Findings of the RIBA ethics and sustainable development commission*. Retrieved from https://riba-prd-assets.azureedge.net/-/media/GatherContent/Work-with-Us/Additional-Documents/Ethics-and-Sustainable-Development-Commission%2D%2DFull-findingspdf.pdf?rev=6b098490558e4aaeb79cb5216cc17700, 30 October 2023.

RIBA. (2019). *RIBA's Ethics and Sustainable Development Commission: final report*. Retrieved from https://www.architecture.com/knowledge-and-resources/resources-landing-page/ribas-ethics-and-sustainable-development-commission-final-report, 30 October 2023.

RIBA. (2023). *Royal Gold Medal 2023 recipient: Yasmeen Lari*. Retrieved from https://www.architecture.com/awards-and-competitions-landing-page/awards/royal-gold-medal/Royal-Gold-Medal-2023, 30 October 2023.

Schwab, H. (2012). *Meet The Winners Of The 27th Annual Crystal Award*. Retrieved from https://www.forbes.com/sites/worldeconomicforum/2021/01/18/meet-the-winners-of-the-27th-annual-crystal-award/?sh=5e31c1445417, 25 April 2024.

Sisak, M.R. (2024). *New York hush money case*. Retrieved from https://apnews.com/projects/trump-investigations-civil-criminal-tracker/, 24 June 2024.

Slessor, C. (2023). *"What do such dutifully dull shortlists say about the wider state of the Stirling?"*. Retrieved from https://www.dezeen.com/2023/10/20/riba-stirling-prize-catherine-slessor-opinion/, 30 October 2023.

Spero, J. & Raval, A. (2023) *Sir David Adjaye: The celebrated architect accused of sexual misconduct*. Financial Times. Retrieved from https://www.ft.com/content/1a03a13c-a0b3-41c9-9a7b-702719ac6d58, 12 October 2023.

Spocchia, G. (2023) *Adjaye no longer working on the line, NEOM confirms*. Architects' Journal. Retrieved from https://www.architectsjournal.co.UK/news/adjaye-no-longer-working-on-the-line-neom-confirms, 12 October 2023.

Tonnelier, A. (2022). *The abolition of France's wealth tax still has no proven effect on the economy*. Le Monde. Retrieved from https://www.lemonde.fr/en/politics/article/2022/10/24/the-abolition-of-france-s-wealth-tax-still-has-no-proven-effect-on-the-economy_6001505_5.html, 15 November 2023.

USGBC. (2018). *Case Study: National Museum of African American History & Culture*. Retrieved from https://www.usgbc.org/sites/default/files/NMAAHC%20Case%20Study_November%202018_0.pdf, 8 February 2023.

Vale, B, and Vale, R. (1975). *The Autononmous House*. Thames and Hudson.

Wainwright, O. (2020). *The despot dilemma: Should architects work for repressive regimes?*. The Guardian. Retrieved from https://www.theguardian.com/artanddesign/2020/jan/27/despot-dilemma-should-architects-work-for-repressive-regimes-bjarke-ingels, 19 October 2023.

Wainwright, O. (2023). *'A polite beige consensus': Are these really Britain's best new buildings?*. Retrieved from https://www.theguardian.com/artanddesign/2023/sep/06/britain-best-new-buildings-riba-stirling-prize-shortlist, 30 October 2023.

Waite, R. (2023). *RIBA National Awards: Regional 'best buildings' miss out—again*. Retrieved from https://www.architectsjournal.co.uk/news/riba-national-awards-regional-best-buildings-miss-out-again, 30 October 2023.

Wilson, R. (2013). Not doing/overdoing: 'Omission' and 'Excess'—Lacaton & Vassal's place Léon Aucoc, Bordeaux, and Construire's Le Channel, Scène Nationale de Calais, Calais. *Architectural Design, 83*(6), 44–51.

Worland, J. (2019). *Donald Trump called climate change a hoax. Now He's awkwardly boasting about fighting it*. Time. Retrieved from https://time.com/5622374/donald-trump-climate-change-hoax-event/, 28 September 2023.

Part III

Unlearning the Goal of Wealth for Educating Future Designers

The final three chapters of the book are dedicated to how architects have been trained and the role wealth has played in this, looking at what might be a quality education (SDG 4) for those involved in creating the built environment. The book ends by discussing the importance of ethics and resilience in creating an alternative concept for designers based on non-wealth-related creativity as the core of the education of future architects.

10 The Changing Nature of Built Environment Education and Its Effect on What Is Built

Introduction

SDG 4 aims at an inclusive and equitable education for all. This raises the issue of whether education has ever been equitable or whether having an education depends on personal wealth. A recent study in India found that the need to fulfil domestic duties still affected the education of girls (Ramanaik et al., 2018). In fact, the education of children in the developed world is still relatively new.

> In the first half of the nineteenth century, child labour continued in agriculture; in new works associated with the industrial revolution, such as mining, ore-processing, potteries, and the infamous cotton-textile manufactories; in domestic industries such as weaving; in miscellaneous trades such as tobacco processing, and in low-skilled services such as errand running. Children were, finally, widely employed in unpaid domestic work, including child-minding. (Smelser, 1991:257)

If you were wealthy in Victorian England and a male, then you would receive an education, probably at one of the all-male public schools. To delve into the connection between being wealthy and receiving a good education, this chapter looks at the history of architectural education. The aim is to see the extent to which wealth and privilege have played a role in success as an architect. As noted earlier, the education of architects has been around for centuries more than that of planners and urban designers. For many years, the "architect" was the person involved in these various disciplines, and architects like Palladio and Alberti wrote on how to plan cities. Town planning emerged as a separate educational subject much later: "The term town planning was first used in Britain in 1906 and was then embodied in the 1909 Housing and Town Planning Act" (Royal Town Planning Institute, n.d.). Planning education began in 1909 in the UK with a course at the University of Liverpool and in 1924 in the USA at Harvard (Fainstein, 2024). Germany had embraced planning education earlier, with the first course in 1880 in Aachen (Campbell. 2022). Given the relative newness of planning education, what follows is a history of architectural education, much of which, therefore, includes planning education. Apart from looking to the distant past, there is then a brief case study of UK architectural education because of its influence in many parts of the world through the former British Empire, recognising that pursuing the Beaux Arts tradition of France and its influence in places like the USA might lead to a different conclusion. This journey through history has a focus on who was likely to become an architect because of their background. The second part of the chapter then ponders on how the type of education received might affect what is built.

E. J. García, B. Vale, *Wealth and the Built Environment*, Sustainable Development Goals Series,
https://doi.org/10.1007/978-3-031-81954-4_10

The Distant Past

Hemiunu (2625–2500 B.C.E), the architect credited with the design of the Great Pyramid of Gizeh, the tomb of King Khufu, and its other associated pyramids and tombs, had a privileged background. He was the nephew of King Khufu and the grandson of King Sneferu. The inscription on Hemiunu's statue, which survives, states that not only was he a priest but he was "Overseer of All [these] Construction Projects" (Anon, 2005). This little about an architect of Ancient Egypt fails to explain how the pyramids were either laid out or who designed either the permanent or temporary housing that was constructed for those working on the pyramids (Redding, 1992). Rossi (2020) is clearer on who was responsible for what in Ancient Egypt.

> The overall impression is that part of the planning and building process of monuments was envisaged in advance by the officers in charge, but that then a substantial part of the practical implementation of the project was entrusted to the actual builders and workmen, including the final dimensions of several architectural elements.

This association between the architect and the large and impressive building continues in Ancient Greece, where Weir (2016) argues that Aristotle saw the building of everyday houses the province of the householder and that architects designed the major buildings of society. This sentiment was echoed by the Roman, Vitruvius, who began "his book on residential architecture with the polite reminder that the best communities produce great public architecture, not great houses" (Weir, 2016). Although Vitruvius saw the architect as the overseer of such work, is also clear from his writing that building knowledge was crucial for the architect (Oksanish, 2019:11–12).

> Not all temples will be built on the same principles for every deity; nor will the same principles be useful for the building of siege cranes and ladders in different places. Crucially, such an arrangement presumes that the practitioner of the art possesses both a certain level of generalized cultural capital and discipline-specific technical knowledge, some of which will have been learned through experience. Accordingly, Vitruvius stipulates a broad, literary education for his architect in various disciplines (which includes but is not limited to general "culture"), as well as an experiential, manual component...Only with both of these components in hand will the ideal architect and his products achieve auctoritas. [authority]

From these fragments, it seems that the architect in the ancient world was associated with the design of prestigious buildings and also civil engineering structures, and although not a builder was expected to have building knowledge. The architect was also expected to be an educated person. In Rome, a wealthy family would hire a tutor to educate the sons of the family and those who could not afford this would send them to primary school at age seven and so learn "to read and write, to count, weigh, measure and calculate" (Bonner, 2012:34), all useful skills for the potential architect. What Vitruvius suggests is that detailed learning of construction would be gained on the job, presumably by working alongside some with these skills. This was to be the general pattern of architectural education for many centuries to come.

Medieval Architects

In his 1899 book *The Cathedral Builders,* on the influence of the guild of masons that stemmed from Como in Italy, Scott (1899:15) noted: "No really qualified Comacine architect is ever mentioned...without the prefix Magister [while] the subordinate ranks of stone-cutters or masons were called by the generic name of operarius." He goes on to note a similar definition is given by Alessandro Da Morrona whereby an operator (operarius) works with materials while the architect who designs and commands is a Magister. In his discussion of wages in the fourteenth century, Mortimer (2009:100) states an average mason could earn £7 1s 6d annually while a master mason could earn much more. "At Wallingford in 1365, the master mason is paid 4s a week, and at Windsor Castle the master mason in charge of the grand rebuilding is paid 7s a week...an [annual] earning bracket of £10-£7, similar to highly educated lawyers and physicians." Additionally, the master mason "has by his completed education risen to the stage of perfection, when he may teach others" (Scott, 1899:15). To be an architect, therefore, meant having an education and learning the skills of the trade on the job from his master.

As a UK example, William of Wykeham was born c.1324 into a yeoman family in Wickham, Hampshire, but was sponsored by the Sheriff of Hampshire to attend grammar school in his teenage years where apart from grammar, he learned "geometry, French, arithmetic and dialectic—a complete secular education for those days" (Moberly, 1887:7). This, with an emphasis on geometry, was the educational path to becoming a master mason. Wykeham rose to become "surveyor of the works at Windsor castle and park" (Bennett, 2008). Later he became Bishop of Winchester and amassed a considerable fortune, which enabled him to establish both Winchester College and New College, Oxford.

William of Wykeham came of a good family and this with his education allowed him to rise in his profession. Mortimer (2009:99) makes the further point that unless you had the money, you could not buy your way into a craft guild. To become an apprentice in a craft also required money at a level that would be difficult to save for, being some £1–£2. This was a time when a labourer was paid 3¼d per day for a 6-day working week noting there were 40–50 non-working holy-days (holidays) a year (Mortimer, 2009:99–100), making approximately £3 14s. per annum. This makes clear the extreme difficulty of a labourer saving enough to pay for an apprenticeship as a step up in the world. Clearly, inequality has always been with us when it comes to creating the built environment and raises the issue of how SDG 4 (Quality Education) and SDG 10 (Reduced Inequalities) can both be achieved simultaneously. SDG 5 (Gender Equity) is another issue as so far, all people in the role of architect in ancient and medieval times have been men, although as noted in Chap. 6 medieval women labourers on the building site were paid the same as men.

Architectural Education in Tudor (1485–1603) and Stuart (1603–1711) Times

The architectural historian John Summerson distinguishes between UK designers of buildings of this period who had come up through the trade route and those who had been educated. For him, Inigo Jones (1573–1652) was the educated architect, notwithstanding the fact Jones had come from a trade background, as his father was a dealer in cloth and Jones had started by being apprenticed to a joiner (Summerson, 1966:25–26).

> At the time when Inigo Jones first turned his attention to architecture many surveyors with strong architectural inclinations were at work in and outside London…These men came, for the most part, out of the building trades, but could draw and write well and often knew some Latin. Their culture was, however, strictly limited; they rarely travelled abroad and possessed no intellectual authority. The role of designer as a man of intellectual status was left to Jones.

However, Jones had travelled in Europe and on his return had secured a job at court as a designer of stage sets and costumes for the court dramas or masques (Summerson, 1966:15–19). Jones went on to remodel parliament, being made an MP in 1621 "more for his architectural expertise and technical skills" (Hawkyard, 2013), although he is better known as the architect of the Banqueting Hall at Whitehall (completed 1622), and the Queen's House at Greenwich (c1636) (Fig. 10.1), both in the classical tradition.

An architect of the time who came up through the trade route was Robert Stickells (?-1620). In his will, Robert Stickells described himself as a freemason (Summerson, 1959). Stickells had worked on Dover harbour 1584–1585 and was responsible for the design of the earlier Banqueting Hall in Whitehall, scene of many masques and plays, that burned down in 1619 due to some "clumsy workmen with a lamp" (Astington, 2019:203). King James was said to see this as a blessing, as the interior wooden columns, painted to look like stone, interrupted the view (Anon, 2016). Later Stickells was to supply designs to Sir Thomas Tresham. The confusion between a gentleman who had a keen interest in

Fig. 10.1 Queen's House, Greenwich. (Source: Creative Commons (n.d.-a))

architecture, in this case Sir Thomas Tresham, and the architect who had learned a craft, in this case Robert Stickells, illustrates the changing status of the architect at this time. Circa 1604, it seems Stickells was involved in the designs for a new type of lodge for Tresham, the never completed Lyveden New Building, which was a cross-shaped structure, the centre of which probably had a dome. Thomas Tresham (1543–1605) was a wealthy landowner who owned many books on architecture and who must have had practical knowledge as he emphasised the need for adequate drainage of sunken walks in his garden design at Lyveden (Eburne, 2008) but who would also call on another designer as necessary, such as Stickells at Lyveden New Building.

However, it is another person from this period who is often claimed as the first English architect and he came through the building trade route. Robert Smythson (1534/5–1614) was a stone mason and, in 1568, was leading a team of masons when working on the rebuilding of Longleat House, so was, therefore, a master mason. Smythson went on to design a number of large Elizabethan mansions including Wollaton Hall (Fig. 10.2). The fact his son joined the family business (Beckett, 2014) shows, with exceptions like Tresham, how the training of architects then was both grounded in craft and linked to family, but still associated with the design of significant buildings.

The Royal Works

The most famous UK architect of this period is Christopher Wren (1632–1723). His father was a rector and his uncle a Bishop, so the Wren family had standing. Wren went up to Wadham College, Oxford in 1649 as a "gentleman commoner" (Tinniswood, 2001:27). Notwithstanding Wren's considerable intellectual abilities, it must have been a help that his first architectural commis-

Fig. 10.2 Wollaton Hall, Nottingham, 1580–1588. (Source: Creative Commons (n.d.-b))

sion—Pembroke College Chapel—came from his uncle, then Bishop of Ely (Pembroke College, n.d.) (Fig. 10.3).

Wren was a founding member of the Royal Society, the remit of which was to encourage and publish research, so although the Royal Society had members who practised architecture it was not a body for educating architects. However, when Wren was appointed Surveyor-General to the Royal Works in 1669, his organisation of this establishment through the incorporation of "designers, draughtsmen, clerks and craftsmen" made it a "flexible ladder of learning that could be scaled by an aspirant architect" (Crinson and Lubbock: 1994:13–14). Perhaps the most famous architect to follow this path was Nicholas Hawksmoor (1661–1736), son of a yeoman farmer and described as "a commoner of lowly origins" (Hopkins, 2016:25).

At this stage, therefore, the architect could be a gentleman who had been given a liberal education or someone who had come up through the trade route, which often ran in families. It was also advantageous to be well connected, as Wren had been. Hopkins (2016:25) notes that Wren was: "A well-connected gentlemen with impeccable royalist credentials, he was fortunate to have the ear of the monarch at various moments in his career as both scientist and architect." For those stemming from more lowly origins, like Hawksmoor, a degree of luck was needed for someone of influence to notice and nurture the potential talent.

Only for Gentlemen: Museum Minervae

Within this period in the UK, there was an institution that purported to teach architecture. The 1636 Museum Minervae was a liberal arts school for gentlemen in London, where the subjects included "modern languages, mathematics, painting, architecture, riding, fortifications, antiquities and "the science of medals"" (Ayres, 2014:423). The important issue here is that this

Fig. 10.3 Pembroke College Chapel, street elevation. (Source: Creative Commons (n.d.-c))

was for gentlemen, and hence men of substance and Ayres (2014:424) further notes these institutions "had little relevance to an ordinary apprentice or journeyman – indeed they would have been excluded from them. The great value of these aristocratic educational institutions was that they had the capacity to establish a visually educated elite, the potential patrons of the arts." The Museum Minervae foundered after the civil war and the rise of the commonwealth.

Formal architectural training at this time was very much for the gentleman architect, who would have money and leisure to pursue study and no need to make a living from the profession. Those who did need to make a living from architecture often found it a struggle. These were those generally trained through apprenticeship in building craft rather than in a drawing school. The successful architect James Paine (1717–1789), who was the son of a carpenter, took the building route to becoming an architect (Crinson and Lubbock, 1994:22). In turn, he articled eight pupils between 1756 and 1758 as a means of training others. This apprenticeship route became known as architectural pupillage (see below).

At this time, architects also combined the design of buildings with other money-making ventures such as land surveying and rent collecting. Architects also became developers such as the prolific John Nash (1752–1835), whose father was a millwright. "His business practices were a little shady, but he had a remarkable creative flair. A brilliant planner, he could adapt a wide variety of styles to suit the aspirations of his clients" (Country Life, 2009) (Fig. 10.4).

Fig. 10.4 Regent Street laid out under direction of John Nash. (Source: Creative Commons (n.d.-d))

Pupillage

Pupillage had its roots in the eighteenth century at a time when smaller professional groups "such as architects and surveyors" were growing in numbers (Earle, 1989:72). Whittick (1842:11), in his fascinating contemporary advice to Victorian parents covering all possible trades and vocations, is clear that architecture is not for the faint-hearted or those of restricted means.

> The youth desirous of becoming an Architect, should be liberally educated, and in addition to the Latin language, he should be master of French and Italian; have some knowledge of mathematics, geometry and drawing. The premium required with a pupil by a respectable master, is from two to five hundred pounds: the youth will also require a considerable sum for the purchase of books, instruments, and drawing materials.

The training a pupil received entirely depended on the dedication of the pupil master to this task (Crinson & Lubbock, 1994:45). At one end of the spectrum was Pecksniff, who appeared in Dickens' 1844 novel *Martin Chuzzlewit.* Pecksniff was an architect who made a living from collecting rents and the premiums of young gentlemen who then boarded in his establishment, where the young gentleman "improved himself, for three or five years, according to his articles, in making elevations of Salisbury Cathedral from every possible point of sight; and in constructing in the air a vast quantity of Castles, Houses of Parliament, and other Public Buildings" (Dickens, 1844; 1968:13).

At the other extreme were architects like Sir John Soane who believed in education and took his responsibility as a pupil master very seriously (Saint 1983:54). In all Soane had 64 pupils who "worked twelve hours a day for an average of five years, with six weeks' holiday" (Mordaunt Crook, 1969). Fees for pupillage with Soane were initially £50 but increased to 175 guineas as his fame rose (Crinson & Lubbock, 1994:25). For a date of 1830, £50 is equivalent to £3390 in 2017 and 175 guineas to £12,459 (National Archives, n.d.). Although somewhat later, these figures are in line with those quoted by Mordaunt Crook (1969) who stated that in the second half of the eighteenth century premiums varied from £10 to £210, normally tied to a seven-year contract. Of Soane's pupils, George Basevi (1794–1845) was the son of a city of London merchant and cousin of Benjamin Disraeli (Jacobs & Lipkind, 2021), and Robert Smirke (1780–1867) was the son of a painter and book illustrator (Mordaunt Crook, 1972:73), although his pupillage only lasted a year (Bolton, 1927:15). Soane himself was the "son of a bricklayer, or small builder" (Bolton, 1927:3).

Early Institutions for the Study of Architecture

Alongside articled pupillage were institutions that supplemented the study of architecture, normally on a part-time basis. The history of these and the efforts made to create an architectural profession have been well documented elsewhere (Crinson & Lubbock, 1994:38–88; Mordaunt Crook, 1969). The aim here is to look at two institutions and ponder on who was likely to study architecture because of the cost.

The Royal Academy Schools

The Royal Academy Schools were established in 1768 to teach painting, sculpture and architecture, but not through a knowledge of the crafts associated with making buildings. Ayres (2014:436) notes the Royal Academy Schools were to establish "a well regulated School of Design". From this, it is evident that "Design" was to take precedence over the processes of "making". Of the 77 students in 1768, only 3 were architects (Royal Academy. n.d.). The architecture students were assumed to be articled and therefore were exempt from daytime drawing classes but "the Academy library was opened three evenings a week. The Academy library was well stocked, early on, with books on architecture" (Royal Academy, n.d.). However, from the start, architecture was a poor relation in the Royal Academy Schools, which were "weighted in favour of painting" (Mordaunt Crook, 1969).

In general, the study of architecture at the Royal Academy involved composition, critical review of famous buildings, critical reading and critical review of structures, but lectures given by teaching professors were often erratic and of inconsistent quality (Royal Academy, n.d.; Morgan, 1973). In 1838, a London architect complained about the Royal Academy course (Mordaunt Crook, 1969):

> 'The science of construction,' adds Blyth, 'is of the first importance, but in the Royal Academy the student has no guide; practical details form no portion of his studies; his works not uncommonly, are judged of, not by the real standard of excellence, purity, propriety, and practicability of execution, but by their power of pleasing the fancy. It is thus that young architects, knowing that their designs will for the most part be judged of by artists, not unfrequently aim at producing pictorial and picturesque effects.

Pleasing those who were to judge the work was still uppermost in the mind of the student, as it is today both in modern education and in architectural competitions (see Chap. 8).

The Government Schools

Unlike the Royal Academy, the government schools of design were set up on a more practical basis. The government schools of design, which were established in 1836 following the 1835–1836 Parliamentary Commissions on Art and Design (Rifkin, 1988), were meant to educate designers who would be capable of leading the new manufacturers of "stuff" that had proliferated with the industrial revolution (Anon, 1843).

The first new school was the 1837 Central School of Design in London, which in 1896 became the Royal College of Art (Royal College of Art Archives, n.d.). When first established the curriculum did not include architecture so as not to upset the pupillage system, given that at least some architects relied on the premiums of their pupils (Crinson & Lubbock, 1994:54). By 1849, there were 21 schools in various parts of the UK in response to the need for the "diffusion of a knowledge of the arts and the principles of design amongst the manufacturing population" (Rifkin,1988). Rifkin describes the 1841 Manchester school as having 141 part-time pupils, who:

> Roughly speaking and including apprentices, or 'boys intended for' one or another trade, they break down into thirty-one categories. The largest was draughtsmen for calico printers (25) [Manchester was a centre for calico printing] and their apprentices (I3), wood engravers (II), then copper engravers (9), house-painters (8), architects (8), and a veritable hotch-potch of others.

Although initially not central to the training of architects, these seeds of architectural education in government design schools were, in some places, to grow into substantial and well-known UK schools of architecture; "Canterbury, Hull, Birmingham and Heriot-Watt, for example" (Crinson & Lubbock, 1994:55–56). These same government schools also spread architectural training to the colonies. In India, in the middle of the nineteenth century, government-sponsored design schools with a focus on industrial design were established in Bombay, Madras, Calcutta and Lahore (Fujita, 2017). As in the UK, these Indian schools over the years developed architectural courses, and these in turn became schools of architecture.

Two Routes to Becoming an Architect

The RIBA was established in 1834, and from about this time offered voluntary examinations in architecture. The RIBA exams began in 1882, when most aspiring architects learned by being articled pupils, and these exams had to be passed before a person could apply to be an associate of the RIBA (Howarth, 1959), and hence be seen to have achieved a level of architectural standing. The exams later evolved into a sequence of three separate assessments, the forerunners of the current three-part educational structure (Lester, 2022). The RIBA examinations were simply there to ensure the profession maintained a standard of entry. For a person in articled pupillage, taking the RIBA exams was the route into the profession. Those studying full-time in the schools of architecture would take exams that had been accredited by the RIBA. There were thus two routes, one of which—pupillage—meant that the student would be earning something, rather than paying out fees to study. Articled pupils could also attend night school to prepare them for the exams. Discovering the costs of these various routes in the early twentieth century is difficult. For those trained in the architecture schools, Anderson (2016) notes:

> State support for universities has a longer history than is often supposed, and allowed student fees to be kept low. Even before 1914, outside Oxford and Cambridge, fees usually only made up well below half of university income. Making students pay the full cost of their education is a radical innovation.

However, if you were a student in the 1930s, anecdotal evidence suggests the costs were not insubstantial. "My father studied Engineering at Liverpool University about 1930. He was in the minority of students who weren't the sons of nobility. He came from a fairly wealthy middle-class background and even then had sponsorship because his family couldn't have afforded it by themselves" (Oddy, n.d.). Phyllis Hudson who in 1932–1936 studied at Manchester University for an Arts degree said: "Although the work was hard and the fees were expensive, she felt lucky to be there and she loved her experience" (Pierce, 2013). As an example of the costs of tertiary study in the first half of the twentieth century, albeit not from the UK, the University of Pennsylvania (University Archives and Records Centre, 2022a) gives the annual fees for studying architecture in 1930 as US$400, and the total cost of studying including board, food and text books, as US$955. This has the equivalent buying power of US$7097 and US$16,945 in 2022. In 1920, fees for an architecture course were US$320 and for fees, board, food and books US$750, equating to a buying power of US$4741 and US$11,112, respectively. Along with the School of Engineering and Applied Science, architecture fees were the highest in the university at that time. The school of architecture at University of Pennsylvania opened in 1890, although courses in architecture were being offered earlier in 1869 (University Archives and Records Centre, 2022b).

The higher cost of architecture fees in 1920 compared to other courses in the institution was also noted, when Cardiff Technical College founded its school of architecture that year (Powell, 2009:1,4). The history of the Welsh School of Architecture is revealing in a number of other aspects. The first is the struggle to have this new course, which could be taken full-time or part-time, recognised by the university as a degree course, something that did not happen until 1939 (Powell, 2009:17). The second concerns the views on architectural education of W.S. Purchon, the first head of the school and at the start its only teacher (from 1907 to 1920 he had been "Lecturer in charge of the Department of Architecture in the University of Sheffield") (Chatterton, 1923:201). Purchon, with years of architectural experience gained in part during WWI, believed in life-long learning and that: "Construction teaching should be integrated with design; working drawings were important, as were full-size details, laboratory work, and visits to materials producers and sites" (Powell, 2009:9). Powell (2009:5) also tells the story of David Williams, an early part-time student at the school who became a miner at 14 and worked his way through the school by taking night shifts in the mine. This prize-winning student went on to become Chief Assistant Architect at Essex County Council, clearly demonstrating that whatever the cost and whatever your background, if architecture is your chosen vocation, it is possible to become an architect.

The cost of pupillage at this time is even harder to discover. In discussing the entry of women to the Architectural Association school of architecture in London in 1920, McEwan (2018) notes that at that time, even women from the "comfortable classes" might have hesitated in asking the family for the 300 guineas or more it would cost for them to become an articled pupil. This equates to a 2017 buying power of £9589 (National Archives, n.d.-b). The 1920 cost of sitting the RIBA final examination, known as the Special Examination, was £6 6 s (6 guineas) in Australia (Anon, 1920). This is approximately £336 in 2022 but that is only the cost of the final examination that would give the title ARIBA (Associate of the Royal Institute of British Architects) if passed, not the tuition required in preparing for it.

The Oxford Conference

In the UK, entry to the architectural profession was to change with the 1958 Oxford Conference. Before the Oxford Conference those with five years of successful secondary schooling could work in an architect's practice, take part-time or evening courses at local schools of art or technical colleges and so work towards taking the RIBA external examinations which effectively gave them entrance to the profession. This was useful for those living away from large urban centres where the schools of architecture were housed. Whatever the wages, such students would be earning from the age of 16, however many years it might take to become fully qualified.

According to the Oxford Conference chairman's (Leslie Martin) post-conference report (Martin, 1958), the idea of holding a conference came from the RIBA Council as means of exploring all aspects of architectural education. Amongst other things, the Oxford Conference recommended raising the entry level into some form of architectural education from five years of secondary schooling followed by being an articled pupil to seven years of secondary schooling followed by a degree course undertaken in a university or polytechnic.

To be an architect now meant, after school, undertaking a course of study at a university or polytechnic school of architecture. This was normally another five years of formal full-time instruction sandwiched with two years working in an architect's office before registration as an architect. This is still the norm in UK-based courses today, such as those found in commonwealth countries. At the time of the Oxford Conference, pupillage, the cheaper and less exclusive route to becoming an architect, still formed 46% of those successful in the examinations in 1957, and Fig. 10.5 shows the dominance of pupillage in this period. The five-year architecture course proposed by the Oxford Conference changed all this and effectively meant five years when the student needed financial support before they could be working full-time.

The conference felt that some of the current tuition students received was not adequate, but however worthy the aim of improving the overall standard of instruction, the change to the university or polytechnic entry requirement must have closed off the possibility of becoming an architect for those who needed to work and study simultaneously.

What a student would have had to pay if entering a recognised school of architecture in 1958 is hard to determine beyond the fact that some form of fees would have been paid, although this was to change soon afterwards (Anderson, 2016).

> Free higher education…[was] introduced in 1962 following the report of the Anderson committee. This was intended to simplify what had become a jungle of grants and scholarships, and had two aspects. First, though fees were not formally abolished, full-time domestic students now had them paid by the state. Second, students were entitled to a maintenance grant, whether at their local university or away from home. The maintenance grants were means tested (dependent on parental income), as were fee grants until 1977, but both were outright payments, not loans. Provided parents paid their share (if any), students were free of financial burdens.

What is more relevant is that Anderson (2016) goes on to make the point that this move to free university education came before comprehensive schools were established and when only 20–25% of children were in grammar schools leading to the chance of university entrance. An exam at the end of primary education (age 11) would either send a child to a grammar school, where the focus was on academic studies with a view to going to university, or to a secondary modern school, where the focus was on educating children to go into a trade (Richardson, 2016). The UK fee paying private (but known as "public") schools would provide an education geared to university entrance. To be an architect, therefore, meant you either had to be clever or privileged. For children bright enough to qualify to attend a grammar school, the fact that university education was free enabled some from poorer backgrounds to become architects (Iddon, 2015),

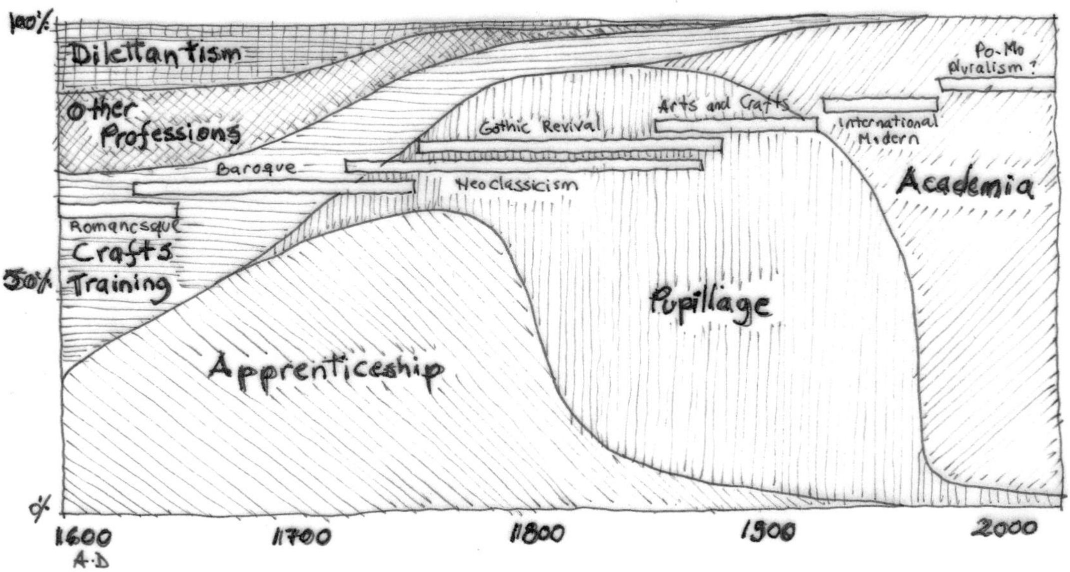

Fig. 10.5 Four hundred years of architectural education. (Based on McCormack (2015))

although many surveys and reports on education in the 1960s repeated that the home background was critical to children's success in education and that those from the middle classes were more likely to succeed than those from poor, working-class families (Gillard, 2018).

Writing at the end of the twentieth century, Cunningham (1999) saw the loss of pupillage as a dangerous legacy of the Oxford Conference. He also noted that architecture had an uneasy place within the university system, being neither part of the humanities nor part of science. For him, design was something that architects did, but he viewed design as a tool, not a university discipline like law or medicine. Design was something that a builder or a home handy-man might do when making a building without having undertaken a course in architecture. Cunningham was very critical of the effects of the Oxford Conference.

> The discouragement of part-time education and the 'burden of pupillage' eliminated any obligation for practitioners to serve as masters to guide aspirants across the threshold 'when play becomes real'—as do lawyers, doctors, musicians, sports people and fine artists—and therefore placed the onus on academics to produce 'complete' architects, a fatuous expectation which endures.

The Oxford Conference, and more importantly the acceptance by the RIBA of its recommendations of making entry into the profession via tertiary education, made architecture a more exclusive profession as to enter it a person had to be clever at passing exams and somehow find sufficient funds to undertake the five non-earning years of study. It is fortunate that 1962 in the UK saw the era of potentially free education based on means testing, so at least one of these barriers was removed for a while. However, by the 1990s, the maintenance grants became loans "with some outright payments retained for poorer students (until abolished in 2015)", with top-up fees introduced in 1998 (under a Labour government) and full tuition fees in 2010 (Anderson, 2016). Embarking on becoming an architect in the UK was now not for the faint-hearted, intellectually challenged or poor.

Recent Architectural Education in the UK

The move to full tuition fees has made architecture an expensive course of study. A survey of architectural students undertaken by the *Architects' Journal* reported in 2015 that the average annual cost to an architecture student was almost £13,000, and nearly 25% expected to owe more than £50,000 by the end of the course (Mark, 2015). In 2021, as in 2020, the average salary of registered architects of less than five year's standing was £34,000, rising to £40,000 for those registered for five years and above (RIBA, 2022a). It is, therefore, no surprise that nearly a third of students in the 2015 survey thought they would never repay all the loans incurred for their education. The fact that students are now paying heavily for their architectural education suggests they may have an influence on what is being taught. The problem with this is that students do not necessarily know what they need to know to become architects, beyond the need to find a job to pay for their education. What they see as students in the magazines and in lectures are the works of successful architects, normally shown without any notion of the budget spent to make these praiseworthy buildings. For many years the *Architects' Journal* published cost information on its featured buildings, although recently this has been replaced by a self-reported carbon footprint. If architecture is presented without costs, it is no wonder that those students who see the reality of how difficult and chancy it is to become well known, and hence wealthy, might feel they will never be able to repay their loans.

Also hidden is the fact the much cheaper pupillage route to becoming an architect is still available, although confusingly the RIBA now call this an "apprenticeship" (RIBA, 2022b). The fact that in the UK it is still possible to learn (and earn) on the job (RIBA, 2022c) through these routes seems a well-kept secret as in 2015 only 0.5% of each year's cohort of those studying to be architects were not in full-time education (McCormack, 2015).

Problems with Current Architectural Education

In a 1981 article on teaching, Kroll and Mikellides made the point that contrary to how architecture is normally taught, "architecture is not made by the architect" but rather results from the co-operative efforts of many people, including the client, users, builders, materials suppliers, quantity surveyors, engineers, planners, law makers and those checking that laws are met. In architecture school, the student is normally the architect for an imaginary client and seldom are any of the other parties that make architecture involved. This makes it hard for students to question what they need to know. Argentinian architect Rodolfo Livingston points out that much work for architects comes as renovation projects, something seldom dealt with in schools of architecture: "One hundred percent of architects have never seen a renovation in college, never trained to deal with a real client. It is the same as training pilots by showing drawings of airplanes, without doing flight practice" (Jurado, 2016). In 2019, Patrik Schumacher, the head of Zaha Hadid Architects, made the same points as Kroll and Mikellides, claiming architecture schools were like curriculum-less art schools: "Accordingly architectural education is detached from the profession and from societal realities and needs as expressed in real (public or private) client briefs". Schumacher also noted that "that too many architectural educators do not have sufficient experience working as architects" (Ravenscroft, 2019). This seems strange as the point made in 1958 when architecture in the UK became a degree course, and thus set the pattern for the current schools, was that the teachers in the schools of architecture should be skilled practitioners who, on occasion, would give up time to teaching (Martin, 1958).

The Role of the Architect

Apart from those teaching in schools of architecture not necessarily being "skilled practitioners", the other problem with what is taught is different views of the role of the architect. A 1969 first year University of Cambridge *Theory of Architecture* exam paper consisted of one question: "Commit to paper, as concisely as you can, your own ideal concerning the architect, his position and his product." Setting aside the gender bias, given the current problems that face humanity—climate change, inequality, resource limitations—it would be easy, were this exam question to be posed now, to structure an answer that saw the architect as the enabler in his/her local corner of society. The job would be to help households, and commercial clients make the best use of the budget and the resources available to them to create homes and buildings with a minimum environmental impact that were also weather tight, easy to repair and easy to take apart, so that the resources in them could be reclaimed. There are architects who do fulfil such a role but in the developed world perhaps not as many as there should be by a significant margin.

It could be argued that twentieth-century modernism did espouse using architecture as a tool with which to give everyone decent housing and good working conditions but without the complete commitment to respecting the environment that is necessary today. Since the potential problem of climate change from fossil fuel consumption was noted as long ago as the end of the nineteenth century (Vale & Vale, 1991:21), it seems hard to understand why all new buildings have not aspired to use as little energy as possible, whether in hot or cold climates, given the all-glass tower continues to dominate the commercial building world whatever the climate. Until this absurdity stops, it is difficult to imagine how the profession can add anything to SDG 11 (Sustainable Cities and Communities).

Architecture as an Interconnected Process

The child who writes his or her address as the number, street, suburb, city, country, world, universe, big bang has a clear idea of the interconnectedness of things and how a relationship (where you live) can be perceived at a number of scales, and even over time since the big bang was the start of everything. Such interconnectedness is how eco-systems work, where scale is equally important. Architecture can be thought of in the same way as an interconnected process that leads from the desire to have a building to the building itself via a design idea that is explored in depth, so that third parties can see that the building meets legal expectations and that the depth is sufficient for the building to be constructed, handed over and occupied. Other connections exist, as noted above, such that the building has been designed to last a long time because it is easy to maintain and has been detailed to preserve its integrity, whereas inside-out buildings like Lloyds of London where the services are exposed have been problematic (Kollewe, 2015).

In the modern world, a building will normally need energy to make it run, as people are no longer satisfied to stop working when it gets dark. The materials from which it is made will also stem from the wider environment, some of which may never be replaced naturally (some plastics are made from oil which is fossilised animal and plant remains), and some of which may never be reusable (plastic shrink wrap round materials). The once complete cycle of buildings that were no longer needed being moved or mined for all the materials that could be reused has been lost, at least at some levels of architecture, including that taught in many architecture schools. Waite (2022) noted that 12.5% of UK architecture schools never even deal with retrofitting existing buildings, which is unfortunate as reusing existing structures can use fewer resources than building new ones. In some societies without architects, something more like a complete cycle in terms of materials used for buildings still exists (Fig. 10.6). Here, what is built depends on what is available that the owner/builder can afford, so that all old materials have value rather than incurring a cost for their disposal as waste, as happens in the developed world (Ministry for the Environment, 2022). This may look haphazard and untidy but is an example of the use of "good building knowledge" (see Chap. 7).

Architecture Stems from Nature

The point of this discussion is to consider how and what we teach students about the nature of architecture, such that it will lead students to see the architect and his/her product as part of the interconnectedness of everything. There is a zero point in architecture before there is any idea of what a building might be, or a dream of what a space might look like, where architecture is already part of something bigger—nature. In this sense, nature is the total physical environment that people draw on to make things. A computer is made of nature, a bed is nature, a house is nature, and the built environment also comes from nature. The materials used and the way of shaping them might differ, but everything comes from and exists on the same planet. Viewed in this way it becomes obvious that all architecture, planning, and urban design need, is both to respect this fact and minimise any disturbance to nature. However, architecture has seldom been viewed in this way, and probably not since it was included in the fine arts or identified with large and monumental buildings.

The Architect as Saviour of the Environment

Saint's (1983) seven images (or ages) of the architect (The Architect as Hero and Genius; Myth and the Medieval Architect; The Architect as Professional; The Architect as Businessman; The Architect as Gentleman; The Battle of the Bauhaus and The Architect as Entrepreneur) miss what might be termed the Architect as Saviour of the Environment. Perhaps the rather homespun looking espousers of alternative technology and design that were intended to work with rather than against the environment, as appeared in the 1970s, were not seen as proper architects, or at least not then. Graham Caine's 1972 Eco-house (Fig. 10.7) was built with the help of a £2000 grant from the Architectural Association where Caine was a fourth-year student, although many of the materials were scavenged from waste.

Fig. 10.6 Self-built informal housing. (Source: Creative Commons (n.d.-e))

The house, which was lived in and run by Caine and his wife with their baby daughter, was demolished when its two-year temporary planning permission lapsed. It was designed on the principle of interconnectedness, as wastes were recycled to grow food, as well as making use of insulation, solar energy collection, phase change materials and rainwater collection. Although Caine presented the monitoring results as his final year diploma submission, he failed the exam so never became an architect, although he was immediately employed by the AA (Kallipoliti, 2012). The same reference also notes:

> The story of the Eco-House in South London raises a significant disciplinary paradox. It was an experimental laboratory sponsored by the Environmental Council of London and a house that compelled the public's imagination as well as the interest of the popular press; it raised hopes for an alternative sustainable occupation of the urban sphere. At the same time, it was rejected and notoriously criticized by the architectural community for its lack of canonical references to core disciplinary conventions: investigations of form, proportions, and spatial syntax.

The moment when architectural education could have embraced environmental issues whole heartedly was lost. This sends a caveat, when it comes to SDG 4 (Quality Education) in that what is taught is as important as the fact that something is being taught, and what is taught is a harder thing to measure.

The Architect as Problem-Solver

This idea that not everything new is good forms a further caveat for architecture. Basic human needs do not change, however many new smart phone models are produced. In education, much is made of the problem-based approach and solving design problems. However, "solving" a problem inevitably creates a new problem. At the start

Fig. 10.7 Graham Caine's 1972 Eco-house, with living quarters behind and greenhouse in front. (Based on Twentieth Century Society (2023))

of the twentieth century, a shortage of rural housing for rent at a price the average labourer could afford led architects to design cottages (Weaver, 1919:6–11). Such an affordable architect-designed cottage at that time might well incorporate a services core with downstairs toilet, or even a toilet accessed from outside. This was not a problem when bedroom chamber pots were common and accepted. However, this same dwelling becomes a problem for an architect who is asked to bring it up to what are now viewed as acceptable modern standards with an internal, upstairs bathroom. All designs are temporary solutions to present problems and the starting point of future problems (Garcia et al., 2021:99–134).

Another example of creating future problems comes with the move to neoliberal economics introduced in the UK under Thatcher. Her reign as UK prime minister starting in 1979 saw the abolition of local authority departments of architecture, the loosening of planning regulations that allowed out-of-town shopping centres, the deregulation of the finance industry that changed the skyline of London with the new office buildings, and the abolition of the mandatory fee scales for architects (Fulcher, 2013). The latter meant that the time for research built into fee scales that could have produced buildings more aligned with environmental goals was no longer there. Making money was the goal. This has left a legacy of buildings whose environmental performance needs to be upgraded. Solving one problem—trying to make cheaper buildings through low fees—produces another—poor environmental performance.

What was current in architectural journals also changed post-Thatcher—at least in the case of the UK journal *Architectural Design* (AD). From the end of 1971, it ran an almost regular "Recycling" column featuring items such as the Clivus composting toilet (Moorcraft, 1971) and wind power (Stabb, 1972). Late in 1972, this morphed into an almost regular column entitled Ecotech, which ran until the following year. The name with its fusion of architect and ecology is important, as it suggested that architecture had a role to play in environmental issues. The journal continued with articles on such issues as Alternative Technology and autonomous houses until 1976, when an index was included of all AD articles on environmental issues. This moment also saw a change in editor at which point there was an intermittent column entitled "Energy" that lasted until 1978 and after that environmental issues were no longer seen as relevant to the readers, whether architects or students. The chance for a revival of interest in environmental issues in the built environment had to wait until a certain urgency was perceived, when it came to climate change and the need for more sustainable built environments.

The Teaching of Sustainability in Architecture

With the warning from the IPCC (2022) that the window of preventing major change to the climate of the planet is shrinking fast, and realising that in 2018 buildings accounted for almost 40% of global energy and consequent greenhouse gas emissions, and that over 80% of global primary energy was from non-renewable sources (Ritchie et al., 2020), it seems it would make sense now, if not even in the 1970s, for every building designed in a School of Architecture to be zero energy in use, made of low-energy materials, and working with rather than against nature. In this way, the next generation of architects would be equipped to cope with the enormous problem for humanity of the contribution current buildings are making to climate change. Experience shows that this is far from true in all schools. Álvarez et al. (2016) state:

> Architecture has an essential role in sustainable development, and it is necessary to educate professionals who have a well-balanced and integrated knowledge of local issues and global standards. In order to achieve that, an appropriate implementation of sustainable architectural education curricula is needed in the schools.

The same authors looked at sustainability teaching in 20 schools representing the leading universities in 11 Asian countries (including Hong Kong and Singapore). In terms of the required credits for graduation, only one school had 25% sustainability-related credits with four more around 20%. Three schools had 5% or less, three more under 10%, and the rest varied between 10% and 15%. Most courses that dealt with sustainability issues were technology related. Four schools had no required design courses based on sustainability. A study of the websites of architectural schools in the USA and UK (Ismail et al., 2017) found that there was no consistent method of delivering sustainability as part of architectural education and no way to gauge how the knowledge learned in various courses transferred "into practice as sustainable design". A more recent study of three university schools of architecture has been undertaken (University of Auckland, University of Texas at Saint Antonio and CEU Cardenal Herrera University) (Boarin et al., 2020), partly based on a student survey. In the first two institutions, sustainability courses were integrated through the years but appeared less in the design studios, whereas in the Spanish school sustainability was incorporated into all design studios. In Auckland and Cardenal Herrera universities, 26% and 30%, respectively, of students thought the subject was dispensable, whereas only 9% held this view in Texas, which, as the authors note, "is remarkable given the great majority of respondents were from the undergraduate degree where the subject finds little space in both lecture-based courses and design studios". Perhaps, it is the delivery of sustainability knowledge through lectures and seminars rather than in the design studio that is the issue here. Unless dealing with environmental problems is seen as part of the design of buildings, as Graham Caine did in 1972, then why should it be perceived as important for architecture? In the section of the survey where students could comment, common to Auckland and Texas, students felt "sustainability should be more integrated into design studios and not be a topic for courses or seminars only." But even in the 1970s, the work that Graham Caine did was not perceived as "proper" architecture to the extent that designing, constructing and monitoring a building was not sufficient to satisfy the diploma year exam at the AA.

One of the problems that emerged in the study of architecture at Auckland was that architect tutors were brought in to teach studio who may never have been exposed to sustainability teaching or have contradictory ideas of what it means. In effect, students were taught either what the tutors had learned in architecture school before the need to respond to climate change became an issue or what they had learned in practice. The formal lecture and seminar courses were delivered by permanent staff members with research experience in sustainability. These issues with the teaching of sustainability in architecture happened despite, in 2009, the University of Auckland signing up to the statement on sustainability of Universitas 21, which includes the idea that: "Appreciation for and general knowledge of sustainability must be actively promoted" (Universitas 21, 2009). The University of Auckland's own statement includes the words that "staff members are attracted to the University to research, teach and practice sustainability [and] students are attracted to the University to learn, research and contribute to sustainability" (University of Auckland, 2021). Given the lack of embedded sustainability in design teaching at the Auckland school of architecture brings to mind the seventeenth-century proverb "fine words butter no parsnips".

The Study of Vernacular Architecture: Local Plus Repetitive Equals Unique

The making of buildings has not always been divorced from environmental concerns. The study of vernacular architecture is almost exclusively the study of how to make buildings that use local resources and that respond to environmental conditions. These can range from an enclosed thick-walled mud building in the hot desert, designed to keep out the heat but be warm in the cold desert night, to a thick-walled mud building with a thatched roof and small windows designed to keep the heat in (Fig. 10.8).

The buildings in Fig. 10.8 represent a tradition that evolved through initial experiment, refinement and then repetition. This does not mean that every building is the same, but the way of making a building for a locality is passed on through the generations, such that differences emerge from repetition (Deleuze, 1968). This is what happens in the natural world, where each leaf of the same plant is both a repetition and a difference from every other leaf. This difference through repetition is what inspired Christopher Alexander and colleagues to produce *A Pattern Language* (1977), which collected together repeated aspects of the built environment, whether a town, neighbourhood, house, garden or room. These patterns are viewed as important just because they are repeated, meaning that they satisfy some inner human need when it comes to the built environment.

> The more living patterns there are in a place—a room, a building, or a town—the more it comes to life as an entirety, the more it glows, the more it has that self-maintaining fire which is the quality without a name. And when a building has this fire, then it becomes a part of nature. Like ocean waves, or blades of grass, its parts are governed by the endless play of repetition and variety created in the presence of the fact that all things pass. (Alexander, 1979:x–xi)

However, these are the buildings made by people who understood where they lived and the materials available to them. These are not the trained architects as outlined in this brief history who produced large and apparently important buildings. In the past, however, architects have embraced difference through repetition. Vitruvius may, as an architect, have laid down the ideal house form for making the best use of solar energy whether in the north or south of the empire (Gelder, 2018), but not every Roman house built on such principles was the same as its neighbour in a particular location. Nor is every villa by Palladio the same, though they share characteristics of symmetry and a porticoed entrance as well as simplicity of geometric forms and being sparing with ornament, so differences can emerge through repetition even within the oeuvre of one architect. These points challenge the preconception that SDG 11 (Sustainable Cities and Communities) will be the result of the emphasis on innovation in SDG 9 (Industry, Innovation and Infrastructure).

Fig. 10.8 (left) Beehive houses Syria. Source: Creative Commons (n.d.-f); (right) Cob Cottage. Source: Creative Commons (n.d.-g).

Building on tradition and repetition is very different from the way architecture is now taught, where the emphasis seems to be on creating something not seen before and then having to puzzle out how it might be constructed. Why has the nature of architecture changed from the repetitive tradition to the quest for the new? This may have been an inheritance of Modernism, as Habraken referred to this when he stated that the built environment was a design problem in which everything had to be designed anew in response to the problem (Montaner, 2015:9). The current global problems as outlined in the SDGs need a built environment that is both robust and flexible, as well as one that minimises environmental impact. Those training the designers of such an environment need to change the view of the architect from that of the prima donna artist to that of the servant, not only of humanity but of the planet.

References

Alexander, C. (1979). *The timeless way of building*. Oxford University Press.

Alexander, C., Ishikawa, S., & Silverstein, M. (1977). *A pattern language*. Oxford University Press.

Álvarez, S. P., Lee, K., Park, J., & Rieh, S.-Y. (2016). A comparative study on sustainability in architectural education in Asia—With a focus on professional degree curricula. *Sustainability, 8*. 32pp.

Anderson, R. (2016). *University fees in historical perspective.*. Retrieved from https://www.historyandpolicy.org/policy-papers/papers/university-fees-in-historical-perspective, 7 July 2022.

Anon. (1843). The Government School of Design. *Illustrated London News, 2*, 375–376.

Anon. (1920). For intending candidates in New South Wales. *Architecture, 7*(3), 76–77.

Anon. (2005). Hemiunu, Son of Nefermaat. In E. I. Bleiberg et al. (Eds.), *Arts and humanities through the eras* (vol. 1: Ancient Egypt 2675–332 B.C.E) (pp. 59–60). Gale eBooks.

Anon. (2016). *The Lost Palace of Whitehall.* Retrieved from https://tudortimes.co.uk/guest-articles/the-lost-palace-of-whitehall/a-modern-banqueting-house, 9 August 2022.

Astington, J. H. (2019). The Jacobean banqueting house as a performance space. In S. Chiari & J. Mucciolo (Eds.), *Performances at court in the age of Shakespeare*. Cambridge University Press.

Ayres, J. (2014). *Art, artisans and apprentices*. Oxbow Books.

Beckett, M. (2014). *The first architect? 400 years on from the death of Robert Smythson.* Retrieved from https://thecountryseat.org.uk/2014/12/05/the-first-architect-400-years-on-from-death-of-robert-smythson/, 4 August 2022.

Bennett, M. (2008). *A review of William Wykeham: A life.* Retrieved from https://reviews.history.ac.uk/review/670, 13 June 2022.

Boarin, P., Martinez-Molina, A. and Juan-Ferruses, I. (2020). Understanding students' perception of sustainability in architectural education: A comparison among universities in three different continents. *Journal of Cleaner Production 248*. 16pp.

Bolton, A.T. (1927). The portrait of Sir John Soane, R.A. (1753–1837). London: Sir John Soane's Museum Publication.

Bonner, S. (2012). *Education in Ancient Rome: From the elder Cato to the younger Pliny*. Routledge.

Campbell, S. (2022). *Planning history timeline: A selected chronology of events.* Retrieved from https://websites.umich.edu/~sdcamp/timeline.html, 2 May 2024.

Chatterton, F. (Ed.). (1923). *Who's who in architecture*. Architectural Press.

Country Life. (2009). *Great British Architects: John Nash*. Retrieved from https://www.countrylife.co.uk/news/great-british-architects-john-nash-27592, 13 July 2022.

Creative Commons. (n.d.-a). "Greenwich Park - Blackheath Avenue - View NNW on Queen's House 1617 Inigo Jones & Greenwich Hospital 1712 Sir Christopher Wren" by Txllxt TxllxT is licensed under CC BY-SA 4.0.

Creative Commons. (n.d.-b). "Wollaton Hall and Park 8-3-2014" by Martin Pettitt is licensed under CC BY 2.0.

Creative Commons. (n.d.-c) "Pembroke College Chapel" by alasdair massie is licensed under CC BY-NC-SA 2.0.

Creative Commons. (n.d.-d). "London - Regent Street" by roger4336 is licensed under CC BY-SA 2.0.

Creative Commons. (n.d.-e). "Slums alongside the railway" by wallygrom is licensed under CC BY-SA 2.0.

Creative Commons. (n.d.-f). "Beehive houses" by upyernoz is licensed under CC BY 2.0.

Creative Commons. (n.d.-g). Woolfardisworthy, Glebe Cottage - geograph.org.uk - 255043" by Martin Bodman is licensed under CC BY-SA 2.0.

Crinson, M., & Lubbock, J. (1994). *Architecture: Art of profession?* Manchester University Press.

Cunningham, A. (1999). Getting other, not better. *Architects' Journal*, 23rd *September*, 55–56.

Deleuze, G. (1968). *Différence et Répétition*. Presses Universitaires de France.

Dickens, C, (1844; 1968). *Martin Chuzzlewit*. Dent.

Earle, P. (1989). *The making of the English middle classes*. Methuen.

Eburne, A. (2008). The Passion of Sir Thomas Tresham: New Light on the Gardens and Lodge at Lyveden. *Garden History, 36*(1), 114–134.

Fainstein, S.S. (2024). *Urban planning*. Retrieved from https://www.britannica.com/topic/urban-planning, 2 May 2024.

Fujita, H. (2017). Art and design education in nineteenth century India: British background and development in South Asia. In *The Second Asian Conference of Design History and Theory—Design Education beyond Boundaries*. Tokyo, 1–2 September, pp. 109–120.

Fulcher, M. (2013). *The Thatcher years: Architects reflect on the legacy of the Iron Lady.* Retrieved from https://www.architectsjournal.co.uk/archive/the-thatcher-years-architects-reflect-on-the-legacy-of-the-iron-lady, 7 July 2023.

Garcia, E., Vale, B., & Vale, R. (2021). *Collapsing gracefully: Making a built environment that is fit for the future*. Springer.

Gelder, J. (2018). Design with climate in Ancient Rome: Vitruvius meets Olgyay. In P. Rajagopalan & M. M. Andamon (Eds.), *Engaging architectural science: Meeting the challenges of higher density* (52nd International Conference of the Architectural Science Association) (pp. 745–752).

Gillard, D. (2018). *Education in England.* Retrieved from http://www.educationengland.org.uk/history/chapter12.html, 25 July 2022.

Hawkyard, S. (2013). Inigo Jones, the surveyors of the works and the 'parliament house'. *Parliamentary History, 32*(1), 16–59.

Hopkins, O. (2016). *From the shadows: The architecture and afterlife of Nicholas Hawksmoor*. Reaktion Books Ltd.

Howarth, T. (1959). Background to architectural education. *Journal of Architectural Education, 14*(2), 25–30.

Iddon, P. (2015). *The shocking cost of becoming an Architect - is it worth it?* Retrieved from https://www.linkedin.com/pulse/shocking-cost-becoming-architect-worth-paul-iddon/, 27 August 2024.

IPCC. (2022). *Climate change 2022: Impacts, adaptation, and vulnerability*. Cambridge University Press.

Ismail, M. A., Keumala, N., & Dabdoob, R. M. (2017). Review on integrating sustainability knowledge into architectural education: Practice in the UK and the USA. *Journal of Cleaner Production, 140*, 1542–1552.

Jacobs, J., & Lipkind, G. (2021). *George, Basevi (Joshua)*. Retrieved https://www.jewishencyclopedia.com/articles/2616-basevi-ge, 28 July 2022.

Jurado, M. (2016). *Three dimensions.* Retrieved from https://www.clarin.com/ciudades/Dios-arquitecto-perjudica_0_NJkm9-TAg.html, 1 August 2022.

Kallipoliti, L. (2012). From shit to food: Graham Caine's eco-house in South London, 1972-1975. *Buildings and Landscapes, 19*(1), 87–106.

Kollewe, J. (2015). *Cheesegrater tower loses third bolt in three months.* Retrieved from https://www.theguardian.com/business/2015/jan/14/cheesegrater-leadenhall-building-bolt-city-london, 27 June 2022.

Kroll, L., & Mikellides, B. (1981). Can architecture be taught? *Journal of Architectural Education, 35*(1), 36–39.

Lester, S. (2022). Architecture in the UK: A study in professional entry-routes and entry-gates. *Higher Education, Skills and Work-Based Learning, 12*(1), 1–12.

Mark, L. (2015). *Revealed: The AJ student survey results 2015.* Retrieved from https://www.architectsjournal.co.uk/archive/revealed-the-aj-student-survey-results-2015, 8 July 2022.

Martin, L. (1958). Conference on architectural education. *Royal Institute of British Architects Journal, 65*(8), 279–282.

McCormack, K. (2015). *How do we learn to be architects?* Retrieved from https://www.ribaj.com/intelligence/how-do-we-learn-to-be-architects, 11 July 2022.

Ministry for the Environment. (2022). *Waste disposal levy expansion.* Retrieved from https://environment.govt.nz/what-government-is-doing/areas-of-work/waste/waste-disposal-levy/expansion/, 31 July 2022.

Moberly, G. H. (1887). *Life of William of Wykeham.* Warren and Son/Castle and Lamb.

Montaner, J. M. (2015). *La arquitectura de la vivienda colectiva*. Reverte.

Moorcraft, C. (1971, September). Clivus Multrum. *Architectural Design, 41*, 589.

Mordaunt Crook, J. (1969). The pre-Victorian architect: Professionalism & patronage. *Architectural History, 12*, 62–78.

Mordaunt Crook, J. (1972). *The British Museum: A case-study in architectural politics*. Pelican Books.

Morgan, H. C. (1973). The schools of the Royal Academy. *British Journal of Educational Studies, 21*(1), 88–103.

Mortimer, I. (2009). *The Time Traveller's Guide to Medieval England*. London: Vintage Books.

Oddy, L. (n.d.). *Quora: How much did the whole undergraduate course cost in Oxford in the 19th century?* Retrieved from https://www.quora.com/How-much-did-the-whole-undergraduate-course-cost-in-Oxford-in-the-19th-Century-Could-children-from-proletariat-families-attend-such-universities-at-that-time, 27 July 2022.

Oksanish, J. (2019). *Vitruvian man: Rome under construction*. Oxford University Press.

Pembroke College. (n.d.). *Chapel History.* Retrieved from https://www.pem.cam.ac.uk/college/about-pembroke/chapel/chapel-history, 3 August 2022.

Pierce, C. (2013). *What was it like to be a university student in the 1930s?* Retrieved from https://www.theguardian.com/education/mortarboard/2013/sep/27/university-life-1930s-1960s, 27 July 2022.

Powell, C. (2009). *The Welsh School of Architecture 1920–2008: A history*. The Welsh School of Architecture.

Ramanaik, S., Collumbien, M., Prakash, R., Howard-Merrill, L., Thalinja, R., Javalkar, P., Murthy, S., Cislaghi, B., Beattie, T., Isac, S., Moses, S., Heise, L. and Bhattacharjee, P. (2018). Education, poverty and "purity" in the context of adolescent girls' secondary school retention and dropout: A qualitative study from Karnataka, southern India. *PLoS One 13*(9). 22pp.

Ravenscroft, T. (2019). *Architecture education is in crisis and detached from the profession, says Schumacher.* Retrieved from Patrik Schumacher outlines the crisis in architectural education (dezeen.com), 11 July 2022.

Redding, R. W. (1992). Egyptian old kingdom patterns of animal use and the value of Fauna data in modelling socioeconomic systems. *Paléorient, 18*(2), 99–107.

RIBA (Royal Institute of British Architects). (2022a). *Salary guide 2022—architects and architecture.* Retrieved from https://jobs.architecture.com/staticpages/10290/salary-guide-2022-architects-and-architecture/#:~:text=Like%20last%20year%2C%20the%20average,the%20same%20as%20last%20year. 11 July 2022.

RIBA (Royal Institute of British Architects). (2022b). *Training an architecture apprentice.* Retrieved from https://www.architecture.com/education-cpd-and-careers/apprenticeships/training-an-architecture-apprentice, 27 July 2022.

RIBA (Royal Institute of British Architects). (2022c). *Alternative routes to professional qualification.* Retrieved from https://www.architecture.com/education-cpd-and-careers/studying-architecture/riba-studio, 27 July 2022.

Richardson, H. (2016). Grammar schools: What are they and why are they controversial? Retrieved from https://www.bbc.com/news/education-34538222, 19 July 2022.

Rifkin, A. (1988). Success disavowed: The Schools of Design in Mid-Nineteenth-Century Britain. (An allegory). *Journal of Design History, 1*(2), 89–102.

Ritchie, H., Roser, M. and Rosado, P. (2020). *Renewable energy*. Retrieved from https://ourworldindata.org/renewable-energy, 30 August 2024.

Rossi, C. (2020). On measuring ancient Egyptian architecture. *Journal of Egyptian Archaeology, 106*(1–2), 229–238.

Royal Academy. (n.d.). *Records of the school of architecture.* Retrieved from https://www.royalacademy.org.uk/art-artists/archive/records-of-the-school-of-architecture, 5 July 2022.

Royal College of Art Archive. (n.d.). *Administrative/Biographical History.* Retrieved from https://archiveshub.jisc.ac.uk/search/archives/9953bc62-bc90-3d1d-93d4-90100158881e#:~:text=The%20Royal%20College%20of%20Art%20was%20founded%20in%201837%20as,for%20the%20Exhibition%20of%201851. 2 August 2022.

Royal Town Planning Institute. (n.d.). History. Retrieved from https://www.rtpi.org.uk/about-the-rtpi/history/. 2 May 2024.

Saint, A. (1983). *The image of the architect.* Yale University Press.

Scott, L. (1899). *The Cathedral Builders.* Retrieved from The Project Gutenberg eBook of The Cathedral Builders, by Leader Scott, 13 June 2022.

Smelser, N. J. (1991). *Social paralysis and social change: British working-class education in the nineteenth century*. University of California Press.

Stabb, D. (1972, April). Wind. *Architectural Design, 43*, 253–254.

Summerson, J. (1959). *Three Elizabethan Architects.* Retrieved from https://www.escholar.manchester.ac.uk/api/datastream?publicationPid=uk-ac-man-scw:1m1942&datastreamId=POST-PEER-REVIEW-PUBLISHERS-DOCUMENT.PDF, 8 August 2022.

Summerson, J. (1966). *Inigo Jones*. Penguin.

The National Archives (n.d.). *Currency converter: 1270–2017.* Retrieved from https://www.nationalarchives.gov.uk/currency-converter/#currency-result, 14 June 2022.

Tinniswood, A. (2001). *His invention so fertile: A life of Christopher Wren*. Oxford University Press.

Twentieth Century Society. (2023). *Building of the month: June 2023—Street Farmhouse, Eltham.* Retrieved from https://c20society.org.uk/building-of-the-month/street-farmhouse-eltham, 30 October 2024.

Universitas 21. (2009). *Universitas 21 Statement on Sustainability.* Retrieved from https://cdn.auckland.ac.nz/assets/auckland/about-us/the-university/sustainability-and-environment/universitas-21-statement-on-sustainability.pdf, 31 July 2022.

University Archives and Records Centre. (2022a). *Tuition and Mandated Fees, Room and Board, and other Educational Costs at Penn: 1930–1939.* Retrieved from https://archives.upenn.edu/exhibits/penn-history/tuition/tuition-1930-1939/, 28 July 2022.

University Archives and Records Centre. (2022b). *School of design: A brief history.* Retrieved from https://archives.upenn.edu/exhibits/penn-history/school-histories/design/#:~:text=While%201990%20marked%20the%20centennial,program%20in%20the%20United%20States, 28 July 2022.

University of Auckland. (2021). *Sustainability policy.* Retrieved from https://www.auckland.ac.nz/en/about-us/about-the-university/policy-hub/strategic-engagement/sustainability-policy.html, 31 July 2022.

Vale, B., & Vale, R. (1991). *Green architecture*. Thames and Hudson.

Waite, R. (2022). *'Students told to ignore existing building'—survey reveals retrofit teaching gap.* Retrieved from https://www.architectsjournal.co.uk/news/students-told-to-ignore-existing-building-survey-reveals-retrofit-teaching-gaps, 19 June 2023.

Weaver, L. (1919). *The 'Country Life' book of cottages*. Country Life Ltd.

Weir, S. (2016). On the origin of the architect: Architects and *xenía* in the ancient Greek theatre. *Interstices, 17*, 9–15.

Whittick, N. (1842). *The complete book of trades: Or the Parent's guide and youths' instructor*. Thomas Tegg.

11 Is the Current Education of Architects Only Geared to Wealth?

Costs and Salaries

As mentioned in the previous chapter, a 2015 survey found the average annual cost to an architecture student was almost £13,000, and nearly 25% of students expected to owe more than £50,000 by the end of the course (Mark, 2015), with nearly a third of students thinking they would never be able to pay back the money they had borrowed. Salaries for registered architects are also falling below the rate of inflation, rising only 3–4% (RIBA, 2023) compared to a 2022 inflation rate of around 5.5% (House of Commons Library, 2023). However, it is female architects and architecture students who should be worried: "The gender pay gap has been widening since 2020, when it stood at 13 per cent. Last year it was 15 per cent and it has now grown to 17 per cent. Female staff continue to be under-represented in the highest quartile of incomes and over-represented in the lowest" (RIBA, 2023). Given that in 2021, in the UK, there were more first year female architecture students (2900) than male (2730) (Waite, 2021), the prospect for women in architecture is not good. Equity in architectural education has a way to go.

Currently (2023), 31% of UK registered architects are female (ARB, 2023), with the implication that these women will be in lower paid jobs with little influence on what is designed and built. This is a pity, as at the end of the last century Davidson and Freudenburg (1996) showed that women tend to have more concerns for the environment and the impact of technology than men. In a more recent Pew Research Center survey of 11 developed nations, and arguably developed nations have to change the most in serious climate change mitigation, "women are more likely to consider climate change a serious problem, be concerned it will harm them personally and say that major lifestyle changes are needed to solve the problem" (Zainulbhai, 2015). It, therefore, seems a great pity that women have so little power within the architectural profession, or at least in that of the UK, to try and steer design towards climate change mitigation. The situation is worse in the USA where, although in 2020 nearly half of all architectural students were female, only 17% of registered architects were women (AIA, 2020). In a 2016 survey (Arch20, n.d.), along with Japan, the USA and the UK were the top countries for a poor ratio of female to male architects. The best countries were Vietnam with 59% female architects, Turkey 56%, Norway 50%, Sweden 50%, Germany 43%, France 43% and Spain 42% (Arch20. n.d.). The annual Pritzker Price for architecture was first awarded in 1979, and of the 46 recipients only three awards have ever gone to women. The first was to Zaha Hadid in 2004, the only sole woman recipient (and not noted for her concern about climate change), the second to Kazuyo Sejima who with Ryue Nishizawa received the award in 2010, and the third was the joint 2020

E. J. García, B. Vale, *Wealth and the Built Environment*, Sustainable Development Goals Series,
https://doi.org/10.1007/978-3-031-81954-4_11

award of Yvonne Farrell and Shelley McNamara (The Hyatt Foundation, 2023). None of the women recipients came from a country with a high proportion of female registered architects. If it is easier to gain the award coming from a wealthy country where there is more money to spend on buildings, where does that leave those coming from a less privileged background who may be equally talented, or women, or both?

These few figures raise the issue not just of the poor prospect of a female architecture student in terms of being able to pay off an education debt, but of what type of work female architects, or indeed most male architects, are likely to be doing and whether this bears any relationship to the design challenges faced during their education. The fact 80% of architects work in small practices with fewer than 10 people and doing relatively small building work in a traditional way has been noted in Chap. 3.

How Do Architects Earn Money?

Table 11.1 sets out the fee structure for architects from the New Zealand Institute of Architects (NZIA) and how the fees received relate to stages in the design process. The design process is similar in the UK and other countries. The total fee for a small building like a house would be around 10% of the final cost. Fees earned in this way are for a project where there is an existing budget, and the fees are negotiated as a percentage of the total. To explain the terminology in Table 11.1, resource consent in New Zealand enables a project to go ahead in terms of the use of land, fitting in with existing district plans and ensuring there is no adverse environmental impact. This is the equivalent of obtaining planning consent in the UK. Building consent concerns the technical details of compliance with regulations regarding all aspects of construction.

The first thing to note is that the creative part of the design process, which is developing the concept, is unpaid labour. This might well explain why conceptual drawings are often "off-the-cuff", such as Paxton's famous sketch for the Crystal Palace, which was reputedly done on a table napkin (Guney, 2018).

Unfortunately, it is the unpaid conceptual design phase that often forms the focus of much studio time in architectural education, whereas the working drawing stage, where most of the fee is earned, is given much less, if any, attention.

Architects can also work for a fixed fee or lump sum where the project is such that a realistic estimate of the time involved can be made. An architect might also charge an hourly rate, especially for feasibility studies. Hourly rates for 2018 in the UK ranged between £80 and £100, with the higher figure being charged around London (The Fees Bureau, 2018:44). Earning fees as a percentage of the total job worked well, when architects were responsible for the whole job. With the advent of contract administrators and project managers, these parts of traditional fee income are often lost. How much an architect can earn depends on being asked to do a job. This

Table 11.1 NZIA breakdown of architects' fees

NZIA contract stage	What the architect does	Fee for stage as % final fee
Pre-design	Brief from client/project context analysis/research	2–10%
Concept design	Concept design	-
Preliminary design	Preparing design for client approval	4–18%
Developed design	Responding to client and preparing design/drawings for resource consent	5–20%
Resource consent		
Detailed design and documentation	Working drawings/building consent	30–50%
Procurement	Procurement of services for undertaking the job	1–20%
Contract administration	Contact administration/overseeing works/overseeing completion	5–33%

raises the issue of how architectural practice starts, especially the practices of the famous. After the first successful job, recommendations often lead to others but the first job is the big hurdle.

How to Get the First Job

To be a famous architect obviously takes more than just talent. It also requires clients prepared to pay for extraordinary architecture. How do yet to be famous architects find these clients? Chap. 3 notes that Gensler, the largest practice in the world, started from modest beginnings by looking for the work other architects did not want to do, in Gensler's case this was interior design. Money also plays a role in finding the first job. The relatively humble early life of Lord Foster, the world's richest architect, has been well documented (Sudjic, 2010). Having worked to support himself through architecture school as he did not receive a maintenance grant, Foster gained a scholarship to Yale where he met Richard Rogers. The ensuing friendship led to setting up Team 4 in 1963 with their partners Su Brumwell who married Rogers and Wendy Cheeseman who married Foster, all meeting at Yale and establishing the practice before any had qualified (Icon, 2023). The first commission was a house, Creek Vean, for Brumwell's parents and their art collection, which was reputedly funded by the sale of a Mondrian painting. Su Brumwell's father had been a founder of the Misha Black Design Research Unit, "and the connection was to provide useful commissions later" (Sudjic, 1986:18). The commission that foreshadowed the architecture that was to come was the Reliance Controls factory, completed in 1967, the year Team 4 was dissolved, after which Foster set up Foster Associates with Wendy Cheeseman. This commission resulted from the recommendation of another architect Jim Stirling (Sudjic, 1986:19), who had taught Rogers at the AA and who had also met up with Rogers and Foster at Yale.

Whether having wealthy relatives leads to the first commission or it is more a case of having the right connections as well as talent is explored further in the following sections. These look at a number of well-known architects and how they found their first architectural commission, starting with the four greats of modernism—Le Corbusier, Mies van der Rohe, Walter Gropius and Frank Lloyd Wright—then moving to more recent architects, Gehry, Hadid and Ingles, all of whom appear in the list of the richest architects.

Le Corbusier

Le Corbusier or, when born, Charles-Edouard Jeanneret, grew up in the Swiss town of La Chaux-de-Fonds, which was then and still is famous for watchmaking. His father was an engraver and his mother a musician who gave piano lessons, coming from a musical family (Curtis, 1987:13). Le Corbusier trained as an engraver and was mentored at the local art school by Charles L'Eplattenier. This connection led to his first commission for a villa in 1905 for another master at the art school, Louis Fallet (Pardo, 1971:3). The design of the Villa Fallet, built in 1906, was aided by a local architect (Curtis, 1987:13) with Le Corbusier normally being credited with the overall organisation of the house (Sisson, 2015). His first true commissions stemmed from this house, being a house for Albert Stozer, who was married to a member of the Fallet family, designed in 1907 and completed the following year, and the 1908 Villa Jacquemet for Fallet's brother-in-law (Fig. 11.1). These two villas were designed when Le Corbusier had left La Chaux-de-Fonds and was in Vienna. Le Corbusier claimed this journey was made possible by the money he earned for the Villa Fallet (Jenks, 1973:30).

Other commissions followed in the same locality, such as the 1912 Villa Favre-Jacot and the house for his parents in the same year. Of this period, Jenks (1973:40) notes that the clients "were all well-to-do and mostly self-made industrialists. It was this type of client Le Corbusier would later search for, mostly in vain", while "Jeanneret himself came from the bourgeoisie, and although he was often destitute and lived like

Fig. 11.1 1908 Villa Jacquemet. (Based on Foundation Le Corbusier (n.d.))

a monk, he kept a certain middle-class style throughout his life".

Mies van der Rohe

Born in 1886 and named Ludwig Mies, he later added his mother's surname to become Ludwig Mies van de Rohe. His father was a stone mason from a family that had been stone masons for generations, recently specialising in marble, so Mies is part of a tradition of stone mason families producing architects (see Chap. 10). His family has been described as "craftsman/middle class" (Schulze & Windhorst, 2012:8). In 1905, when Mies was 19, he left his birthplace of Aachen, where he had been apprenticed to two different architects "attending Evening and Sunday vocational school while working full-time and living at home" (Schulze & Windhorst, 2012:12) to work for an architect in Berlin. This job did not last long, and he left and started work for Bruno Paul, who was a furniture designer and founding member of the Deutscher Werkbund (Von Eckhardt, 2023). It was through Bruno Paul that Mies gained his first job, a weekend and holiday house in Neubabelsberg, Postdam for Professor Alois Riehl and his wife Sophie (Fig. 11.2). "The couple did not want to hire any architect already established but instead preferred to favor a beginner with talent" (Architectuul, 2023). This made sense, as Riehl, born in 1844, was the son of an innkeeper and farm owner. From such relatively humble beginnings, he had worked his way through the university system until he became a full professor at the University of Graz in 1878, succeeding to the chair in philosophy at the University of Berlin in 1905 (Martin-Luther-Universität, 2023).

Mies went to work for Peter Behrens in 1908, where he worked alongside Walter Gropius and later met Le Corbusier, though he continued to build villas in a neoclassical style during this period, such as the Hugo Perls house (1910–1913), the Villa Urbig in 1915 and as late as 1924–1926 the Mosler House. However, in 1925, Erich Wolf commissioned a house and gave Mies a free hand, producing what is recognised as Mies' first modernist house, unfortunately bombed in World War II (Fig. 11.3). The modernist Villa Tugendhat

Fig. 11.2 Riehl House 1906–1907. (Based on Architectuul (2023))

Fig. 11.3 Wolf House 1926. (Based on Metalocus (2016))

in Brno was to appear two years later in 1928–1930.

In terms of a first commission, the story of Mies van der Rohe is that of a talented young architect being in the right place at the right time. It does, however, show that Mies was prepared to build what his clients wanted and had to wait until the right client arrived in order to be able to build in the modern idiom that was to make him famous.

Walter Gropius

Born in 1883 in Berlin, Gropius was the son of an architect although it was his uncle Martin Gropius who had a significant architectural practice in Berlin (MacCarthy, 2019:13). Walter Gropius received formal architectural training in Munich (1903–1904) and Berlin-Charlottenburg (1905–1907) with a break for military service in between. His ancestors came from Prussia and the family has been described as "solidly respectable high bourgeoisie" (MacCarthy, 2019:12).

His first commission came from another uncle on his father's side who had an estate in Janikow in Pomerania and was interested in farming. Uncle Erich was keen to have ideal farm buildings. "This commission came in autumn 1905. Gropius was twenty-three and was only just beginning his training at the Hochschule. But the opportunity to design real working farm buildings, including a smithy and a laundry, was too tempting to resist" (MacCarthy, 2019:21). After the successful completion of these structures, Uncle Erich commissioned a granary from his nephew (Fig. 11.4). About this time, Gropius also designed a house for family friends and four cottages for farm workers for his uncle. His studies having been interrupted by this architectural work and with a legacy he had been left, he stopped his studies and instead travelled to Spain where he spent a year (MacCarthy, 2019:22). In 1908, he joined Behrens' office in Berlin, where he met Mies van der Rohe (Windsor, 1981:94) and in 1910 Le Corbusier, on the occasion of the latter's visit to the office (Glancey, 2019).

Frank Lloyd Wright

Frank Lloyd Wright's father was a church minister, while his mother came from a farming family of Welsh immigrants with many church connections. The first building over which Frank Lloyd Wright had complete artistic control was the 1889 house and studio he built for himself and his family in Oak Park, a village suburb adjacent to Chicago (Fig. 11.5). At the time he was working for Adler and Sullivan, and particularly with the latter. After agreeing to a five-year contract so Wright could get married, Sullivan loaned him the money ($5000 plus another $1200 to complete the project) to buy the land in Oak Park and build the marital house, with Frank Lloyd Wright gradually paying this back. The money to do this came in part from the houses Wright designed out of office hours, since Adler and Sullivan did not have time for such projects. Wright stepped in to do these, at least until the break from Adler and Sullivan came just before the expiry of the contract, although by then the loan had been repaid (Gill, 1988:88–101). Such confidence is in part what made the reputation of Frank Lloyd Wright as an architect, though Gill also notes that even in the early years of his career Wright was living beyond his means (Gill, 1988:102). Throughout his life, Frank Lloyd Wright earned considerable money but was also very good at spending, especially on luxuries such as fast cars or oriental rugs, meaning he was often short of funds (Alexander, 2021).

Wright's first commission came from his own belief in himself both as a man and an architect. However, once the first house appeared in Oak Park, other residential commissions followed such that by 1914 there were 24 Wright-designed houses in the village (Ahlfeldt & Mastro, 2012).

Frank Gehry

Frank Gehry was born in 1929 in Toronto. His grandfather owned a hardware store, and the story is that his grandmother helped the young Frank build cities on the floor using off-cuts of wood (archisoup, 2023). His family moved to

Fig. 11.4 Granary at Janikow. (Based on Warsaw Bauhaus (n.d.))

Los Angeles when he was 17, and he then drove a truck to help support the family while attending night classes. Having aborted various careers, Gehry went on to study architecture at the University of Southern California (Friedmann, 2009:60). After attending the city planning course at Harvard in 1956–1957, he worked for architects in Los Angeles and later in Paris, setting up his own practice with Greg Walsh in 1962 (Zelazko, 2022). However, before this, his first commission was jointly with Greg Walsh, then a fellow student, in 1957. This was for a retreat house in the mountains, known as the David Cabin, for a neighbour of the family (Morey, 2018) (Fig. 11.6).

This was followed by the 1959 Steeves house in Los Angeles and the 1960 renovations and recording studio for another Los Angeles house (USModernist, 2023). The Hillcrest apartments of 1962 in Santa Monica, also with Walsh, must have established the new practice. These had been designed as a speculative venture with help from the family and also from a colleague at Victor Gruen, the architectural firm where Gehry was then working (Apeles, n.d.). The Kenmore Apartments followed in 1964 along with the Kline house and many other commissions in this area of California (USModernist, 2023). Perhaps only with the Gehry House of 1977 (Fig. 11.7), where an existing Californian bungalow was

Fig. 11.5 Home and Studio at Oak Park. (Source: Creative Commons (n.d.-a))

wrapped in everyday materials such as corrugated steel and chain link, did the architect have free rein within a limited budget to produce the type of architecture that was later to make him both famous and rich.

Zaha Hadid

Hadid was born in 1950 in Baghdad into a wealthy Iraqi family. Her father was an industrialist and her mother was an artist. Hadid attended boarding schools in Europe, and after studying mathematics at a university in Beirut, she entered the Architectural Association (AA) in 1972 (Fontana-Giusti, 2016), graduating in 1977 (Celant, 2006:17). The London-based AA was a full fee-paying school of architecture as in July 1972, in a written answer, the then UK prime minister Mrs. Thatcher announced that education authorities would no longer make awards to new students entering the school as: "Authorities had for several years objected to the high level of the tuition fee" (Hansard, n.d.). At the AA under Alvin Boyarsky, Hadid was tutored by Rem Koolhaas and Elia Zenghalis. Cook (2004:11) notes: "The inevitable student success and symbiosis with Zenghalis and Koolhaas led to her inclusion in the embryonic Office for Metropolitan Architecture (OMA). It centred around their project for the Dutch Parliament extension." Hadid established Zaha Hadid Architects in 1979 (Zukowsky, 2023), but her early designs were mostly on paper, including the 1982–83 competition-winning design for the Peak in Hong Kong. Cook (2004:13) also notes that 1981–82 saw a house conversion project for her brother at 59 Eaton Place that was never realised. Zukowsky (2023) emphasises this point.

> Hadid began to be known as a "paper architect," meaning her designs were too avant-garde to move beyond the sketch phase and actually be built. This impression of her was heightened when her beautifully rendered designs—often in the form of exquisitely detailed coloured paintings—were exhibited as works of art in major museums.

Fig. 11.6 1957 David Cabin. (Source: Creative Commons (n.d.-b))

The first building was the (1989–90) interior of the Monsoon Restaurant in Sapporo, while the first free-standing building was the 1990–1994 Vitra Fire Station, Weil am Rhine, Germany (Fig. 11.8).

During these early years of practice, Hadid must have had the money to support both it and herself, although she was teaching at the AA in the early 1980s. Cook (2004:11) talks of her "Kensington Mews house of the 1980s", an expensive even if small residence and perhaps an address in London not normally associated with an architect struggling to establish herself. This wealthy background does not deny the extraordinary talent of Hadid and her architectural achievements. She was also noted for her hard work: "This drive towards the inventive and experimental matched with a deep-seated willingness to see her projects constructed: the reason why Hadid went through the pain of arduous global travelling and project management" (Fontana-Giusti, 2016).

Bjarke Ingels

Born in Copenhagen in 1974, his mother was a dentist, and his father was an engineer working with fibre optic cables (Parker, 2012). Ingles originally had ambitions of becoming a cartoonist and, in 1993, began to study architecture at the Royal Danish Academy of Fine Arts to improve his drawing skills (Famous Architects, 2023). While still studying, he was part of a competition-winning team for phase one of the new buildings at Copenhagen University (KUA) in Ørestad Nord (Danish Architecture Center, 2009). However, the first KUA building (KUA 1) was designed by KHR Arkitekter and built in 2000 (Anon, 2013). Ingels also studied at Escola Tecnica Superior de Arquitectura in Barcelona but, by 1998, was working for Rem Koolhaas at the Office for Metropolitan Architecture (OMA) in Rotterdam. "Here he met Julien de Smedt, with whom he founded PLOT in 2001" (Danish Architecture Center, 2009). This practice based

Fig. 11.7 Gehry House, 1977. (Source: Creative Commons (n.d.-c))

in Copenhagen lasted for five years until the partners split, and Ingels founded BIG while de Smedt founded JDS Architects. By the time Ingels was 33, he is claimed to have "inaugurated" his fourth building, although not all were proceeding (Farnese, 2006). In 2001, PLOT won a "theoretical competition for a Water Culture House" sponsored by the Sports Facility Fund, and in 2003, the latter gave them a real commission. This resulted in the Copenhagen Maritime Youth House (Betsky, 2004). To avoid the cost of dealing with ground polluted with heavy metals, the structure sits on a timber platform on top of the ground, creating a public space, a yacht club and the youth centre (Fig. 11.9).

At the same time, PLOT were throwing out architectural ideas, such as the 2003 off-shore Danish Superharbour that would free up onshore land and, in 2004, a scheme for the new Stavanger Concert Hall that won a prize at the Venice Biennale for that year (JDS Architects, n.d.). PLOT's first major project was the 2005 VM housing complex in Ørestad (Fig. 11.10).

> A few years ago, Ingels met the developer Per Høpfner, who had joined up with the Danish Oil Company to create housing on land purchased from the city-and-state corporation planning Ørestad. Høpfner decided PLOT should design VM Houses, even though Ingels and De Smedt hadn't built anything on that scale before. Per's son Peter Høpfner, who also works in the business, explains that Ingels has a very important trait (besides creativity): "He isn't stubborn. When we would tell him that something was too expensive, he would say, 'Give me two days,' and then come up with something even better." Høpfner adds, "VM may look crazy, but it came in at budget". (Stephens, 2009)

Parker (2012) noted a disconnect between age and success when it came to Bjarke Ingels given the scale of projects undertaken internationally by BIG, and Ingels being only 37 at that time. The fact that he is already among the list of the richest architects suggests that his wealth will only increase with time.

Fig. 11.8 Vitra Fire Station. (Source: Creative Commons (n.d.-d))

Fig. 11.9 Maritime Youth House. (Source: Creative Commons (n.d.-e))

Fig. 11.10 VM Housing. Source Emilio Garcia

The First Step to Architectural Fame (and Wealth)

The first steps to becoming a famous and possibly wealthy architect are certainly talent and a penchant for hard work. Standing out from the crowd can lead to first work, as in the case of Le Corbusier and Mies van der Rohe who were both recommended, Le Corbusier by a teacher and Mies by an employer. Having relatives sufficiently wealthy to commission a building is a sure way to the first job, and this statement building can lead to other commissions. Frank Lloyd Wright's career start is perhaps the most unusual among those considered here, given that he talked his way into obtaining the money for his first job, a house for himself. So, although privilege and wealth are helpful in becoming a famous architect, they are far from being the only ingredients. Nevertheless, it does feel as if initial fame does attract in a "rich get richer" way, since popularity tends to follow the process of preferential attraction. "Popularity is a phenomenon characterized by extreme imbalances: while almost everyone goes through life known only to people in their immediate social circles, a few people achieve wider visibility, and a very, very few attain global name recognition" (Easley & Kleinberg, 2010:479). The same authors (2010:485) also note that the first step in becoming popular, in this case a popular architect, does depend on luck and being in the right place at the right time. Many architects have had the talent and work ethic of the very famous but not that first push or piece of luck that set them on the path to fame and fortune.

Education Only for Wealth and Fame?

The real point of this chapter, however, is to consider the state of the education of the majority of qualified architects, some of whom are also highly talented, and who never have the chance to set up practice on their own but spend their career as salaried professionals. Is current education serving the majority well, or is it only aimed at helping the next star architect? Is too much pressure put on students to be original rather than building on what has gone before? The common thread in the brief discussions of the careers of Gehry, Hadid and Ingels is, whatever the influences they came under, they did not build on what had gone before but broke with tradition, only to create a new tradition. There are limits to how far this can be taken. As in Tainter's theory of collapse (Tainter, 1988), solving one problem leads to an increase in complexity and a step nearer to collapse. Breaking with tradition to create a new tradition means that buildings, which are what architecture is, have somehow to be radically different from what has gone before. It is this difference that makes the star architect. Unlike science, where knowledge builds on knowledge, the cry in architectural education is always for the outrageously creative and novel, whether it works as a building or not, rather than learning from what has gone before and using this knowledge to create new architecture. Given how few star architects there are, an education grounded in learning what works and how to be creative with that knowledge would seem to benefit the majority. Harper (2017) notes: "In theatre, a culture of covers is arguably the bedrock of the industry, with directors constantly reworking centuries-old scripts in innovative ways", and he points out that this holds true for music and other arts.

Most architects will earn enough but will not be able to accumulate wealth, because, at present, wealth largely goes to those who already have it. Unfortunately, most current architectural education is based on the idea that money is no object when it comes to design projects. Additionally, projects are geared to the idea of creating the star architect, rather than simply trying to do the ordinary better. Since people like Foster and Ingels will succeed as architects whatever their education, it would seem more fitting to provide an education geared to those providing the buildings that form the backdrop to most people's lives. Most suburban streets and high streets are not composed of mini Bilbao Museums. This leaves a question for SDG 4—Quality Education. There needs to be serious discussion about what "quality" means in this context, so that a relevant education is delivered such that all can benefit from it.

Let the Teachers Teach

Wealth gets in between everything that has honest intentions and transforms it into something else that does not resemble the real thing. This also interferes with and limits education. In architectural education, external limitations are linked to a wider understanding of what architecture and architects are and do. The environmental crisis should be helping to narrow down what architects ought to be doing, but it is not just the fixation on wealth that is preventing this. Those teaching architecture within the university environment are also under stress. We cannot expect most teachers to research and produce original and ground-breaking content, deliver lectures and tutorials, supervise postgraduate student theses, write articles and books, apply for research grants and lobby to promote their research to stakeholders while remembering to keep an active presence in social media, and still find time to engage with design for climate mitigation. Moreover, promotional possibilities are limited by the money a staff member can bring into the university through research grants. These are seen as important research outcomes that prove that the research is relevant and can impact society. However, this is a market logic applied to the education sector, rather than directing research in the built environment to what needs to be done in the best interest of human society.

Other problems come from applying the market model to education. At university, it is students who judge the performance of their teachers simply because they are paying. This sounds logical if you think in terms of relationships between machines and users. For example, who would be the best judge of the efficiency of a domestic stand mixer other than its users? Such a situation rarely applies to people's work relationships and performance. Asking students to quantify their teacher's performance is as biased as asking the divorce lawyer of your ex how things should be divided. The situation with students is similar. Those who do not like their results will blame the teachers. The students are pushed into an uncomfortable situation and ordered to do a task they are not trained to do. This whole idea of evaluation is based on the idea that people do not want to work and, therefore, they should be controlled and evaluated regularly to be more productive. The problem here is that education is not a production line. Education is about drawing out from students their innate ability to learn. It is not about how much stuff a teacher can put in. Before SDG 4 can be truly implemented, there needs to be a serious consideration of how education should be delivered.

The fear of being disapproved of by students has created a type of likeability mania, aggravated by the worst qualities of social media habits. This is detrimental for everyone. Staff ease the student workload, bump up grades, put more energy into being entertaining than being relevant and truly challenging and, above all, try to fail no one. Something challenging for the wealthiest civilisation ever is that education is not a spectacle. It is a guided practice for deliberation and learning essential content to help someone keep learning for the rest of their life. It does not need to follow the predominant way of communication developed by the market, like social media, nor the market itself.

If education follows wealth and works for it, it will always be dated and irrelevant. Researching or developing a product for a stakeholder to bring money into an institution is not the purpose of universities. According to Universities New Zealand (2024), "…universities don't just train, they educate. Universities also enhance society through their contribution to our understanding of social issues and our achievement of social, economic and physical well-being." Universities are not there to serve the goal of wealth.

Resilience Through Diversity

In resilience theory, more diverse systems tend to be more resilient (Garcia & Vale, 2017:63). This should also apply to education. In the not-so-distant past, if you happened to live in a small town, only a few people did international travel, probably once in their lifetime. So when someone was visiting from another city or country, everyone was interested in inviting him or her to listen to their stories and share food. We were curious about what they would think about our habits and traditions. It was a very unique opportunity to learn for everyone. However, the same enthusiasm is only sometimes shared in architecture schools. In the architecture schools of developed countries, it is common to find professionals from different countries on the staff. Rather than using the different experiences of these staff members, especially those from developing countries, to enrich the curriculum, architecture is taught as if there is only one viewpoint and one set of valid design solutions. The diversity clause in academic job offers may state that everyone is welcome, especially people from overseas with different cultures and beliefs, but in education these experiences are normally ignored. When it comes to the SDGs, the experiences of such people now working in developed countries will have much to offer.

Built Environment Education for Meeting SDG Goals

Much in this book has been written about the problems of modern built environment education, particularly the education of architects. It seems fitting, therefore, to end with a brief discussion of what built environment education might need to be to meet the SDG goals. The first step would seem to be to learn the craft of building before learning how to use that craft in designing something new. This links back to the Dewey decimal system of knowledge classification (see Chap. 7), where each class builds on knowledge in the previous division, so building comes before architecture and fine arts. The easiest way to learn the craft of building would be to watch what someone experienced is doing, as in the old pupillage system. Building craft could become a lecture subject but needs to be integrated with design teaching. Thinking back to the early days of the Welsh School of Architecture (see Chap. 10), when construction and working drawings were given equal importance to design, it seems like a way of teaching that must be revived. Alongside this is the pressing need to address climate change mitigation and the unnecessary wasting of resources so as to preserve land and sea. These address the knowledge base of becoming a designer of the built environment. On top of this is considering what to design to ensure more equality and an end to poverty. This suggests teaching how to make humble buildings that work well for everyone and making use of the knowledge of staff members with direct experience in making humble buildings. Only in this way can the goal of making sustainable communities (SDG 11) be achieved.

References

Ahlfeldt, G., & Mastro, A. (2012). Valuing iconic design: Frank Lloyd Wright architecture in Oak Park, Illinois. *Housing Studies, 27*(8), 1079–1099.

AIA (American Institute of Architects). (2020). *Women in architecture.* Retrieved from https://www.aia.org/articles/6252982-women-in-architecture, 10 May 2023.

Alexander, P. C. (2021). *Insufficient funds: The financial life of Frank Lloyd Wright.* Dorrance Publishing Co.

Anon. (2013). *Københavns Universitet Åbner sig forørestad.* Retrieved from https://byoghavn.dk/koebenhavns-universitet-aabner-sig-for-oerestad/#:~:text=kaffe%20i%20caf%C3%A9en.-,%C3%98restad%20Nord%20som%20vidensby,omkring%20Universitetskanalen%2C%20taget%20i%20brug., 10 June 2023.

Apeles, T. (n.d.). *KCRW's Frances Anderton's Gehry-designed apartment is far-out.* Retrieved from https://www.hunker.com/13717849/kcrws-frances-andertons-gehry-designed-apartment-is-far-out, 12 June 2023.

ARB (Architects Registration Board). (2023). *New data from ARB highlights underrepresented groups in the architects' profession.* Retrieved from https://arb.org.uk/new-data-from-arb-highlights-underrepresented-groups-in-the-architects-profession/#:~:text=Gender%3A,joining%20the%20register%20were%20female, 10 May 2023.

Arch20. (n.d.). *Survey shows best and worst countries for female architects.* Retrieved from https://www.arch2o.com/survey-shows-best-and-worst-countries-for-female-architects/, 10 May 2023.

archisoup. (2023). *The life and architectural career of Frank Gehry.* Retrieved from https://www.archisoup.com/frank-gehry, 12 June 2023.

Architectuul. (2023). *Riehl House.* Retrieved from https://architectuul.com/architecture/riehl-house, 24 May 2023.

Betsky, A. (2004). *Following the PLOT.* Retrieved from https://www.domusweb.it/en/architecture/2004/10/06/following-the-plot.html, 12 June 2023.

Celant, G. (2006). In G. Celant & Ramirez-Montagut (Eds.), *Zaha Hadid adventure in architecture* (pp. 17–22). Guggenheim Museum Publications.

Cook, P. (2004). The emergence of Zaha Hadid. In P. Schumacher & G. Fontana-Giusti (Eds.), *Zaha Hadid complete works: Texts and references* (pp. 6–17). Thames and Hudson.

Creative Commons. (n.d.-a). "Frank Lloyd Wright Home and Studio, Oak Park, IL" by naotakem is licensed under CC BY 2.0.

Creative Commons. (n.d.-b). "Frank Gehry - David Cabin, Idyllwild CA. 1957" by Mmenorris is licensed under CC BY-SA 4.0.

Creative Commons. (n.d.-c) "Facciata principale della Gehry House a Santa Monica" by André Corboz is licensed under CC BY-SA 4.0.

Creative Commons. (n.d.-d). "Zaha Hadid- Vitra Fire Station, 1993" by roryrory is licensed under CC BY-SA 2.0.

Creative Commons. (n.d.-e). "PLOT - Maritime Youth House - Photo 06" by 準建築人手札網站 Forgemind ArchiMedia is licensed under CC BY 2.0.

Curtis, W. (1987). Le Corbusier: Nature and tradition. In M. Raeburn & V. Wilson (Eds.), *Le Corbusier architect of the century* (pp. 13–23). Arts Council of Great Britain.

Danish Architecture Center. (2009). *Bjarke Ingles.* Retrieved from Blue Book: Bjarke Ingels (archive.org), 10 June 2023.

Davidson, D. J., & Freudenburg, W. R. (1996). Gender and environmental risk concerns: A review and analysis of available research. *Environment and Behavior, 28*(3), 302–339.

Easley, D., & Kleinberg, J. (2010). *Networks, crowds, and markets: Reasoning about a highly connected world.* Cambridge University Press.

Famous Architects. (2023). *Bjarke Ingels.* Retrieved from https://www.famous-architects.org/bjarke-ingels/, 10 June 2023.

Farnese, A. (2006). The PLOT thickens. *Women's Wear Daily.* 20th November, p. 72.

Fontana-Giusti, G. (2016). Zaha Hadid 1950-2016. *Architectural Research Quarterly, 20*(2), 95–98.

Foundation Le Corbusier. (n.d.). *Villa Jacquemet.* Retrieved from https://www.fondationlecorbusier.fr/en/work-architecture/achievements-villa-jaquemet-la-chaux-de-fonds-switzerland-1907-1908/, 30 October 2024.

Friedmann, M. S. (2009). *Frank Gehry: The houses.* Rizzoli.

Garcia, E., & Vale, B. (2017). *Unravelling sustainability and resilience in the built environment.* Routledge.

Gill, B. (1988). *Many masks: A life of Frank Lloyd Wright.* Heinemann.

Glancey, J. (2019). The metamorphosis of Charles-Édouard Jeanneret. *Architectural Design, 89*(6), 41–47.

Guney, Y. I. (2018). Embracing today's economic and technological reality: What it means for design professionals. *Journal of Contemporary Urban Affairs, 2*(3), 107–111.

Hansard. (n.d.). *Architectural Association School of Architecture.* Retrieved from https://api.parliament.uk/historic-hansard/written-answers/1972/jul/25/architectural-association-school-of, 7 June 2023.

Harper, P. (2017). *Call us dull, call us sellouts, call us gentrifiers—just don't call us copycats.* Retrieved from https://www.dezeen.com/2017/07/18/phineas-harper-opinion-copying-originality-architecture-assemble-cineroleum/, 14 June 2023.

House of Commons Library. (2023). *Rising cost of living in the UK.* Retrieved from https://commonslibrary.parliament.uk/research-briefings/cbp-9428/#:~:text=The%20cost%20of%20living%20increased,month%20of%20double%2Ddigit%20inflation. 10 May 2023.

Icon. (2023). *Team 4: and architecture supergroup.* Retrieved from https://www.iconeye.com/architecture/team-4-an-architecture-supergroup, 14 June 2023.

JDS Architects. (n.d.). *HAV/Superharbour*, Retrieved from http://jdsa.eu/hav/, and *STA/Stavanger Concert Hall.* Retrieved from http://jdsa.eu/sta/, 12 June 2023.

Jenks, C. (1973). *Le Corbusier and the tragic view of architecture.* Allen Lane.

MacCarthy, F. (2019). *The man who built the Bauhaus.* Harvard University Press.

Mark, L. (2015). *Revealed: The AJ student survey results 2015.* Retrieved from https://www.architectsjournal.co.uk/archive/revealed-the-aj-student-survey-results-2015, 8 July 2022.

Martin-Luther-Universität. (2023). *Alios Riehl.* Retrieved from https://www.catalogus-professorum-halensis.de/riehlalois.html, 24 May 2023.

Metalocus. (2016). *The debate around Mies van der Rohe's Wolf House.* Retrieved from https://www.metalocus.es/en/news/debate-around-mies-van-der-rohes-wolf-house, 24 May 2023.

Morey, A. G. (2018). *From the ground up: Frank Gehry.* Retrieved from https://archinect.com/features/article/150063907/from-the-ground-up-frank-gehry, 12 June 2023.

Pardo, V. F. (1971). *Le Corbusier.* Thames and Hudson.

Parker, I. (2012). *High Rise: A bold Danish architect charms his way to the top.* Retrieved from High Rise | The New Yorker, 8 June 2023.

RIBA. (2023). *What does our Business Benchmarking Report tell us about architects' salaries.* Retrieved from https://jobs.architecture.com/staticpages/10290/what-does-our-business-benchmarking-report-tell-us-about-architects-salaries/, 10 May 2023.

Schulze, F., & Windhorst, E. (2012). *Mies van der Rohe.* University of Chicago Press.

Sisson, P. (2015). *21 First Drafts: Le Corbusier's Villa Fallet.* Retrieved from https://archive.curbed.com/2015/8/10/9932192/21-first-drafts-le-corbusiers-villa-fallet, 5 February 2025.

Stephens, S. (2009). BIG Bjarke Ingels Group: A young Danish architectural firm is living up to its acronym with bold projects home and abroad. *Architectural Record, 197*(12), 35.

Sudjic, D. (1986). *New directions in British Architecture: Norman Foster, Richard Rogers, James Stirling.* Thames and Hudson.

Sudjic, D. (2010). *Norman Foster: A life in architecture.* Orion Publishing Group.

Tainter, J. (1988). *The collapse of complex societies.* Cambridge University Press.

The Fees Bureau. (2018). *RIBA business benchmarking 2018.* RIBA.

The Hyatt Foundation. (2023). *The Pritzker Architecture Prize.* Retrieved from https://www.pritzkerprize.com/, 10 May 2023.

Universities New Zealand. (2024). *About the university sector.* Retrieved from https://www.universitiesnz.ac.nz/about-university-sector#:~:text=That%20is%2C%20universities%20don't,economic%20and%20physical%20well%2Dbeing. 8 July 2024.

USModernist. (2023). *Frank Owen Gehry FAIA (1929-).* Retrieved from https://usmodernist.org/gehry.htm, 12 June 2023.

Von Eckhardt, W. (2023). *Ludwig Mies van der Rohe.* Retrieved from https://www.britannica.com/biography/Ludwig-Mies-van-der-Rohe, 24 May 2023.

Waite, R. (2021). *UCAS data: More women than men start architecture courses in 2021.* Retrieved from https://www.architectsjournal.co.uk/news/ucas-data-more-women-than-men-start-architecture-courses-in-2021, 10 May 2023.

Warsaw Bauhaus. (n.d.). *The future of the Granary.* Retrieved from https://warsawbauhaus.org/projekt-spichlerz-gropiusa/, 24 May 2023.

Windsor, A. (1981). *Peter Behrens: Architect and Designer 1868–1940.* The Architectural Press.

Zainulbhai, H. (2015). *Women, more than men, say climate change will harm them personally.* Retrieved from https://www.pewresearch.org/short-reads/2015/12/02/women-more-than-men-say-climate-change-will-harm-them-personally/, 10 May 2023.

Zelazko, A. (2022). *Frank Gehry.* Retrieved from https://www.britannica.com/biography/Frank-Gehry, 12 June 2023.

Zukowsky, J. (2023). *Zaha Hadid: British Architect.* Retrieved from https://www.britannica.com/biography/Zaha-Hadid, 7 June 2023.

12 The Chicken and the Egg

Breaking the Vicious Cycle of SDGs: From Development Goals to Sustainable Goals

The application of SDGs to the built environment shows that sustainability and development are very different and not always compatible. As shown in Chap. 1, the path to achieving most of the Sustainable Development Goals is assessed through economic indicators. The SDGs work under two assumptions: development means getting wealthier, and sustainability means developing whatever area needs development. As a result, getting wealthier is the only path to sustainable development. These ideas are influenced by the Environmental Kuznets Curve (EKC), a green version of the economic Kuznets Curve.

Awarded the Nobel Prize in Economics, Kuznets (1955) formulated a hypothesis in the 1950s that concerned the connection between economic growth and income inequality. His hypothesis suggested that income inequality tends to rise at lower income levels and then decline as income levels increase, resulting in an inverted U-shaped relationship between income inequality and GDP per capita, commonly referred to as the Kuznets Curve (EKC) (Bo, 2011). The theory is that as wealthy people get wealthier, they will reinvest that money in growing their businesses and opening new factories, creating more job opportunities for everyone. Doing this would increase the incomes of the employed workers and shorten inequality gaps, a tide that lifts all the boats. Bo (2011) and Grossman and Krueger (1995) found the same inverted U-shaped curve for the relationship between income and environment when assessing the environmental effect of the North American Free Trade Agreement (NAFTA). Their idea was supported by the World Development Report (1992). Environmental issues would first worsen but would then improve with economic growth, as inequality did in the Kuznets economic study using pre-war data. If this relationship were real, it would be a very convenient path. It would mean that societies, by getting wealthier, would eventually overcome environmental problems (if they do not destroy the world while trying). As mentioned in Chap. 1, current environmental problems have been caused by the consumption patterns of wealthy countries. Following the idea derived from the EKC implies that developing countries will need to travel the same path of damaging the environment before being able to improve it. If developing countries must impact the environment first to develop later, can the world survive that development process before they can fix it?

Using these ideas, wealthy countries could opt to invest in the SDGs through getting wealthier to finally overcome environmental problems. However, before doing this, it might be wise to check if the EKC hypothesis and even the idea of green growth, as proposed by Macron in France

E. J. García, B. Vale, *Wealth and the Built Environment*, Sustainable Development Goals Series,
https://doi.org/10.1007/978-3-031-81954-4_12

(see Chap. 9), work. Jefim Vogel, a scientist from the Sustainability Research Institute in Leeds, and Jason Hickel, from the Institute of Environmental Science and Technology in Barcelona, have raised concerns about the possibility of high-income countries reducing their CO_2 emissions to keep global warming at the 1.5 °C Paris Agreement level while increasing GDP (Vogel and Hickel, 2023:e759).

> The emission reductions that high-income countries achieved through absolute decoupling fall far short of Paris-compliant rates. At the achieved rates, these countries would on average take more than 220 years to reduce their emissions by 95%, emitting 27 times their remaining 1.5 °C fair-shares in the process. To meet their 1.5 °C fair-shares alongside continued economic growth, decoupling rates would on average need to increase by a factor of ten by 2025.

Historically, many of the wealthy countries built their wealth and economic growth by profiting from wars, slavery, colonisation (Piketty, 2020) and the development of fossil fuel resources. The development suggested by the SDGs is just a peaceful and greener version of what developed countries did to be wealthy. The transition proposed by the SDGs is only focused on development through economic growth, assuming sustainability as an outcome without acknowledging the environmental cost of economic development. Chapter 4 shows that industrial outputs based on innovation, like cars and road infrastructure, produced economic growth and wealth for a few but also debts, inequalities, pollution and sprawl that had a direct negative impact on cities, working against SDG11—Sustainable Cities and Communities. Cars have also made no contribution to reducing poverty or hunger. This is a chicken and egg situation. More economic growth, based on industry innovation and infrastructure, may not achieve sustainable development goals, but can add more problems in achieving these goals, which will require more investment to alleviate the same problems. How can this vicious cycle be broken using the SDGs?

Countries suffering from high levels of hunger, poverty and inequality do not have the means to make the same investments as wealthy countries. Although they contain most of the world's population, they do not have these same options. However, their lower incomes are also their potential. Even though they do not have much, and they need to reduce poverty and hunger, they have less impact on the planet. Without investing and sacrificing resources to get wealthier and later to try to fix what they did, many low-income countries already have low emissions per capita. Why bother following the wealthy path suggested by the EKC or SDGs?

The SDGs can be very useful for remembering what countries should be trying to aim at, but they must be radically reviewed and changed. The first step is to separate their sustainability goals from the development goals. The rationale is logical and ethical. The sustainability goals within the SDGs encompass the rest of the goals, but they are mixed and confused, and there is no hierarchy. The interrelationship between the SDGs has been analysed here to unravel the hidden hierarchy. The analysis was done by looking at how relevant each SDG is for achieving other SDGs. Figure 12.1 shows that SDG 13 (Climate Action) is related to achieving 15 out of the 17 goals, while SDG 17 (Partnerships for the Goals) is the least relevant for achieving all the other SDGs. SDG 8 (Economic Growth) is relevant to achieving 11 goals, and SDG 9 (Industry and Innovation) is relevant for only 10 of the SDGs. The chart shows that more than half (9 out of 17) of SDGs are more relevant and connected to achieving other goals than economic growth (SDG 8, Decent Work and Economic Growth), and two-thirds are more relevant for achieving goals other than Industry, Innovation and Infrastructure (SDG 9).

Without the environment, there is no society or economy (Chap. 1). From an ethical point of view, all SDGs, with the possible exception of SDG 17 (Partnerships for the Goals), are good in themselves. Achieving SDG 1 (No Poverty) or SDG 2 (Zero Hunger) worldwide benefits all communities. However, as seen across the book, there are three SDGs that can be positive, negative or make no difference, and these are SDG 8 (Decent Work and Economic Growth), SDG 9 (Industry, Innovation and Infrastructure) and SDG 17 (Partnerships for the Goals). These are

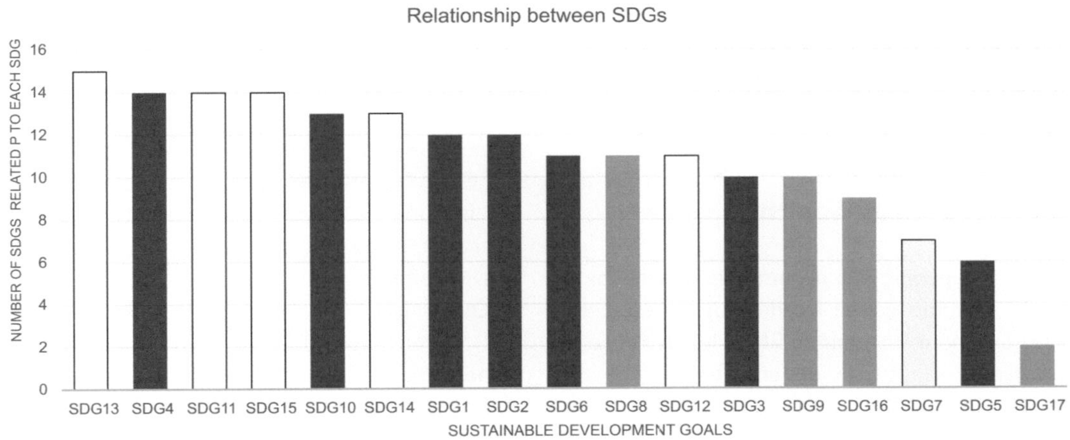

Fig. 12.1 Relationships between SDGs. White bars represent SDGs linked to predominantly environmental issues, like Climate Action. Black colour bars are SDGs related to social issues, like ending poverty, hunger and inequality, among others. Grey colour bars are related to economic and political variables, like economic growth and industry development

Fig. 12.2 SDGs grouped by their social, environmental and economic interests

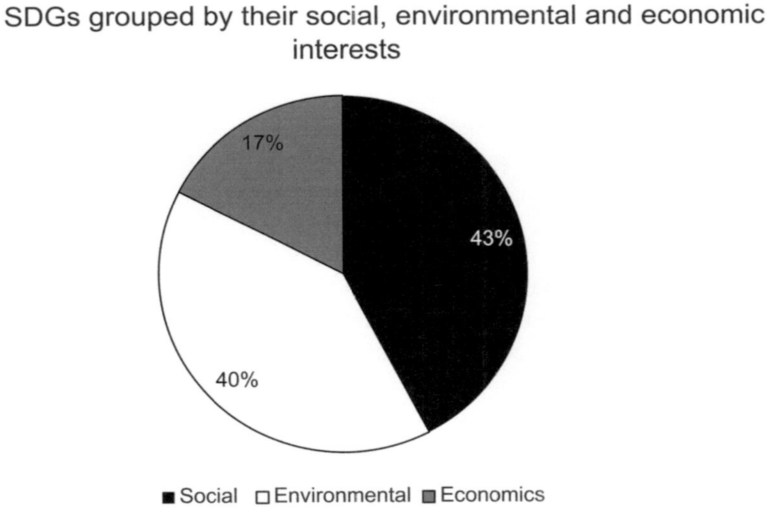

susceptible to affecting the achievement of all the other goals negatively. If countries could make good use of SDG 8 and SDG 9 to end poverty or hunger, they are in fact achieving SDG 1 and SDG 2. SDGs 1 and 2 are ends, but economic growth (SDG 8) and innovation (SDG 9) are, in the best-case scenario, a means to an end.

Figure 12.2 shows that if the relationships between SDGs illustrated in Fig. 12.1 are grouped according to their interest in solving environmental, social and economic issues (white, black and grey bars, respectively, in Fig. 12.1), there are more SDGs related to social (43%) and environmental (40%) matters than the 17% of SDGs related to economic and political goals.

Arguably, there are too many SDGs. The urgency is to solve environmental and social issues first, in other words, climate action and ending poverty. Without climate action, there will not be the resources to end poverty, and without ending poverty, there is no reason to have SDGs. There is only one sustainability goal: ending poverty through climate action. The real challenge is committing wealthy people and societies to doing this if they choose not to collaborate. Are there any limits to the freedom of wealth?

The Freedom of Wealth

When few people can accumulate unsustainable wealth, it removes the freedom of all the rest to choose a more sustainable lifestyle. As proposed in Chap. 2, wealth provides an individual sense of entitlement and freedom that goes against collective interests. Limiting freedom is an issue for architects and planners, especially when dealing with wealthy clients and big projects, as shown in Chaps. 3 and 9. Can architects limit the sense of freedom of their wealthy clients? From churches in the past to expensive single-family housing in the present, this has been historically challenging. Architects have been serving the wealthy because they can pay. Within the context of climate change and evidence of the impact of wealth on the environment and society, are not these arguments enough evidence to question the freedom of the wealthy to use their wealth without limitations?

An understanding of the meaning of freedom is essential for understanding the impact of wealth and moving towards equity, less hunger and sustainability for everyone. Without understanding freedom, there is no room for discovering what we want as a global society or why we want it. The key problem is that freedom is understood differently depending on how wealthy a person is. The impact of wealth is proportional to the freedom of the wealthy to increase inequalities at all scales. Two researchers at the London School of Economics suggest the UK tax system has issues, when it comes to the wealthy paying a fair tax share: "Where you get your money from (or at least how you package it) matters because investment income and capital gains are taxed at lower rates than income from work" (Summers & Advani, 2021). Everyone has to work and pay income tax, but it cannot be fair that this is taxed at a higher rate than sources of income associated only with the wealthy. Without a fair distribution of wealth, there is no way of dealing with climate change issues, helping the poor and creating a society where merit and work matter. The more money someone has, the more responsibilities for the environment they have acquired. However, this is a non-wealth-related viewpoint, an ethical point. The responsibilities of wealth seem to be considered less important than the freedom to use it. Otherwise, how can building a linear city in the desert (see Chap. 3) accept these responsibilities?

For the wealthy, freedom means limitless power to act, regardless of how it might interfere with someone else's freedom to do the same or whether it is the best for everyone. Any governmental intervention, for example, to tax wealth and promote a fair redistribution of wealth in society, would be understood as coercion by the taxed wealthy. However, freedom, even market freedom, means something because it has limitations. Anyone can drive wherever they want if they respect the traffic lights and road code, and no one feels coerced by these because they agree it is the way to keep everyone safe. This is important for architects and planners. There is a need to reconsider the limits of a client's freedom to demand whatever they want and the limits of the designer to do whatever pleases their egos. It is the misunderstanding of freedom as limitless power to act that leads to the excess of luxury that is damaging the environment and goes against the achievement of any SDGs other than economic growth and the development of industry.

The Nobel Prize winner in economics, Joseph Stiglitz (2024: xxii), claimed the meaning of freedom is key to moving on as a society, and that freedom has been subverted and ideologically twisted. In his view, it is used as a synonym for neoliberal market freedom, namely, markets without the rules or regulations that condition the "freedom" of banks and wealthy people to exploit other people. Without rules imposed by a government, trust would be low, making trade impossible, and everyone would be free to cheat, challenging the existence of any market. Stiglitz is concerned about this unethical economic freedom that, for example, led to the 2008 economic crisis, whose consequences have been paid for by taxpayers. The current form of capitalism does not promote or nurture freedom in societies. It only means "freeing trade to accelerate deindustrialization; and freeing corporations to exploit consumers, workers, and the environment alike" (Stiglitz, 2024: xvi). Stiglitz, paraphrasing the philosopher Isaiah Berlin, proposes that neoliberal market freedom is "freedom of a few at the expense of the many. Freedom for the wolves; death to the sheep" (Stiglitz, 2024: xxii).

As long as limitations on clients' wealth remain untouchable and frameworks like the SDGs are optional and easily overwritten by economic priorities, the architect or planner is a sheep at the mercy of the market wolves. There is no achievable sustainability under these circumstances. However, it is possible to break the loop. The power of wealth is useful if it can be exercised to achieve what people want. It depends on other people saying yes. The good news is that the new generations are starting to do this and are open to change.

Democracy, Wealth and Its Paradoxes

Many centuries ago, the freedom for some offered by wealth, at the expense of the many, was a topic of concern for Plato. In his book *The Republic*, Plato adverts that if democracies allow an excess of freedom, the rich become too rich and the poor too poor and enslaved by the rich. For this reason, "the chances are that democracy is the ideal place to find the origin of tyranny—the harshest and most complete slavery arising, I guess, from the most extreme freedom" (Plato, 2000:564a). Plato further asserts that "when we look at the growth of a tyrant, I said, 'one thing at least is clear. This position of champion is the sole root from which the tyrant springs'" (Plato, 2000:565d).

For Plato, tyranny can arise from a democracy that concentrates in one person the same luxuries and power of the displaced oligarchy. Plato raised awareness about how the context from which freedom emerges is susceptible to turning its realisation into a negative output. Even freedom has limitations and should be framed to avoid excess. On one side, freedom allows political systems like democracy to be possible. On the other side, it creates positions of power, like being president of a country or the governor of a state or province, that can be sought after and reached by people who are not fit for such office or whose intentions go against democracy or the environment.

The philosopher Karl Popper viewed this as a "paradox of freedom", where "freedom in the sense of absence of any restraining control must lead to very great restraint, since it makes the bully free to enslave the meek" (Popper & Gombrich, 1994:581). Considering the rise of extreme political ideas in Europe and the USA, Plato's concern about the limits of democracy should make us think about the consequences of excessive wealth for a few and the inequalities thus produced, especially because oligarchs, tyrants and extreme parties have little interest in fixing climate change or inequalities.

What to Do with Wealthy People?

Wealth and the wealthy have been scrutinised for a couple of thousand years. However, such scrutiny is more relevant than ever before, since everyone is not equally responsible for the current environmental and inequality crisis (Chap. 9). With the rise of climate change victims, what is the threshold of tolerance for the freedom of the wealthy to destroy the planet and its people?

In his analysis of the global rich, the British philosopher Thomas Wells, who teaches ethics and political philosophy at Leiden University, asserts that the rich have accumulated so much wealth to the point of having a level of "independence and command over others" (Wells, 2016: 19) that is detrimental for both democracy and capitalism. Wells proposes that "if we believe our democracy is worth preserving, we should offer the rich a choice: give up your money or give up your citizenship" (Wells, 2016: 19). Wells (2016:27) admits his proposal is utopian, but his aim is to turn the discussion about wealth into a political problem: "If you do not need us, we can ask, why do we need you?" However, moving businesses from one city or country to another to avoid or to pay lower taxes, as seen in the case of Ford, has happened before without giving up money. Would trying to get rid of wealth or the wealthy ever work to fix the impact of wealth?

The French Revolution tried to kill all the rich, but this did not work. The bloodbath caused by the revolution only paved the way for the upper-middle class to replace the nobility, all of which resulted in increased inequality after the revolution (Piketty, 2014:342). Changes in political and economic systems do not seem to end wealth inequalities. After going through the Cultural Revolution and, later, the economic transformation of the 1980s, China became wealthier, but inequality increased almost to the same levels as in the USA (Piketty, 2020:618). Therefore, the impact of wealth has historically not been fixed by eliminating wealth or the wealthy.

As proposed by Plato, the excessive freedom of some does not end well because the rest feel entitled to be intolerant of them, which leads to a divided society. However, the relationship between freedom and tolerance was approached differently by Karl Popper in his "paradox of tolerance". Popper proposed that "…unlimited tolerance must lead to the disappearance of tolerance. If we extend unlimited tolerance even to those who are intolerant, if we are not prepared to defend a tolerant society against the onslaught of the intolerant, then the tolerant will be destroyed, and tolerance with them" (Popper & Gombrich, 1994:581).

The limit of freedom is found in the tolerance to freedom. Tolerance is the right of people to tolerate others' ideologies and beliefs, as long as they are reasonable and do not go against the principle of tolerance. This idea is also applicable to the design of the built environment. What to do with the intolerant? If SDG 13 (Climate Action) means something, how can it be optional? What should be done about designing buildings that are intolerant to the limitations imposed by the environmental crisis?

Towards a Pedagogy of the Oppressed by Wealth

Freedom and justice can be at risk when one population sector oppresses another. In his book *Pedagogy of the Oppressed*, the Brazilian educational theorist Paulo Freire analyses the relationship between oppressors and the oppressed in a way that might be helpful in dealing with the paradox of tolerance. Freire (1996:18) explains that the oppressor dehumanises the oppressed, something that also happens when people are turned into mere consumers in a market that pushes them to want things they cannot afford and that are detrimental to the environment: "…sooner or later being less human leads the oppressed to struggle against those who made them so. The oppressed must not, in seeking to regain their humanity (which is a way to create it) become in turn, oppressors of the oppressors, but rather restorers of the humanity of both."

The aim of this book is to raise awareness about the impact of wealth and the challenges this imposes for accomplishing sustainability within the built environment. It is not to expose wealthy people as targets to be hated, envied or discriminated against. The climate change crisis pushes us to think of alternative solutions that do not involve the tyranny of tyrants or the oppression of the oppressors. People who feel entitled to use the freedom given by their wealth to bully and oppress others must be stopped. Architects and planners must be prepared to work with those wealthy people who want to use their wealth differently. This has been done before in critical times. During World War II, in a few months, factories were turned to manufacture completely different products. The president of the USA, Franklin Roosevelt, promoted the creation of the War Production Board (WPB) in January 1942. The objectives were to convert factories into manufacturing plants for military equipment and to ration materials demanded for wartime, like metal and gasoline. Ford Motor Company (see Chap. 4) moved from producing cars to assembling B-24 Liberator bombers, the Lionel toy train company switched to making compasses for warships and The Mattatuck Manufacturing Company, which made upholstery nails, switched to making rifle cartridge clips (Vergun, 2020). To give a sense of the task demanded by the government of firms like Ford, automobiles had 15,000 parts, and a B-24 had "450,000 parts and 360,000 rivets in 550 sizes, and it weighed 18 tons" (Trainor, 2019). However, in mid-1944, Ford produced one B-24 per hour.

Other social changes happen with crises. During the great depression in the USA, a wealth tax was introduced to pay for Roosevelt's New Deal programmes. "Those making more than $5 million a year were taxed up to 75 percent" (IRS, n.d.). This tax was not popular. Perhaps more relevant is the tax change in the USA caused by WWII. The Victory tax of 1942 was meant to raise those paying income tax from 5% in 1939 to 75% of all workers (IRS, n.d.). Taxes also rose in the UK to support the war effort. In 1938, income tax was increased, and a surtax of 41% was introduced for those earning over £50,000. Income tax rates were also raised further during the war (UK Parliament, 2024). This was accepted because people were prepared to contribute to the common goal of defeating Hitler's Germany. The UK war effort has also been associated with a decrease in inequality: "…the War created a coalition among voters for social reform which was then implemented by the Labour Party, leading to a reduction in (wealth) inequality in England" (Heldring et al., 2022). The point is that if it has been possible to change not only what a country produces, but also the willingness to accept additional taxation, it can be done again. For those in the forefront of the effects of climate change, such as increased flooding and loss of crops through drought, the threat to lives and livelihoods is as real as in wartime. When threats are perceived as real, as in the public response to COVID-19, people can change what they do and what they accept.

World War II also affected the building industry and hence the built environment. One effect was a change in the types of building now required: "Factories, factory extensions, temporary hostels and a host of temporary buildings for both civil and military use…" (Bowley, 1966:83). Of more pertinence to the SDGs and the aim of sustainable human settlements, the war also saw the need to economise on the materials that went into buildings, such as the use of welding to reduce the use of steel and the use of standardised designs and factory-made products. This was achieved by drawing on the new techniques that had been developed but not exploited (Bowley, 1966:83–84). Reducing the materials that go into buildings can reduce the energy or carbon embodied in them, and hence their overall environmental impact. Although these ideas are known, they have not yet been applied in regulations in developed countries (Rasmussen et al., 2024), but to meet the requirements of SDG 11 (Sustainable Cities and Communities), they should be.

What to Do with the Buildings of the Wealthy?

Across the book, we have seen that if there is something architects have been exposed to, it is dealing with wealth and wealthy clients. Architects and planners could play a key role in the climate change era if education is geared towards learning how to work for and with the poor. It is essential legally to commit the wealthy to reducing poverty without destroying the planet. All the SDGs should only be about making this happen. That said, the built environment is a legacy from the past and even if attitudes change, we are left with many buildings that reflect wealth in some form.

To envision what future societies can do with the buildings of the wealthy, history can be used as a reference to learn what happened in the past with the ones that disappeared (Gracia et al., 2021: 59). The American anthropologist Joseph Tainter (1988:4) proposes that big buildings and monumental architecture were linked to the peak of a civilisation's development and subsequent collapse. After societies collapsed, people tended to move to the periphery to live in humble houses or remain in what was left of the political centre (Tainter, 1988:20).

> What populations remain in urban or other political centers reuse existing architecture in a characteristic manner. There is little new construction, and that which is attempted concentrates on adapting existing buildings. Great rooms will be subdivided, flimsy facades are built, and public space will be converted to private…People may reside in upper-story rooms as lower ones deteriorate. Monuments are often mined as easy sources of building materials. When a building begins to collapse, the residents simply move to another.

The practice of reusing big buildings by converting them is very old and still relevant. Another possible destiny for big buildings, using the collapse of past civilisations from Meso-America to the late Roman Empire as examples, is for them to be abandoned or mined for their materials. Big buildings embody extracted raw materials adjusted for building purposes, which makes it easier for people to reuse them to build new. Tainter (1988:55), using the work of another anthropologist, Philips (1979:138–140), proposed that monumental buildings are abandoned or dismantled because, after a collapse, society needs to be more efficient in its use of resources, and monumental buildings are not essential for rebuilding a society since they demand a lot of care and resources and give very little in return.

Buildings that are very expensive to build, maintain and repair could burden the future generations dealing with the impacts of climate change. To make possible the emergence of a post-climate change society, it is essential to limit and tax big wealthy buildings proportionally to the problems they might cause future generations. The more difficult it will be for future generations to adapt, change, reuse or dismantle them, the more economically inviable they must become in the present.

False and True Generosity in Buildings

In the past, important buildings have been constructed through the consistent help of generous donations from people who do not always have a lot (see Chap. 5). However, generous donations do not always produce generous buildings. In the same trend as the Louis Vuitton Foundation in Paris and the Guggenheim in Bilbao (see Chap. 6), both by Frank Gehry, the Soumaya Museum in Mexico City (Fig. 12.3) is big, shiny and has a complicated shape, forming an envelope that competes with the content of the museum. Entry to the Soumaya Museum is free, and it has been promoted as a gift to the city made possible by the charity of the Mexican billionaire Carlos Salim, whose wealth came through investments, including in the largest mobile telecom company in Latin America. His architect son-in-law, Fernando Romero, designed the building. The museum is named after Salim's late wife, Soumaya, and displays a wide range of works of art, including many pieces that Carlos Salim collected throughout his life.

The design of the museum does not contribute to enhancing public space. It is not a building that gives too much; it is closed and isolated, like a silo. However, it takes a lot of space, since it needs to be detached from everything to showcase its shape. The design decision to limit natural daylight in a museum is usually justified by the need to preserve the works of art. However, in the Soumaya Museum, the façade is mostly next to the circular ramps that run peripheral to exhibition areas at the core of each level, making the closed nature of the façade difficult to explain other than for aesthetic preferences and visual impact. The façade has multiple layers, and its final skin is modulated to overcome the complications posed by the form, which adds very little to the visitor experience of the interior quality but makes the building difficult to modify in the future. Even though the building attracts tourists, its Instagrammable quality mostly entertains them. While tourists take pictures of themselves, local workers queue in the corner of the block where the building is located, waiting for a small, old bus to take them back home (Fig. 12.4). The glass barrier tries to suggest there is no barrier between the pavement and the museum but fails. The opportunity to provide a public space outside the building that could accommodate those waiting for the bus has been lost.

Paulo Freire (1996: 18–19) proposes that fighting for true generosity consists in "destroy-

Fig. 12.3 Soumaya Museum, Mexico City. (Source Emilio Garcia)

ing the causes which nourish false charity", namely, fighting against acts that pretend to be generous but only consolidate unfairness. True generosity consists of extending a hand to the ones that need it the most, and, for this reason, it constitutes an "act of love". These ideas can be used as guidelines in the design of the built environment. A generous public space at the bus stop could have been much appreciated by local workers and their families, who might never visit the museum, but at least would feel a bit better while waiting for the bus. However, such things probably only matter for designers who use the bus. The Museum Soumaya contrasts sharply with the generosity and simplicity of the building in the Sahara described by Lacaton and Vassal, which is only a roof sustained by six posts and a television that uses solar energy and serves to educate nomadic communities (see Chap. 9). More museums do not mean more education, less inequality or less poverty. If there is a commitment to the SDGs, whatever is built in the landscape has to be an act of love for the people who need it the most.

Transparency

To achieve a sustainable built environment, we should know how much each building costs and its environmental impact; in the same way food packaging displays what the contents are made of and how many calories each portion contains. This information has to be published.

If the aim is to make buildings that can contribute to future generations, then it is vital to have a decommissioning plan at the outset. Whereas from the start materials were reclaimed from cars at the end of their useful life, and today cars are dismantled by specialists for what can be reused, from parts like transmission systems, to materials like steel and plastic and even to components like the refrigerant in air conditioning

Fig. 12.4 Bus stop in front of Soumaya Museum, Mexico City. (Source Emilio Garcia)

systems (Cassells, 2004:48–50), no such system exists for buildings, at least in developed societies. As an example, in New Zealand, 40–50% of all waste in landfills comes from construction waste and from demolished buildings (BRANZ, 2024). To avoid this means planning how a building could be deconstructed and which materials might be easily recycled. A building containing 300,000 standard bricks might be recycled to make 40 standard houses, whereas a building with 300,000 special bricks, like the UTS Business School (see Chap. 3), might not. The real innovation in design for future generations is to plan for building material reuse, not to make weird shapes.

Wealth Is the Sponsor of Innovation and Its Cause of Demise

In the same way that supporters of market unfairness have appropriated the meaning of freedom, the meaning of innovation has been totally twisted and appropriated by industry and the market. This situation has also affected the understanding of sustainability. The first misconception is that innovation means technological change, which is assumed to be essential to achieving sustainability. If you have money, you can live in a house with central heating produced by photovoltaic panels and move around in an electric car. From this point of view, technological advances can make a difference for the people who can pay for them and even contribute to creating new jobs. In essence, this means that households must spend to consume less. If spending on technological advancements is achieved by not spending the money on items that contribute to environmental degradation, such as flying to holidays abroad, then it less objectionable. This is the assumption behind the Environmental Kuznets Curve and the interest in SDG 8 (Decent Work and Economic Growth) and SDG 9 (Industry, Innovation and Infrastructure), but it does depend on those who are able to make decisions making the right decisions.

However, if you are renting, you probably do not have central heating; you depend on heating one room at a time with inefficient heaters that probably consume more energy, which increases the bill. The more money you have, the greener you can be and the less you expend. This is the result of leaving sustainability in the hands of the market. The person who can buy more gadgets can live better and save more than the poor, who need it but live worse and pay more proportionally. From this point of view, technological and industrial outputs only advance the population sector that produces or can buy them, which is unfair and benefits only a minority. Once again, actions towards achieving SDG 8 (Decent Work and Economic Growth) and SDG 9 (Industry, Innovation and Infrastructure) compromise SDG 1 (No Poverty), SDG 2 (Zero Hunger), SDG 3 (Good Health and Well-Being), SDG 7 (Affordable and Clean Energy), SDG 10 (Reduced Inequalities), SDG 11 (Sustainable Cities and Communities) and SDG 13 (Climate Action). The reality is that no system can be sustainable if being poor is expensive.

The creative field, reserved for genius artists until the Enlightenment (see Chap. 8), is now predominantly identified with jobs similar to those in successful Silicon Valley firms. However, none of the top ten billionaires sell anything new, nor are they making brilliant discoveries to end poverty or climate change. They are not the brightest minds of our generation either, since no Nobel Prize winners are on the Forbes list. Is current economic growth and the perception of technological development producing anything new that breaks with the past in an innovative way? "We find that papers and patents are increasingly less likely to break with the past in ways that push science and technology in new directions" (Park et al., 2023:138).

The contemporary understanding of innovation that has enslaved designers and institutions means succeeding in the market and industry (see Chaps. 3, 6 and 8). Most of the so-called architectural innovations are marketing ideas. As this book has tried to show, part of the history of architecture, even in its most elitist version, is a history of people learning and being inspired by others. Designers must be liberated from the burden of being economically successful to prove relevance to the market, their peers and themselves. This will reduce the social anxiety caused by the continuous pressure of being radical and innovative, increase the quality of their work, help people, impact the environment less and educate new generations of designers to be free of "innovation anxiety".

One of the reasons why architecture is so dependent on its past is its link with humanity. Humans stopped changing biologically 40 to 50 thousand years ago. The scale reference for domestic habitats is still the human body. The relationship between beds and bedrooms is a good example. Most beds are rectangular. From the first vegetable mattresses discovered 77,000 years ago (Wadley et al., 2011:1388) to tatami and futon beds, the most efficient shape used to carry our sleeping body has always been and still is rectangular. The last "innovation" in bedroom design probably occurred during the "Neolithic shift from circular to rectangular building" around 7000 to 8000 years ago, possibly reflecting changes from cooperative to autonomous family households (Duru et al., 2021:19). Even this big innovation did not happen everywhere, all at once. European cities made circular buildings until early medieval times (Bradley, 2020). Materials for mattresses, bed frames and houses have changed. However, innovation in the design of the bedroom has only happened once in thousands of years. Since medieval times, rectangular beds in rectangular houses have been the norm, zero innovation beyond that point. While in architectural magazines and websites, words like new, innovative and radical pollute main headings, the most common bedroom shape is still rectangular. Unsurprisingly, the most common form for houses is also rectangular.

Buildings and cities do not evolve, since they do not have DNA. They are developed by people who inhabit them and whose bodies have evolved in the past. Since buildings tend to last longer than the life of people and consume land, which is finite and very limited in desired locations, inundating the market with innovative architecture is not helpful.

The systematic pressure to be radical, original, creative and innovative has no ethics other than satisfying the market need for consumption. Curiously, words like original or radical are related to the origins and roots of something, to their past. To be original and radical is an enquiry into the start of something that already exists, that is not new.

The architecture of wealth is intended to be unique and recognisable in its aesthetic: to be free, like the market. Under these premises, the architecture of the non-wealthy is traditional, repetitive, forgettable and boring. The architecture of wealth wants to be exceptional, but the final result is the same type of building repeated. The architecture of the non-wealthy starts with what is well-known and popular because it has been historically proven to work. However, in the market, when a product is successful, it becomes more accessible, and everyone can have it. It goes from being a novelty accessible to a few to a mundane and forgotten cheap object in the everyday life of the masses. Therefore, to keep selling innovative products, the market calls the old, new, and even this is not a new idea.

The Real Innovation Would Be to Stop Innovating for the Market

In terms of the built environment, design solutions that can avoid and lessen the impact of climate change already exist. Why must there be something wrong with the repetitive designs from traditional and vernacular architecture that do not fit with the innovation the market demands but promote sustainable lifestyles and are affordable? Should design innovations for the market that are detrimental to society and the environment simply be prohibited?

Future architecture has already happened a hundred years ago. Most of the experimental housing projects built in the inter-war era in Europe have been tested at a larger scale in Africa and South America. As Habraken (Montaner i Martorell, J, 2015:10-11) pointed out, the task is now to use the innovations and make them fit in cities so they can work. There is still a missing step, this being the architecture for the poor and how to build it without destroying the planet. To do this, it is important that the wealthy reduce their consumption levels, so their current emissions can be invested in building for the people who need it the most.

Sustainability and Architectural Design

The gap in knowledge highlighted throughout this book is the lack of consideration given to the impact of wealth on the built environment, on built environment professions and associated education, and how these become an obstacle in achieving a sustainable built environment. A common misunderstanding is that a sustainable building contains innovative technological devices and techniques, without which sustainability cannot be attained. However, as the book shows, these usually lead to buildings that might achieve green certification while still going against sustainability. Therefore, it could be too late if sustainability is only injected into the design process after the agreement between the client and designer, or if consultants patch it in during the development stage. This is a way of making the design more complicated by trying to fix problems that could have been avoided from the start. So when does sustainability start and end in the design process?

As shown in Fig. 12.5, when applied, the domain of sustainable design is normally con-

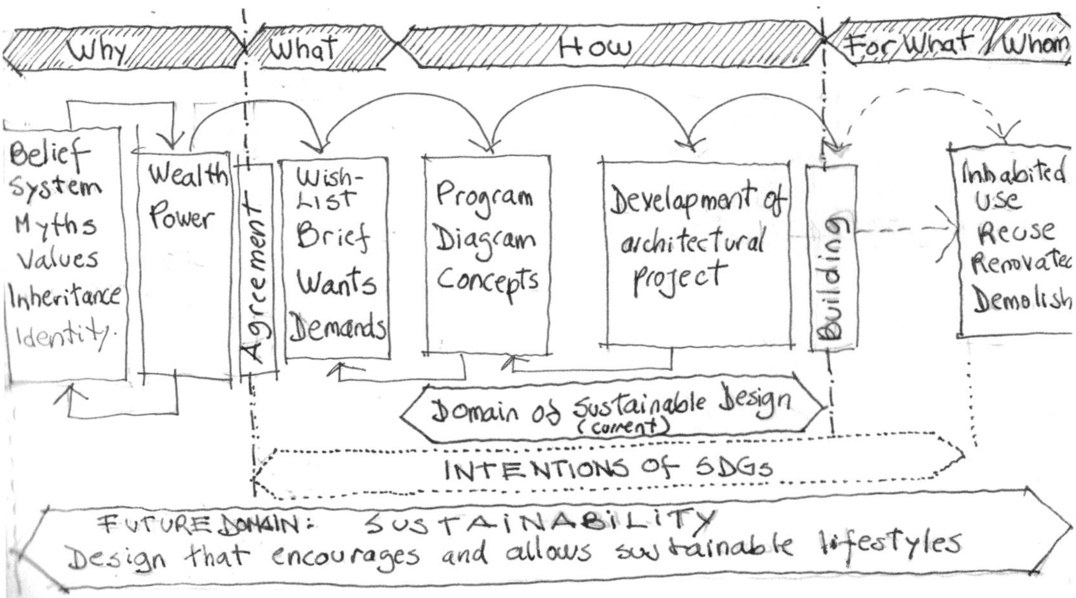

Fig. 12.5 Diagram of the domain of sustainable design within the architectural design process

fined to the development of the architectural project, the “How” phase. Sustainability is considered too late, after all the decisions that define the finality of a project and its impact have been taken. The values and inherited preconceptions of designers and clients are where the drivers for making decisions in the design process exist and originate. All these drivers and preconceived expectations form the “Why” of any design process, which is rarely questioned. Equally unattended is the last stage of the design process, namely, ensuring “What for”—the finality of the design, and most importantly, the beneficiaries, or “for Whom” it is made. The SDGs might be useful in widening the domain of sustainable design but are of very little use if the starting point is a preconception of unlimited resources and wealth.

Throughout this book, there has been an emphasis on proposing that sustainability in architecture and the built environment starts before the conceptual phase of the design process. The link between design and sustainability can be made, but it must start by challenging the “Why” of a project. Why should a project exist? Why is it needed? For whom? These must be the primary questions. “How” that is achieved and “What” building suits the goal coming after the values, conditions and limitations have been agreed upon and set. The link between sustainability and design starts by imagining how people can live in and use a building in a sustainable way. This, in turn, means understanding what a fair share of the resources of the planet is for each person involved. What people feel entitled to and what they can have are two very different things that should be acknowledged before producing a wish list for an architect. Designers and clients would benefit from having clear guidelines about environmental budgets for each building. This is not very different from having speed limits on the street. A few people can buy very expensive sports and racing cars that reach 300 km/h. However, what is the point of it if the speed limit in most city centres ranges between 30 and 50 km/h and rarely exceeds 110 km/h on highways?

However, there are no environmental limits that serve as departure points and references for the built environment, including for its designers and the education of these designers. It is still possible to design buildings that consume as much energy as their owners can afford. In the same way that speed limits exist and are agreed upon for collective safety, there should be energy budgets that help to raise awareness among clients and designers. A clear and regulated understanding of such energy limits could empower designers to make design decisions and create better buildings for everyone. Turning again to the effect of WWII, post-war in the UK: "Those building private houses…were restricted in both the size and cost of any new home so as to avoid unnecessary use of scarce materials." Limits were put on the size of 1000ft^2 (93m^2) for a two-storey and 930ft^2 (86m^2) for a single-storey house (Vale, 1995:158). Such restrictions would save both materials and energy in use. These restrictions do not seem to have limited the creativity of designers or affected the construction industry. On the contrary, the most influential and experimental proposals, at a large scale, in social housing emerged from this period.

Carbon taxes, the equivalent of fines for speed limits, exist but are not implemented in all countries, nor do they cost the same per tonne of CO_2e. They also fail to apply across countries to all sectors and fuels. Major polluters like China and the USA do not have carbon taxes. "Uruguay had the highest carbon tax rate worldwide as of April 2024, at US$167 per tonne of CO_2 equivalent (US$/t$CO_2$e)" (Tiseo, 2024), followed by Liechtenstein (132.12 US$/t$CO_2$e), Switzerland (132.12 US$/t$CO_2$e), Sweden (127.25 US$/tCO_2e), Norway (107.78 US$/t$CO_2$e) and Finland (99.98 US$/tCO_2e), which was the first country to apply them (UNFCCC, 1995:39). One big difference is that in Uruguay, the carbon tax only applies to 5% of their total greenhouse emissions, but in Liechtenstein, 81% are covered (according to data from 2008). Another big difference is that Uruguay has a much lower income per capita (US$17,000) than a European country like Norway (US$108,729), making the tax less equal on a global scale. The key point is that very few countries use carbon taxes, and they do it in their own way. Based on the Annual Utility Decarbonization Report 2023, The National Public Utility Council proposed that "Only 1% of global emissions are priced high enough to meet the Paris Agreement's temperature target in 2024" (Oğuz, 2024). To have a wider impact, carbon taxes must be fair and transparent and cannot be isolated from other levies or be optional.

The French economists Lucas Chancel (2020) and Thomas Piketty (2020) have advocated for a progressive tax on greenhouse gas emissions to bridge the inequality gap between most people emitting 5–10 tonnes a year and wealthy people emitting 100–150 tonnes a year. Progressive taxation means the higher the emissions, the more expensive each tonne of CO_2e emitted becomes. Lucas Chancel (2020:134) estimates that these taxes would help cover the 170 billion per year needed to support the most vulnerable in adapting to the impacts of climate change. Based on historical evidence from nineteenth-century France and the WWII period in the USA and the UK, Piketty (2020:981) adds that carbon taxes must be included within a progressive taxation system composed of a progressive property tax, an inheritance tax and a progressive income tax. This means the higher your salary (as a multiple of average wealth) and the more properties you have or inherit, the higher the taxes you pay. Since consumption and emissions are tightly linked, by grouping them and taxing them, limits can be established at both ends. If a fair climate change-driven taxation system could be implemented, it could change people's lives (Piketty, 2020:983).

> With the system proposed here, every young adult could begin his or her personal and professional life with a fortune equal to 60 percent of the national average, which would open up new possibilities such as purchasing a house or starting a business. Note that this system of public inheritance for all would guarantee every individual a sum of capital at the age of 25.

Piketty's approach would mean starting with a capital sum at the age of 25 instead of a student loan debt. It could also change the volume of what needs to be built and the demand for new housing. This demand has been growing since the post-war era, impacted by the impoverishment of states and the increasing role of private interests. Taxing the wealthy fairly to provide capital for the young could be a game changer. It could alleviate the housing shortage and energise the building industry if properly instrumented. Taxation does not necessarily lead to behaviour change, and it does not redesign people, but a fair system is necessary to deal with the cost of the impacts of climate change and the transition to a post-climate change society.

Complementary to the application of progressive taxation, the reform of the public pension systems could change housing production. One reason people want to own a house in countries like New Zealand is the fear of being asset-poor in retirement and unable to cover the rent and expenses with their pension. Depending on the tax situation of each retiree, the pension ranges between $NZ761 and $NZ660 before tax and is paid fortnightly (Work and Income, 2024). However, the average weekly rent in Auckland for a two-bedroom house is $NZ573 (Barfoot & Thompson, 2024). Retirees could afford to pay a cheap rent with their public pensions, but would have to use their savings for eating and living expenses. This situation impacts the design of the built environment because it makes built-to-rent not a long-term option, which again puts pressure on the housing market, with houses often becoming less and less affordable as a result. If everyone had access to a decent retirement income, owning a property would be less critical in becoming income and asset-poor in retirement. Weak pension systems, like those in New Zealand and the USA, could be avoided. Countries like Denmark and the Netherlands, which rank amongst the best retirement systems in the world (Knox & Hudson, 2023), also have a history of renting with a high percentage of social housing, proving that things can be different if investment in people's future is prioritised. For good reason, "one often hears that a public pension is the patrimony of those without patrimony" (Piketty, 2014:490).

The importance of having a generous pension system and a social housing market majorly owned by the government can impact the attitude towards housing ownership, favouring renting. Austria has a generous pension system and a long tradition of living in rented apartments. In Vienna, renting is almost the norm, and social housing is not only for the poor, which helps with its perception. Many of the 1930s architectural experiments of modernism failed socially and were heavily criticised across European cities, but in Austria, they still work. The Matteotti-Hof in Margareten was built in 1926 and still works for its original purpose. This is also possible because the government of Austria and its local authorities resisted selling their social housing to the private sector, keeping their tradition, which originated in the nineteenth century, of social housing built by cooperatives. After COVID-19, rents have been regulated and increased, but renting in Vienna is three times cheaper than in Paris, London, Oslo or Dublin (Linhart et al., 2023:41). One key to these achievements is the important share of the market that the state has, together with the development of a third sector, a type of cooperative that is not fully private or state-owned. The Limited-Profit Housing Associations (in German, Gemeinnützige Bauvereinigungen or GBVs) own 40% of all rented homes and produce housing for individual ownership, all at building cost (GBV, 2024).

> ...limited profit housing associations are exempt from corporation tax in their main and ancillary areas of business...GBVs calculate on a cost-basis, which means that rents can neither be set above nor below the costs incurred in the production, financing and management of residential buildings ("cost rent") (GVB, 2024:3).

A recent study shows that for each 10% increase in housing by GBVs, there is a 5% decrease in the average rent price. Without building for profit and speculation, everything works better for everyone. It is estimated that by avoiding paying market prices, total household savings are 1.3 billion Euros per year, increasing people's purchasing power and the country's GDP, and contributing to building more houses. "Equity is permanently tied up for limited-profit purposes and surpluses are continuously reinvested. This is guaranteed by a limitation to profit distribution among owners and by an obligation to reinvest any surpluses regularly in housing construction" (GVB, 2024:3).

Economics can be simple when it is intended to help everyone.

What happens if progressive taxation cannot be implemented? What is plan B? This would be to limit certain practices and products. In the case of housing, this could be done by using planning policies established by the government instead of private developers. From building massive outdoor parking areas in housing projects, which increases impervious surfaces and deprives children of space to play, policies can introduce clear boundaries that benefit everyone. This requires breaking the habit of designing houses for cars instead of people. It is important to encourage a generous design for people instead of designing garages for two cars in housing projects where two beds barely fit in a bedroom without a closet. Higher standards that are aimed at achieving true generosity in housing design will deliver better quality with less environmental impact.

Designing to End Poverty

As shown in this book, designers have a significant role in only a small fraction of what is built. However, the challenge for designers and planners is contributing to ending poverty (SDG 1). Housing is an important asset that can change people's lives, especially in countries with no fixed salaries provided by the government after retirement and where owning a house can provide some relief. The demand for affordable housing and shelter will affect 3 billion people by 2030. This means that globally, 96,000 houses need to be built every day. The delivery of housing, especially for the bottom 50% of the population, with high standards and avoidance of design inequalities, is the essence of any design. If governments support the initiative of ending poverty through housing, they must become the main developer to set priorities and standards that are people rather than market focused. If this were possible in the past, it can be done again.

The priorities of built environment professionals will need to change to face the challenge of the large-scale production of affordable housing that is gentle on the environment. This is the opposite of designing a few houses with high environmental impact for those with the means to pay for them. Designing to end poverty means designing for people, not for architects. The design of 96,000 houses per day cannot be done by designing smaller and poorer versions of wealthy houses. This substantial change can be made because every new generation of architecture students is keen to learn how to do things that can make a difference. It just needs tutors to show them what better and good mean in architecture. Defining what better means relies on having a common understanding of a problem and developing a set of criteria that can be used to define a level above which something is better. Throughout this book, we have reviewed the environmental, historical, practical, ethical and aesthetic starting points for facilitating such a mindset change.

Considering the mechanisation processes available for building, a key role of architects is to be enablers of a design that humanises people. Designing large-scale housing projects involves a complex set of trade-offs between what is wanted and needed, what is desired and what is feasible, and mediating these are the political, economic, environmental and social matters inherent in the SDGs. If the Italian architect Giancarlo di Carlo thought that architecture was too important to be left in the hands of architects, it should also be too important to be left at the mercy of algorithms.

Stranger Things: The Ordinary Great

The design of modest houses with limited resources that are great for the people inhabiting them poses the highest level of difficulty, complexity and sophistication to anyone who assumes that designing for poor people is only a matter of making everything smaller and cheaper. The current mindset of designers is tuned into designing something inversely proportional to the client's wealth; the lesser the wealth, the more deprived of humanity the design should be, as if poor people were a smaller version of human beings, with smaller lives and little dreams.

What makes a modest building extraordinary? Nidia Marinaro, the head of Livingston Architects and the widow of the architect Rodolfo Livingston, mentioned in an interview that when they were renovating houses, they used a method for interviewing clients and involving them in the design process. She emphasised that apart from understanding a client's needs it is important to understand their dreams, because everyone should have the right to dream even when the budget is modest. The difference between an ordinary and an extraordinary building is that the ordinary gives you only what you expect to have, while extraordinary buildings can offer their non-wealthy inhabitants something they would never have thought of having, something they do not feel they deserve. Throughout this book, it has been shown how wealth affects the design of the built environment, encouraging the reproduction of expensive, ordinary designs for a few. If the SDGs are to be taken seriously, the next generation must deliver humble buildings using modest resources so there is enough for everyone. Wealth cannot do humble buildings. In scarcity and adversity, creating humble buildings that can deliver hope to people regardless of wealth is the extraordinary design needed to tackle the SDGs.

Delivering the SDGs

Designing for those without much is part of both SDG 1 (No Poverty) and SDG 11 (Sustainable Cities and Communities). It also suggests that wealth is not the starting point for good design. This is a critical idea, as it suggests it would be possible to achieve a sustainable society and habitation once humanity saw wealth accumulation as a curse rather than a goal. What Aristotle and Plato realised centuries ago was that having enough is enough, and anything else would lead to problems like wars over resources. What we know now is that all humanity can only have enough because the planet is small and we have already squandered many of its resources.

The SDGs tell us clearly what global problems humanity faces, but unless we tackle the problem of wealth and its uneven distribution in current society, solving these problems will remain a dream.

References

Barfoot & Thompson. (2024). *July 2024 Rental report.* Retrieved from https://www.barfoot.co.nz/market-reports/2024/july/rental-report, 14 August 2024.

Bo. S. (2011). A literature survey on environmental Kuznets Curve. *Energy Procedia 5*. 1322–1325.

Bowley, M. (1966). *The British building industry.* Cambridge University Press.

Bradley, R. (2020). *The idea of order: The circular archetype in prehistoric Europe*. Oxford University Press.

BRANZ. (2024). *Minimising waste.* Retrieved from https://www.level.org.nz/material-use/minimising-waste/, 22 July 2024.

Cassells, S. M. (2004). *Towards sound management of end-of-life vehicles in New Zealand*, PhD Thesis. Massey University.

Chancel, L. (2020). *Unsustainable inequalities. Social justice and the environment.* Harvard University Press.

Duru, G., Özbaşaran, M., Yelözer, S., Uzdurum, M., & Kuijt, I. (2021). Space making and home making in the world's first villages: Reconsidering the circular to rectangular architectural transition in the Central Anatolian Neolithic. *Journal of Anthropological Archaeology, 64*, 2–23.

Freire, P. (1996). *Pedagogy of the oppressed.* Penguin Books.

GBV. (2024). *Österreichischer Verband Gemeinnütziger Bauvereinigungen—Revisionsverband.* Retrieved from https://www.gbv.at/english/, 14 August 2024.

Gracia, E., Vale B. and Vale, R. (2021). *Collapsing gracefully: Making a built environment that is fit for the future.* Cham: Springer.

Grossman, G.M. and Krueger, A.B. (1995). Economic growth and the environment. *The Quarterly Journal of Economics 110*(2). 353–377.

GVB. (2024). *Limited-profit housing associations in Austria. Overview.* Retrieved from https://cms.gbv.at/repos/files/GBV/Intranet%5fdownloads/PDF/Verbandsstatistik/Studien%5fPublikationen%5fGBV/Summary%20WIFO%20study%5fThe%20economic%20impacts%20of%20limited%20profit%20housing%20associations%2epdf?exp=52039&fps=cfd587f892c524fa0aaab363a0b19ca0dbc8355e, 14 August 2024.

Heldring, L., Robinson, J.A. and Whitfill, P.J. (2022). *The second world war, inequality and the social contract in Britain*. Working Paper No. 2022-18. University of Chicago. Retrieved from https://bfi.uchicago.edu/wp-content/uploads/2022/01/BFI_WP_2022-18.pdf, 22 July 2024.

IRS. (n.d.). *Taxes in U.S. history,* Retrieved from https://apps.irs.gov/app/understandingTaxes/teacher/whys_thm02_les05.jsp, 14 July 2024.

Knox, D., & Hudson, S. (2023). *An analysis and ranking of 47 pension systems around the world.* The Mercer CFA Institute Global Pension Index. Retrieved from https://www.mercer.com/insights/investments/market-outlook-and-trends/mercer-cfa-global-pension-index/#download, 14 August 2024.

Kuznets, S. (1955). Economic growth and incomes inequality. *The American Economic Review 45*(1). 1–28.

Linhart, M., Hána, P., Leško, J., Machula, L., & Marek, D. (2023). *Deloitte property index. Overview of European residential markets*. Retrieved from https://www2.deloitte.com/content/dam/Deloitte/at/Documents/presse/at-deloitte-property-index-2023.pdf, August 14 2024.

Montaner i Martorell, J. (2015). *La arquitectura de la vivienda colectiva*. Reverté.

Oğuz, S. (2024). *Visualized: The price of carbon around the world in 2024.* Visual Capitalist. Retrieved from https://www.visualcapitalist.com/sp/visualized-the-price-of-carbon-around-the-world-in2024/#:~:text=While%20there%20are%20notable%20efforts,Paris%20Agreement%2Daligned%20carbon%20pricing, 12 August 2024.

Park, M., Leahey, E., & Funk, R. J. (2023). Papers and patents are becoming less disruptive over time. *Nature, 613*, 138–144.

Phillips, D. A. (1979). The growth and decline of states in Mesoamerica. *Journal of the Steward Anthropological Society, 10*, 137–159.

Piketty, T. (2014). *Capital in the twenty-first century*. Harvard University Press.

Piketty, T. (2020). *Capital and ideology* [Capital et idéologie]. Harvard University Press.

Plato. (2000). In Ferrari G. (Ed.), *Plato: 'The Republic'. Cambridge texts in the history of political thought The world's popular classics* (T. Griffith Trans.). Cambridge University Press.

Popper, K., & Gombrich, E. H. (1994). *The open society and its enemies; New One-Volume Edition*. Princeton University Press.

Rasmussen, F. N., Birgisdóttir, H., Malmqvist, T., Kuittinen, M., Häkkinen, T., Azari, R., & Moncaster, A. (2024). *The Routledge handbook of embodied carbon in the built environment* (A. Azari, & A. Moncaster, Eds.). Routledge, pp. 85–102.

Stiglitz, J. E. (2024). *The road to freedom: Economics and the good society*. Penguin Books Limited.

Summer, A., & Advani, A. (2021). *How much tax do the rich really pay?* Retrieved from https://www.lse.ac.uk/research/research-for-the-world/economics/how-much-tax-do-the-rich-really-pay, 9 July 2024.

Tainter, J. A. (1988). *The collapse of complex societies*. Cambridge University Press.

The United Nations Framework Convention on Climate Change (FCCC). (1995). *Finland's National Report under the United Nation's Framework Convention on Climate Change*. Retrieved from https://unfccc.int/resource/docs/natc/finnc1.pdf, 12 August 2024.

Tiseo, I. (2024). *Prices of implemented carbon taxes worldwide 2023, by jurisdiction*. Statista. Retrieved from https://www.statista.com/statistics/483590/prices-of-implemented-carbon-pricing-instruments-worldwide-by-select-country/, 12 August 2024.

Trainor, T. (2019). *How Ford's Willow Run Assembly Plant Helped Win World War II*. Retrieved from https://www.assemblymag.com/articles/94614-how-fords-willow-run-assembly-plant-helped-win-world-war-ii, 9 July 2024.

UK Parliament. (2024). *The cost of war.* Retrieved from https://www.parliament.uk/about/living-heritage/transformingsociety/private-lives/taxation/overview/costofwar/, 14 July 2024.

Vale, B. (1995). *Prefabs: A history of the UK Temporary Housing Programme*. E and FN Spon.

Vergun, D. (2020). *During WWII, industries transitioned from peacetime to wartime production*. U.S Department of Defense. Retrieved from https://www.defense.gov/News/Feature-Stories/story/article/2128446/during-wwii-industries-transitioned-from-peacetime-to-wartime-production/, 9 July 2024.

Vogel, J., & Hickel, J. (2023). Is green growth happening? An empirical analysis of achieved versus Paris-compliant CO_2–GDP decoupling in high-income countries. *The Lancet. Planetary Health, 7*(9), e759–e769.

Wadley, L., Sievers, C., Bamford, M., Goldberg, P., Berna, F., & Miller, C. (2011). Middle stone age bedding construction and settlement patterns at Sibudu, South Africa. *Science (American Association for the Advancement of Science), 334*(6061), 1388–1391.

Wells, T. (2016). Exile the Rich. *Krisis (Amsterdam, Netherland: 2000), (1)*.

Work and Income. (2024). *How much you can get for NZ Super.* Ministry of Social Development, New Zealand Government. Retrieved from https://www.workandincome.govt.nz/eligibility/seniors/superannuation/how-much-you-can-get.html, 14 August 2024.

Index

E. J. García, B. Vale, *Wealth and the Built Environment*, Sustainable Development Goals Series,
https://doi.org/10.1007/978-3-031-81954-4

O
Over consumption, 3
Oxford Conference, 202–204

MIX
Papier aus verantwortungsvollen Quellen
Paper from responsible sources
FSC® C105338

If you have any concerns about our products,
you can contact us on
ProductSafety@springernature.com

In case Publisher is established outside the EU,
the EU authorized representative is:
Springer Nature Customer Service Center GmbH
Europaplatz 3, 69115 Heidelberg, Germany

Printed by Libri Plureos GmbH
in Hamburg, Germany